T0202125

Newton's Metaphysics

Newton's Metaphysics

Essays

ERIC SCHLIESSER

OXFORD
UNIVERSITY PRESS

OXFORD
UNIVERSITY PRESS

Oxford University Press is a department of the University of Oxford. It furthers
the University's objective of excellence in research, scholarship, and education
by publishing worldwide. Oxford is a registered trade mark of Oxford University
Press in the UK and certain other countries.

Published in the United States of America by Oxford University Press
198 Madison Avenue, New York, NY 10016, United States of America.

© Oxford University Press 2021

Library of Congress Control Number: 2021934271
ISBN 978-0-19-756769-2

DOI: 10.1093/oso/9780197567692.001.0001

1 3 5 7 9 8 6 4 2

Printed by Integrated Books International, United States of America

Dedicated with admiration and gratitude to Katherine Brading,
Andrew Janiak, Niccolo Guicciardini, and Chris Smeenk.

Contents

Contents

A Note on the Chapters

Chapters 1–7 were previously published in academic journals and scholarly volumes. They have been reproduced without significant alteration, although I corrected a number of typos. I have streamlined the citations (including to the material in this volume), the bibliography, and the section headings; I removed a modest number of duplications in the footnotes and removed some material unnecessary to the argument. In a few places I reworded the text to remove unnecessary polemical tenor (evidence of youthful ambition), corrected some citations, and added some cross references. I have added a small number of footnotes: these are marked with this symbol, "*". In no place did I change the argument.

I did change, however, the formal authorship of chapter 4. When it was originally published it was edited by Mary Domski for a special issue of *Southern Journal of Philosophy*. While editing my paper she also supplied me with the rational reconstruction of an argument by Clarke central to the chapter's overall argument. It was quite clear to me then that this essay was a joint effort between us, and I was disappointed she declined my offer for authorial credit at the time. I am thrilled she has finally agreed with my evaluation of her contribution to the argument and that I can count her among my official coauthors now.

Chapter 8 originates in a conference presentation on the General Scholium in Halifax organized by Steve Snobelen. The chapter was originally intended to appear in a volume edited by Ducheyne, Mandelbrote, and Snobelen, who helped edit it into the current shape. I am very grateful to their generosity.

Chapter 9 provides a hint of a much larger project I have embarked on. Some sections in it overlap with an article edited by Kenney Pearce and Takaharu Oda. However, much of the material in chapter 9 is written especially for the present volume with the assistance and encouragement of Professor David Haig of Harvard University.

The two postscripts, to chapters 2 and 7, in which I respond to criticisms known to me, were written for this volume.

Acknowledgments

I have been unusually lucky to study Newton with three great scholars. First, as an undergraduate, in the fall of 1991, with a great deal of ambivalence about the scientific nature of political science (after the fall of the Berlin Wall), I enrolled in George Smith's by now legendary year-long *Principia* course at Tufts University. Among the many highlights of that year included a free tutorial by George's daughter, Jean Smith, to check out the night sky through a telescope that approximated the magnification of Galileo's telescope; to have the privilege to watch George rediscover Newton's reasoning from one draft of "De Motu" to the next; the thrill to work with microfiche editions of seventeenth-century books in the Boston Library; and the joy of having my college roommates, Matt Murgida and Bryan Taturzyk (both engineering students then), help me figure out Huygens' pendulum experiments.

George has an admirable habit of bringing his undergraduate students into his research projects. Being a Dutch native, I was brought on to help with a translation of work by Huygens. Our collaboration since has now spanned many decades. He taught me how to research and write, and rewrite, many times over.

Second, I attended Howard Stein's year-long history of space-time graduate seminar in my first year as a PhD student at The University of Chicago. I was lucky enough that Erik Curiel, Matt Parker, and Matt Frank were there to help me through that course. Howard also taught Newton in his course on empiricism, where I also learned much from Rachel Zuckert, not in the least the art of when not to ask questions. Bill Wimsatt always found ways to connect everything with everything and continues to shape my approach to scientific metaphysics.

Years later, Jed Buchwald asked me to write a handbook article on Newton's *Principia* "in the spirit of George." I was smart enough to ask Chris Smeenk to co-author with me. Smeenk insisted I learn how to write in LaTeX, which must be destroyed, and to rewrite in LaTeX. While drafting and redrafting our handbook article, Smeenk gave me another detailed graduate level tutorial on each proposition of the *Principia*. Much to my amazement, Smeenk and I have embarked on several large-scale editorial projects together since.

Smith, Stein (who was one of my supervisors in graduate school), and Smeenk are each rigorous and exacting scholars not the least for themselves. They aspire to a level of clarity and precision whose brilliance and insight I admire. While I have learned a lot from them, nobody would confuse my work with their gems.

After I completed my PhD, in part, on the reception of Newton during the eighteenth century—with additional supervision by Dan Garber, Charles Larmore, and the late, and much missed, Ian Mueller—George Smith introduced me to Andrew Janiak so that the three of us could write an entry on Newton and his reception for the *Stanford Encyclopedia of Philosophy* (this ended up being three distinct entries). Andrew has been a great friend, coeditor, and an inspiration for the very possibility of agreeing to disagree amicably about Newton for close to two decades.

In 2007 I hosted a conference at the Museum Boerhaave in Leiden on Newton and/as Philosophy with the help of the museum staff, an NWO grant, some funds from KNAW, the Evert Beth foundation, and the help of my Leiden colleagues, Lies Klumper, and my students, not the least Martine Berenpas and Dagmar Hepp. Since I was splitting my time between Syracuse University and Leiden, I could not have organized this event without such great help. George, Dan, and Andrew together with Michael Friedman and Graciela de Pierris were among the invited keynote speakers. They all presented great papers.

At the Leiden conference, I was amazed to discover an incredible amount of fresh philosophical perspectives on Newton by scholars influenced by the giants, Ted McGuire, Howard Stein, Bill Harper, George Smith, Ernan McMullin, Nico Bertoloni Meli, and Niccolo Guicciardini. I learned that the nature of Newton's metaphysics was not a settled fact, but could generate a huge number of fascinating research questions. In particular, Zvi Biener, Chris Smeenk, Katherine Brading, Lisa Downing, and Ori Belkind taught me that a complex matter theory is inscribed in the *Principia*. Nick Huggett taught me that the distinctions of the scholium to the definitions are worth attending to. Mary Domski and Lynn Joy taught me that even after Stein's and Lisa Downing's work, the relationship between Newton and Locke was still underdetermined. At that conference Dana Jalobeanu, Mary Domski, Karin Verelst, Maarten Van Dyck, Steffen Ducheyne, Lynn Joy, Rob di Salle (and his son), Marco Panza, and Bill Harper all presented scholarship that has transformed the field. The audience included Deborah Brown, Calvin

Normore, Herman Phillipse, and Fred Muller, so the Q&As were often as instructive as the talks.

I was like a kid in a candy store.

The event galvanized me because I recognized I had been mistaken to assume that the true meaning of Newton had been fixed by Stein, McGuire, Harper, Cohen, Feingold, Guicciardini, Smith, and others. I realized I was wrong to assume that all that was to be done was to figure out how Newton was received and contested during the eighteenth and nineteenth centuries through Einstein.

At the time, a junior scholar at Syracuse University, I was surrounded by fantastic metaphysicians who took metaphysics to be an alive and worthy project (the late Andres Gallois and José Benardete, of course, but also Kris McDaniel, Mark Brown, and Mark Heller amongst others). This was transformative for me, and my few semesters there were the happiest of my intellectual life.

In addition, in the same period, at a number of workshops I encountered Michael Della Rocca and his students, Omri Boehm, Sam Newlands, and Yitzhak Melamed among others, as well as Alan Nelson and his students Noa Shein, Hylarie Kochiras, and Lex Newman among others. They inspired me to try to integrate the history and philosophy of science with the history of metaphysics in the early modern period. This was a seed already planted by Steven Horst at Wesleyan, but I had missed the significance of it. Later, at Ghent, Charles Wolfe insisted that our program was to incorporate metaphysics into HOPOS, which we did, in part, by hosting a series of workshops on funky causation. Since then a number of contemporary metaphysicians—not the least Laurie Paul and Andrew Bailey as well as David M. Levy and M. Ali Khan—have encouraged my interest in metaphysics.

The essays in this volume are the product of the satisfying intellectual whirlpool I found myself in. During that period I met and/or intensified my contact with Geoff Gorham, Ed Slowik, Yoram Hazony, Stephen Gaukroger, Tammy Nyden, Dennis Des Chene, Red Watson, Anne-Lise Rey, Siegfried Bodenmann, Sorana Corneanu, Catherine Wilson, Jeffrey McDonough, Allison Simmons, Tad Schmaltz, Ed Curley, Donald Rutherford, Sam Rickless, David Miller, Marcy Lascano, Don Ainslie, Lisa Shapiro, Marcy Lascano, Karolina Hubner, Kerzberg, Michaela Massimi, Andreas Hüttemann, Laura Snyder, Andrea Woody, Abe Stone, Doug Jesseph, Roger Ariew, Peter Anstey, Christia Mercer, Marius Stan, Sorana Corneanu, Alan Shapiro, John Henry, Sarah Hutton, Allison Peterman, Scott Mandelbrote, Marius Stan, Adwait Parker, Steve Snobelen, Dmitri Levitin, Hylarie Kochiras, and quite a few other incredible scholars.

In 2009 I received a generous appointment at Ghent University, funded by BOF. Maarten and Steffen became colleagues, and we developed a research group that attracted incredible talent, including Dan Schneider, Charles Wolfe, Jon Shaheen, Marij Van Strien, Laura Georgescu, Barbaby Hutchins, Jo Van Cauter, and Sylvia Pauw. I learned from each of them. Along the way I incurred many debts of gratitude toward flexible Deans and Department Chairs, especially the late Stewart Thau and Cathy Newton at Syracuse, Pauline Kleingeld at Leiden, and Freddy Mortier as well as Erik Weber at Ghent.

Since 2015 my professional orientation has shifted dramatically, I now work as a "political theorist" as a member of Challenges to Democratic Representation in the political science department, in the stimulating environment of the Amsterdam Institute for Social Science Research at the University of Amsterdam as well as a visiting scholar at the Smith Institute of Philosophy and Political Economy at Chapman University. The latter, in particular, provides me with near perfect circumstances to research and I am grateful to Bart Wilson and Vernon Smith for hosting me; to Bas van der Vossen, Keith Hankins, John Thrasher, and Brennan McDavid for letting me hang out with them; and Carol Campos for organizing it all.

In recent months much, including much nonsense, has been written about Newton's plague year. But when many other plans were postponed as the pandemic unfolded, I had time to reflect on my intellectual commitments. And while the pandemic reinforced the significance of quite a bit of my new scholarly interests in political philosophy, the history of feminism, and philosophy of economics, I found myself, simultaneously and quite unexpectedly, returning to themes of the chapters in the volume.

I am grateful to Peter Ohlin, my marvelous and supportive editor at Oxford University Press, and his three anonymous referees who provide generous and constructive reviews. The referees also suggested the addition of the postscripts in which I respond to my (many!) superb critics. In addition, Sylvia Pauw has generously compiled the indices when Covid felled me. She also helped correct numerous errors that had remained in the final manuscripts. Without her this volume would have been much delayed and more error ridden. I thank Dana Rentenaar for help in completing the name index; and Lisa Mullins and Aviram Rosochotsky who assisted in completing the subject index. Finally, I thank Michael Stein, who is the copyeditor of this book

I could not have had my career without a very supportive family: my mom, Marleen; my late dad, Micha; and my sister, Malka. I want to acknowledge

Wipko Terpstra, who encouraged me after the debacle at the 1995 Huygens conference in Leiden. Jody Joder took me under her wings and introduced me to manuscript scholarship and taught me how to ask librarians and archivists for much needed help. Special thank you to Sarit and Avi for their love and for giving me space to write.

In my scholarly life I have incurred many debts of gratitude. And I increasingly recognize Leon Montes' insight that in scholarship all our achievements are collaborative in nature. Many of my most important debts can be found in the notes to these chapters. Many of these chapters include excerpts from correspondents that have tried to set me straight about some challenging conceptual issues. I am honored that Zvi Biener and Mary Domski have agreed to let me republish chapters 4 and 6 in this volume. They, as much as my four dedicatees, have been true intellectual friends. I dedicate this volume to Katherine Brading, Chris Smeenk, Niccolo Guicciardini, and Andrew Janiak for making scholarship a true joy even in the darkest times.

London, November 3, 2020

Abbreviations

PLA	Principle of Local Action
PSR	Principle of Sufficient Reason
SP	Socratic Problem

Abbreviations Used for Texts

George Berkeley

PHK	*A Treatise Concerning the Principles of Human Knowledge*

Samuel Clarke

Demonstration	*A Demonstration on the Being and Attributes of God. More Particularly in Answer to Mr. Hobbs, Spinoza and Their Followers*

David Hume

EHU	*An Enquiry Concerning Human Understanding*
Treatise	*A Treatise of Human Nature*

Immanuel Kant

UNH	*Universal Natural History and Theory of Heavens*

John Locke

Essay	*An Essay Concerning Human Understanding*

Isaac Newton

De Gravitatione	"De Gravitatione et aequipondio fluidorum et solidorum . . ."
Opticks	*Opticks, Or, A Treatise of the Reflections, Refractions, Inflections & Colours of Light*
Principia	*Mathematical Principles of Natural Philosophy*
Treatise	*A Treatise of the System of the World*

Spinoza

Ethics	*Ethica, ordine geometrico demonstrata*

Newton's Metaphysics

Introduction

Not so long ago the scientific revolution was interpreted in terms of competing metaphysically loaded worldviews. For example, Koyré argued, to simplify, that Huygens and Leibniz rejected Newton's physics because they (correctly) thought it incompatible with the mechanical philosophy, which Newton, in turn, rejected as wedded to the wrong kind of hypotheses (see, e.g., Koyré, 1950: 262). While not strictly false this picture is misleading because the debate between Huygens and Newton was also centered on empirical arguments, which helped settled it (Schliesser and Smith, 1995; Maglo, 2003; Schliesser and Smith, forthcoming).[1] In fact, by building on I.B. Cohen's idea of a "Newtonian style," and by focusing on Newton's evidential arguments, George Smith (2014) and Bill Harper (2011) revolutionized Newton studies (see also Harper and Smith, 1995; for an overview, Smeenk and Schliesser, 2013).

One important casualty of the Smith-Harper reevaluation of Newton's methodology is the idea that Newton embraced hypothetico-deductive reasoning in his 'deduction' of the gravitational law. This is also the lesson of an influential paper by Howard Stein (1990). I applaud this focus on scientific method and the nature of evidence and evidential practices. Even so, one may well wonder why this matters to philosophy in a broader sense.[2]

Leaving aside those philosophers that presented themselves, or were taken to be, 'the Newton' of their field, the strength of Newton's evidential arguments is intrinsic to the significance of Newton to the history of philosophy in four ways. First, because in virtue of the previously unimaginable strength of these evidential practices, Newton and his heirs were capable of transforming many questions central to philosophy into circumscribed

---

[1] Those that objected to this somewhat Hegelian narrative argued, in the spirit of Cassirer (1953), that from this conflict a more modest and simultaneously more general science of law-like or system of quantitative relations was born during the eighteenth century. I discuss the afterlife of this still influential approach in the final section of the Postscript to chapter 2.

[2] Stein (1967, 1977) himself used his work on Newton to shape the philosophy of space-time in the second half of the twentieth century, along the way recasting the debate between substantivalism and relationism. Chapters 5 & 7, in particular, are responses to his work on Newtonian space-time.

Newton's Metaphysics. Eric Schliesser, Oxford University Press. © Oxford University Press 2021.
DOI: 10.1093/oso/9780197567692.003.0001

empirical questions or into significant, forward-looking empirical research projects (Brading, 2017; Brading and Stan, forthcoming).

Second, in so doing, Newton contributed to the already emerging split between philosophy and science. In particular, within science questions that increasingly came to be associated with 'philosophy' could be forestalled. Newton's treatment of the consensus on the collision rules articulated by Huygens, Wallis, and Wren is an important example of this (Schliesser, 2011a; see also chapter 6, with Biener, for more details.).

In addition, Newton helped make prominent a style of argument such that the authority of science can be used to settle debate(s) *within* metaphysics and philosophy more generally. I call this move "Newton's Challenge" to philosophy (see Schliesser, 2011a, including a number of distinct versions of Newton's Challenge, and Schliesser 2012a). I became aware of the strategy because Berkeley diagnoses it *in order* to combat it (Schliesser, 2005a). In fact, such arguments are ubiquitous in contemporary philosophy (often presented as 'indispensability arguments,' or arguments from 'physicalism,' and appeals to 'naturalism'). I offer examples of the presence of "Newton's Challenge" in Newton's works and reception in chapters 1, 4 (with Domski), and 8 below.

Finally, Newton helped stimulate a certain style of argument in philosophy that relied on a progressive conception of science. I do not mean to suggest Newton or Newtonians invented the idea of progress in science. When Halley and Du Châtelet interpret Newton as evidence of progress, they do not just echo Baconian tropes, but even quote Seneca's *Natural Questions* VII:25–31 (Halley, 1705: 2–8).[3] In chapter 9, I explore a set of arguments that made a kind of providential deism and scientific progress seem co-constitutive and Newton's role in this.

For some the tenor of the previous paragraphs will be too anachronistic. While I am a friend of methodological anachronism, the studies that follow draw on practices common among contextual historians of philosophy while not eschewing more contemporary distinctions and some rational reconstruction. I close with a brief defense of this hybrid approach.

In this volume I largely assume the reality and significance of the four claims to significance of Newton's science discussed above. They inform the manner by which I approach Newton's writings and reception from different angles. In

---

[3] Halley seems to have inspired the opening lines of Émilie du Châtelet's chapter 7 (on comets) of her (1749) *Commentary on Newton's* Principia.

so doing I try to be sensitive to various historically salient intellectual contexts in order to create a dense picture and analysis of Newton's metaphysics.

For the present volume assumes, and simultaneously aims to illustrate, that Newton's evidential achievements were made possible, in part, by quite a large number of conceptual innovations and tinkering, some of which was the result of Newton's interests in topics that are now not thought of as especially scientific, not the least of which is theology.[4] The papers collected here track and map some of Newton's evolving metaphysical and conceptual views as they enter into some of his physics or are prompted by his physics. To be sure, my essays are not comprehensive and they are very much themselves the product of and situated in engagement with quite a number of scholarly debates since 2007.

There is, alas, no overarching theme to these essays except that judging by the standard narratives we tell in the history of philosophy (empiricism vs. rationalism; speculative vs. empirical; dualism vs. monism, etc.), Newton's positions are idiosyncratic and conceptually quite fine-grained. If that is right then the significance of the present study in Newton's metaphysics is, in part, to challenge us to reconsider these historical narratives, especially on the character of empiricism and Newton's role in it (Wolfe, 2010a; Biener & Schliesser, 2014; Schliesser, 2018c).

While there is no overarching theme, there is a recurring move: I couple Newton's metaphysics to the polemical reception of Spinoza's metaphysics in Newton's circle. And, I hope that, thereby, I indicate how a reopening of standard narratives might be possible. With the exception of some hints in the final chapter, I leave the larger project of reconceiving our historiographic narratives aside.

As noted above, the work inspired by Smith and Harper managed to dislodge a metaphysical interpretation of Newton. At the end of Postscript to chapter 2, while responding to excellent criticism by Adwait Parker, I offer a diagnosis of a persistent antimetaphysical strain inspired by Kantian and late logical empiricist sensibilities *within* more recent philosophical readings of Newton.[5]

It does not follow all antimetaphysical readings of Newton have such philosophical roots. Dmitri Levitin's polemical (2016) essay very much reads as

---

[4] Once, when the eminent historian of science, Motti Feingold, heard an early version of chapter 8, he queried how many eighteenth-century scholars would have been at ease with the intricate metaphysics detailed therein. I am unsure what I answered then, but, in light of my further study, I now believe that the audience for it was larger and more sophisticated then than it was for much of the mathematical-physical contents of the *Principia*.

[5] Obviously that diagnoses does no justice to Ted McGuire's contributions to Newton scholarship. McGuire's work anticipates most of my own efforts.

a professional historian scolding philosophers for their amateurism. Here I focus exclusively on his claim that Newton "had no time for metaphysics as a discipline" (Levitin, 2016: 68). Levitin offers this claim as part of a much larger argument, rooted in a study of Newton's manuscripts and with repeated appeals to "contextual" evidence that Newton affirmed the "rejection of metaphysics." (Levitin, 2016: 70) Levitin claims that Newton was "far from engaging in an elaborate metaphysical enterprise" (Levitin, 2016: 73). Since this volume has "Newton's metaphysics" in its title, it may be thought useful to respond to Levitin's arguments. In so doing, I make more precise what I mean by suggesting that my essays are about 'Newton's metaphysics.'

First, I agree with Levitin that at some point Newton rejected metaphysics as a discipline, especially as it was still taught in universities. But he offers no evidence that Newton rejects all metaphysics. For example, in De Gravitatione, Newton defends his own treatment of body, which involves an elaborate thought experiment, because "it clearly involves the principal truths of metaphysics and thoroughly confirms and explains them" (Newton, 2004: 31; I discuss this passage in chapter 5). Second, I also agree with his quite salutary criticism of those that have tried to link Newton's use of the analytic/synthetic distinction to the Aristotelian *regressus* tradition (Levitin, 2016: 56–67). After reading Levitin, I am relieved I was never tempted to do so myself. But we ought to learn from each other's mistakes, and there is no shame in trying out new hypotheses.

Third, one strain of Levitin's argument is directed at the claim that for Newton forces are causes, which Levitin calls the "causal reading of Newton on force." If this causal reading can be denied or refuted then, it seems, for Levitin Newton is *ipse facto* antimetaphysical. This is one of the points singled out as a "larger philosophical significance" by Levitin (2016: 65). But in his arguments against the causal reading of Newton on force, Levitin *conflates* treating (i) forces as causes, including the force of gravity as a real cause, and (ii) the causes of forces. There is a further conflation in treating Newton's professed (iii) agnosticism about the cause of the gravitational force as (iv) agnosticism about the causes of all forces (cf. Levitin, 2016: 66). Chapters 1–2 explain why these distinctions matter to understanding Newton.[6]

---

[6] Since Levitin cites the original, published version of chapter 1 as one of the culprits of a bad scholarly tendency it is a bit peculiar he misses this point. He also claims that "Newton only really emphasized the language of 'causes' in . . . Query 31," in the context of certain theological commitments (Levitin, 2016: 66). But this oddly overlooks the wording of Newton's first two Rules of Reasoning (which had been the first two "hypotheses" in the first edition), which are worded in terms of the establishment of natural *causes*. (Newton, 1999: 794–795).

It is only if you think—say because you (perhaps tacitly) embrace a Bradley-esque version of the PSR—that is, in order to treat X as a cause of Y, you must know Z (the cause of X), and so forth—that (i) and (ii) start to collapse into each other. But in the body of the *Principia* there is no sign that Newton conflates (i) and (ii) or embraces this kind of version of the PSR.[7] However it is worth noting here that Newton needs (i) in order to settle or at least provide a new and compelling answer to the Copernican controversy. In the preface to the *Principia*, Newton writes,

> [R]ational mechanics will be the science, expressed in exact propositions and demonstrations, of the motions that result from any forces whatever and of the forces that are required for any motions whatever . . . For the basic problem of philosophy seems to be to discover the forces of nature from the phenomena of motions and then to demonstrate the other phenomena from these forces. (Newton, 1999: 382)

And while it is true that here he does not explicitly treat forces as the causes of the phenomena of motions (that's argued in the book), it does not follow his science is acausal. This is why, in the Scholium to the Definitions, Newton promises in his own voice,

> But in what follows, a fuller explanation will be given of how to determine true motions from their *causes*, effects, and apparent differences, and, conversely, of how to determine from motions, whether true or apparent, *true causes* and effects. For to this was the purpose for which I composed the following treatise (Newton 1999: 415; emphases added).

By contrast, Levitin himself seems to embrace what he calls a "phenomenalist" interpretation of Newton (65–67). In so far as properly categorizing Newton's causal talk as, for example, either phenomenalist, structural realist, instrumentalist, or naïve realist, and drawing the kind of distinctions I have in the previous paragraphs counts as 'metaphysics,' that's metaphysics enough for my present purposes. I allow that what I call 'metaphysics' is not what Newton has in mind when he uses 'metaphysics' according to Levitin.

In fact, lurking in Levitin's argument is a kind of fetishism of actor's categories. Now attention to contemporaneous actor's categories is important

---

[7] In chapters 8–9, I explore the presence of various causal principles in the General Scholium.

in historical studies. But to adopt them as one's own is, as Daniel Schneider first noted to me, often no better than to take sides in the polemics of the past without fully realizing it (Schliesser, 2017: 19). Levitin's strategy is precisely such an instance. Newton is critical of certain kind of metaphysicians and metaphysical practices, but from that one cannot infer he is not doing metaphysics (by 'our' lights) at all (even if he were to say so) or that his is an antimetaphysics.

This is not to deny I agree with some of Levitin's exegetical claims.[8] For example, according to Levitin Newton professes agnosticism about the nature of the interaction between God and the world (Levitin, 2016: 70). That is what Newton explicitly says in the "General Scholium" (Newton, 1999: 942) in language that sounds very Lockean.[9] In my new postscript to chapter 2, I use it as an argument against those that attribute to Newton a Principle of Local Action (PLA). But it does not follow from this that Newton is not doing metaphysics. For, Newton himself asserts that there is a relationship of necessity between God and space as well as God and time (Newton, 1999: 942). To appeal to necessity may not be a species of metaphysics. But to rely on arguments from necessity to settle claims about the nature of, say, space and time, and even the reality of final causes, I consider a potential contribution to metaphysics. I confront such passages throughout the volume, and in chapter 8 I try to explain what Newton meant by this.[10]

There is a broader methodological point to be made here. And this is the use of 'context,' as Levitin does, as a purportedly neutral and decisive arbiter of significance. I am open to Levitin's invitation to "delve into the scholastic textbooks" (73), which he treats as the privileged context to treat many of Newton's more speculative remarks. That this is useful and appropriate to do so, if only for debunking purposes, he shows with great expertise. But he closes his polemic with the claim that "Ralph Bathurst or Nathaniel

---

[8] I also agree with his caution against using De Gravitatione "as some kind of metaphysical" key "to Newton's mature natural philosophy" (Levitin, 2016: 15). I say as much in chapters 2 and 5, also in the versions published several years before Levitin's piece.

[9] Levitin's treatment of Locke is no less unsubtle. He discusses Locke as another antimetaphysician (Levitin, 2016: 75). Here is an instance of the more general problem that if one adopts actor's categories one risks merely reproducing their polemics.

[10] Levitin (2016: 73) dismisses my brief treatment of Newtonian emanation and Spinoza's monism in the predecessor to chapter 1 without bothering to investigate the reference to my much more explicit and elaborate argument now in chapter 5 (but knowable to Levitin). For Levitin an "emanative cause" was, alongside its counterpart the "active cause," one of the subsets of efficient causes. It was a cause that functioned simply though the existence of an object, from the action of which was generated an "emanative effect." (Levitin, 2016: 73) I leave it to the reader to compare this with the four kinds of emanation I discuss in chapter 5 (three of which are in contexts familiar to Newton).

Highmore" are the most important figures (Levitin even uses "heroes") behind the "institutionalized intellectual culture" that Newton inhabited. (77)
The implication is supposed to be that this institutional culture is really the
most fundamental context for understanding Newton's "mature" philosophy.
(Levitin uses variants on "mature" throughout his paper.)

Now, in the chapters that follow, I explore how the *Principia* got embroiled
in the reception of Spinoza and a much broader history of Spinozism, including views most contemporary scholars would not consider Spinozist at
all (such as Epicureanism). This Spinozist system of metaphysics—which
I define in chapter 8—is not Newton's. But as I show, to articulate differences
with it requires not just polemical tropes but some subtle distinctions. And
to mark that fact, and use it fruitfully does not require us to treat it (Spinoza/
Spinozism) as the only or most important context for thinking about
Newton's (to use Levitin's phrase) "mature philosophy."

That Newton's mature philosophy got embroiled in the reception of
Spinozism is clear (since Colie, 1959 and 1963). But it does not follow from
that that Newton was not also responding to, for example, Descartes and
Huygens or very engaged in various priority disputes with (say) Leibniz and
Hooke. These names also appear not infrequently in what follows.

I close by distinguishing three uses of context in order to illuminate my
own. The first approach is this: one should stipulate that a certain *corpus* is
the relevant material one is going to use to establish the historical *meaning*
of a text. (See Laerke [2013] and Smith [2013] for different accounts.) While
judgment is involved here, the boundaries of this corpus is really a matter of
stipulation or legislation.

Second, one uses a certain corpus to rule out causal claims of influence or
significance (by showing that a certain transmission is, say, impossible) As
noted above, this is what Levitin (2016) does in his criticism of Ducheyne.

Now crucially, in these two approaches, one can use only one context
at a time to help fix and make possible other claims. And this is why these
approaches appeal to those who wish to be professional scholars. Once context has been fixed one can make further decisions about which categories
and distinctions to rely on and what disciplinary standards one will use to
explain what one has done. So such contextualism does not conceptually rule
out the use of anachronistic categories or modern distinctions, but the deployment of these goes against the spirit of these two approaches.

Third, one uses real interlocutors of Newton and those that read or edited
Newton to explicate or construct an abstract entity, 'Newton's metaphysics.'

This construct can be projected onto a scholarly community of a time, which read particular and overlapping texts, or onto Newton's own mind (e.g., in a particular period; Levitin's use of 'mature,' engages in such a practice without being explicit about it). Or it can be a construct that it is in some sense unmoored from the lived experience of any historical agent, but rooted in some texts and their evolving reception or in a disciplinary practice today.

In the introduction to chapter 1, I explain why I hesitate about attributing to Newton a "comprehensive" view of, what I call there, his "speculative metaphysics." The barriers I discuss to such a project have not yet been removed. So, instead in each chapter I situate partial features of this comprehensive view in light of particular texts and particular issues. What follows from this is insight into and material for this abstract entity, 'Newton's metaphysics,' even if it is to be doubted that Newton himself would have fully embraced its contents (even when translated back into actor's categories) as his own 'speculative metaphysics' at any given time.

The downside of this third approach is that it looks undisciplined and unprofessional because it risks being eclectic and in principle open-ended. The upside, if there is one, is that one can approach a fairly limited number of Newtonian texts (as I do) from different mutually supporting and refining angles. In so doing I draw on scholarship about the variety of contexts (in the first two senses) that Newton inhabited and in which he was received in as well as more reconstructive scholarship by philosophers about the underlying arguments and concepts. By using these different angles one may build such a construction. Depending on one's claims such a construct can be very robust or fragile.

One may wonder what the benefits of this third kind of use of context is since it seems to fit uneasily with the canons of proper historiography as historians understand it. As I noted above, it is likely that what I construct as 'Newton's metaphysics' made possible evidential arguments that helped completely reshape how to understand science, philosophy, and their interactions. If that is so, then understanding 'Newton's metaphysics' is a way to understand much larger forces that shape our philosophical and scientific scene (Schliesser, 2011a). Such a construct, thus, enters into, if not a kind of historical explanation, then at least a possible philosophical and historical self-understanding.

1

Without God

Gravity as a Relational Quality of Matter in Newton's *Treatise*

1.1 Introduction

In this chapter I interpret Newton's speculative treatment of gravity as a relational, accidental quality of matter that arises through what Newton calls "the shared action" of two bodies.[1] In doing so, I expand and extend on Howard Stein's views (Stein, 2002). However, in developing the details of my interpretation I end up disagreeing with Stein's claim that for Newton a single body can generate a gravity/force field.

I argue that when Newton drafted the first edition of the *Principia* in the mid-1680s, he thought that (at least a part of) the cause of gravity is the disposition inherent in any individual body, but that the force of gravity is the actualization of that disposition; a necessary condition for the actualization of the disposition is the actual obtaining of a relation between two bodies having the disposition. The cause of gravity is not essential to matter because God could have created matter without that disposition. Nevertheless, at least a part of the cause of gravity inheres in individual bodies and were there one body in the universe it would inhere in that body. On the other hand, the force of gravity is neither essential to matter nor inherent in matter, because (to repeat) it is the actualization of a shared disposition. A lone part-less particle would, thus, not generate a gravity field. Seeing this allows us to helpfully distinguish among (i) accepting gravity as causally real; (ii) positing the cause(s) (e.g., the qualities of matter) of the properties of gravity; (iii) making claims about the mechanism or medium by which gravity is transmitted.

---

[1] This chapter first appeared as Eric Schliesser "Without God: Gravity as a relational quality of matter in Newton's treatise." *Vanishing matter and the laws of motion: Descartes and beyond.* Edited by Dana Jalobeanu and Peter R. Anstey. London: Routledge, 2011, 80–102.

Newton's Metaphysics. Eric Schliesser, Oxford University Press. © Oxford University Press 2021.
DOI: 10.1093/oso/9780197567692.003.0002

This clarifies what Newton could have meant when he insisted that gravity is a real force.

The view I attribute to Newton is the view that he held when writing the first edition of the *Principia* in the mid-1680s. My evidence for this is the first draft of Book III of the *Principia*, the *Treatise*.[2] I know of no reason to deny its having been written during the mid-1680s, as Newton was moving from the successive drafts deposited with the Royal Society, known as "De Motu," and the publication of the first edition of the *Principia*. There are two reasons to take it very seriously.

First, it gives us insight into Newton's thinking while he was working out the details of his system. In the published introduction to Book III of the *Principia*, Newton writes that he suppressed his *Treatise* in order to "avoid lengthy disputes" with others' "preconceptions" (Newton, 1999: 793). The prejudices he has in mind are not merely "vulgar," but rather philosophical, that is, those in circulation among the learned (especially Hooke and Huygens). For the *Treatise* is more speculatively metaphysical than the published version of Book III. Nevertheless, because of the timing of its writing and the fact that Newton clearly did not disown it, it is rather surprising that it has been largely neglected in Newton scholarship. Many additions to the *Principia* can be explained as Newton's response to new empirical evidence or corrections to obvious problems. But other changes should be viewed as Newton's evolving response to the concerns expressed by some of his religiously motivated interlocutors and to his evolving arguments with Leibniz and his followers (see chapter 3).

It is a bit strange that the "General Scholium" and, say, the Letter to Bentley have received a lot more attention than the *Treatise* among those who claim to be interested in Newton's metaphysics. Regardless of how one views the relationship between Newton's technical claims and the material he segregated into queries and the "General Scholium," which was added to the second edition almost certainly to stop readers from becoming alarmed over his theological outlook (see chapters 3 and 8), this material deserves more attention. In fact, in uncovering Newton's substantive commitments, we should not focus primarily on the views he expressed for reasons associated with concerns over his religious, moral, and political views.[3] With Newton's

[2] See Newton 1728a and 1728b.
[3] I call this the "Socratic problem" (SP), which is about how social (religious, political, moral) forces can threaten the independence and authority of philosophy. I distinguish among at least five distinct theses: SP1: a philosopher claims that "practical" philosophy takes precedence (in some way) over "theoretical" philosophy; SP2: a philosopher explains how statements of traditional religious/political

WITHOUT GOD 11

general secrecy and flirtations with Arianism, there is no denying his aware-
ness of such constraints; in his unpublished De Gravitatione he calls attention to how Descartes "feared" (Newton, 2004: 25) positions that might be
thought to offer "a path to atheism" (Newton, 2004: 31).

Second, because the *Treatise* was published so shortly after Newton's death,
it is invaluable in helping us understand the reception of Newton in the eighteenth century. While that reception was shaped by the *Opticks*, the Leibniz-
Clarke correspondence, and Newtonian expositors, the *Treatise*'s influence is
much neglected. This is unfortunate. For example, among Newton scholars
now it is unfashionable to read Newton as an instrumentalist or a positivist,
but many probably still read the eighteenth-century British Empiricists as
reading Newton in this (mistaken) way and, thus, as uninformative on
matters of Newton interpretation (Smith, 2001; cf. Schliesser 2011a).

Moreover, scholars are thus also blind to the fact that British Empiricists are
often ambivalent about the new role of Newton's "authority." (See Schliesser,
2009 and 2011a) While here I neither will argue for the importance of the
Treatise to eighteenth-century readers nor hope to rehabilitate eighteenth-
century British thought as a guide to interpreting Newton, it is useful to realize that we do not need to focus exclusively on Huygens, Leibniz, and Kant
in offering philosophical insight into Newton's thought. So I view my project
here as making possible the recovery of a well informed eighteenth-century
view of Newton—one that has no access to De Gravitatione or Newton's alchemical writings, but that can use the *Treatise* to evaluate Newton's original
commitments.

Once we realize that many Empiricists understood Newton as the kind of
realist they were not (EHU 7.1.25, note 16, quoted in chapter 3; Demeter &
Schliesser, 2020), we can also see that part of the great eighteenth-century
debate in philosophy is not between empiricism and rationalism, but between those philosophers who believe in creating complete systems from
the method of inspecting ideas (for instance Descartes, Spinoza, Locke,
and Hume) and those who believe in a more piecemeal mathematical-
experimental approach (a tradition initiated by Galileo and Huygens). The
Newton of De Gravitatione that is now the focus of so much scholarly attention, who unabashedly offers an analysis of the "exceptionally clear idea of

texts can be understood as expressions of his/her philosophical doctrines; SP3: a philosopher appeals
to nonphilosophical (political, religious, social) authorities as justification of doctrine(s); SP4: a philosopher is forced by outside authorities to adjust views; SP5: a philosopher is held accountable for the
impact of teachings on students. For detailed discussion, see Schliesser (2012).

extension" (Newton, 2004: 22, 27) belongs to the first tradition; the Newton of the *Principia* and the *Treatise* is the champion of the second tradition. (Of course, Newton mentions our "ideas of [God's] attributes," in the "General Scholium" (Newton 1999: 942) but his argument is not based on an inspection of those ideas.) If I am right about this, we should date De Gravitatione before the *Treatise*, a claim that is plausible on other grounds too (see chapters 5–6; Smith, 2020).

While the view I attribute to Newton should be a privileged one for present purposes, I provide five methodological/historiographic reasons to remain agnostic about how this view should be fully squared with other, potentially competing proposals that Newton entertained on such matters (e.g., the role and nature of God or a very subtle ether in supplying the mechanism for attraction), as well as about whether or not Newton offered a stable and consistent position throughout his life. First, Newton's manuscripts reveal a man who was willing to entertain and try out many metaphysical ideas, although when such views found their way into print without solid empirical evidence they tended to be segregated to "scholia," "queries," "letters," and the work of his various followers.

Second, without full consideration of his alchemical, political, and religious views at any moment, I despair of discerning a comprehensive view of Newton's speculative metaphysics even at a single point in time. Someday this may be possible. Third, because of the dangerous political and complicated religious context we cannot always take Newton at face value in speculative matters, especially because in some instances Newton hints at his knowledge of an esoteric/exoteric distinction (Snobelen, 1997 and 1999). Fourth, despite a few notable and isolated exceptions, we do not have a comprehensive view of how Newton's views evolved across and among many issues. It is dangerous to treat any of Newton's speculative views in isolation, but it is not always clear how to fit these to a larger evolving view. Finally, Newton is an extraordinarily terse writer, and sometimes lesser mortals could have used further clarification.

In fact, these considerations have hitherto inclined me to restrict my scholarship to the reception of Newton's views. So despite the firm language in what follows, my views are quite provisional. Below, I proceed as follows. First I present Newton's relational account of gravity found in his posthumously published *Treatise*. I often contrast my reading with Andrew Janiak (2007), especially with his influential reading of Newton's famous letter to Bentley as ruling out action at a distance. In the final section I disagree with

Howard Stein's account of how the gravity field is generated by a particle; I argue that a lone particle is not enough to generate a force field. In offering arguments against Stein's view, I analyze the third rule of reasoning and the third law of motion with its corollaries.

1.2 Gravity as a Relational Quality of Matter

In an important and influential paper, Janiak ably demonstrates that Newton thought that a force of gravity "really exists" (Janiak, 2007: 130, 141) because it is one of "the causes which distinguish true motions from relative motions" by way of the "forces impressed upon bodies" (Newton, 1999: 412, quoted in Janiak, 2007: 134, n.17; see also Janiak, 2008: 130–131 and 143–146). Following in the footsteps of Leibniz, Janiak correctly rejects an instrumentalist reading of Newton's views on gravity (Janiak, 2007: 138ff).[4]

Because many philosophers are introduced to Newton's views through Clarke's correspondence with Leibniz, and have found Clarke's arguments wanting on a range of issues, they have been prevented from treating Newton as a serious philosopher. There is, thus, no doubt that Janiak's paper and subsequent 2008 book should encourage a reevaluation of Newton's substantive philosophy on a host of issues. By relying nearly exclusively on Newton's publications and letters Newton wrote to contemporaries, Janiak shows that many of Newton's philosophical views can be gleaned from his published writings if they are read with attention. Newton is often a terse writer, but not obscure. While Newton's unpublished manuscripts on theology and alchemy are fascinating and often provide very helpful context, we need not defer to these to grasp some of the most important strands of Newton's views. This is especially useful if we wish to explore Newton's impact on philosophy, because nearly all these unpublished manuscripts were unavailable to many

---

[4] Janiak (Newton, 2004: 134) quotes Clarke's fifth letter to Leibniz as follows: "by that term [attraction] we do not mean to express the cause of bodies tending toward each other, but barely the effect, or the phenomenon itself, and the laws or proportions of that tendency discovered by experience; whatever be or be not the cause of it." In note 18, Janiak refers to sections 110–116 of Clarke's fifth letter in Leibniz, G VII 437–438. The instrumentalism/realism vocabulary does not do justice to the fact that prior to the *Principia*, Newton seems to have allowed that existence is not univocal; as he writes in De Gravitatione things can have their "own manner of existing which is proper to them" (Newton, 2004: 21). Strictly speaking the quote is only about extension, but context makes clear the doctrine is also applicable to substances and accidents. It is not an isolated occurrence in De Gravitatione; Newton also claims: "[w]hatever has more reality in one space than in another space belongs to body rather than to space" (Newton, 2004: 27).

other influential and insightful readers of Newton in much of the eighteenth and nineteenth centuries (and thereafter).

One of the most important of Janiak's insights is his claim against Leibniz that "if by 'mechanism' one means a natural phenomenon that acts only on the surfaces of other bodies, then Newton rejects the claim that gravity must have some underlying mechanism on the grounds that gravity acts not on the surfaces of bodies, but rather on all the parts of a body" (Newton, 1999: 943; Janiak, 2007: 129, n. 6;). Because "mass is one of the salient variables in the causal chain involving the previously disparate phenomena taken by Newton to be caused by gravity . . . [and therefore] gravity is not a mechanical cause" (Janiak, 2007: 142, 145), Newton rejects a key demand of the mechanical philosophy (Janiak, 2007: 146–147; it is fruitful to treat Newton as redefining the mechanical philosophy—see Kochiras, 2013 and Schliesser, 2009).

Nevertheless, in following Leibniz's lead in reading Newton, Janiak ends up misdiagnosing some very important elements of Newton's metaphysics. In particular, Janiak ignores the *Treatise*. This is important because, without arguing for it, Janiak tacitly attributes to Newton (and the *Principia*) an ahistorical stable position. Focus on the *Treatise* prevents us from ruling out in advance a developmental approach to Newton (see chapter 3 and for a developmental view see, e.g., Cohen's "A Guide to Newton's *Principia*" in Newton [1999] and McGuire [1970a]).* In particular, we should be open to allowing Newton to have developed his views between the first and third editions of the *Principia* (and the intervening editions of the *Opticks*).

In my interpretation, Newton knows that it is a kind of speculative metaphysics or hypothesis that he deplores with increasing vehemence in others as he anticipates and gets embroiled in debates with the mechanical philosophers and later in vituperative, politicized exchanges with Leibniz and his followers. As Newton becomes ever more insistent on the "empirical" and "experimental" nature of his method (Shapiro, 2004), he thus deprives himself of the chance to develop and articulate fully the speculative, metaphysical view that guided the first edition of the *Principia*. Moreover, Newton had a certain amount of self-command in refraining from publicly pursuing certain questions (which we know fascinated him). It follows that he came to devalue the kind of metaphysical interpretation I am about to engage in.

Elsewhere I have called attention to what I call "Newton's Challenge," in which the independent authority of philosophy is challenged by empirical

---

* I thank Karen Verelst for being the first to press this point to me.

science. Newton's intended and unintended role in generating a world in which science and philosophy became competitors is a complex matter, and I cannot do justice to it here (Schliesser, 2011). One symptom of this separation of philosophy and science—as Colin MacLaurin, the leading Newtonian of the first half of the eighteenth century, recognized (MacLaurin 1748, 77–79, 96)—is that against systematizing philosophers, Newton and his followers were willing to pursue more narrow research questions. Newton's achievement is to create a mathematical, theoretical structure that was highly promising and efficient as a research engine for ever more informative measurement; this achievement was won at the expense of settling certain metaphysical matters.

In the *Treatise* Newton offers a distinction between a mathematical and natural point of view:[5] "We may consider one body as attracting, another as attracted: But this distinction is more mathematical than natural" (Newton, 1969: 37). Here, as in the "technical" contrast between the "mathematical" and "physical" of the *Principia* (Janiak, 2007: 131ff; see the postscript to chapter 2 for further discussion), Newton does not deny the reality of the natural point of view. For Newton goes on to write, "the attraction is really common of either [body] to [an]other" (Newton, 1969: 37).

Anticipating Toland, Berkeley, and Hume (Schliesser 2007, 2020), Newton is alerting his readers to the fact that one cannot simply infer ontology from one's mathematical expression. In the *Treatise*'s next paragraph, Newton explains the "natural" perspective more fully, specifically, in terms of the attraction between Jupiter and the Sun. I quote two passages:

> There is a double cause of action, to wit, the disposition of each body. The action is likewise twofold in so far as the action is considered as upon two bodies; But as betwixt two bodies it is but one sole single one. (Newton, 1969: 38)

and

> We are to conceive a single action to be exerted betwixt two Planets, arising from the conspiring nature of both. (Newton, 1969: 39)[6]

---

[5] My discussion of this material is indebted to Howard Stein's translation and interpretation, shared with me in private correspondence. See also Stein, 2002: 287–289.

[6] The Latin is: "*Causa actionis gemina est, nimirum dispositio utriusque corporis; actio item gemina quatenus in bina corpora: at quatenus inter bina corpora simplex est & unica*" (Newton, 1728: 25),

In order to get the conception of the nature of an interaction at work in Newton we must note the difference between (1) the "cause of the action," which is "the disposition of each body" and (2) the "action" (or effect) itself. The action is (i) twofold as it is upon two bodies, and (ii) single as between two bodies. A way to capture this is to say that a body has two dispositions: a "passive" disposition to respond to impressed forces is codified in the second law of motion, whereas an "active" disposition to produce gravitational force is treated as a distinct interaction codified in the third law of motion. (Stein, 2002: 289, discussing the end of Query 31 of the *Opticks* (Newton, 1952 [1730], 397). Of course, this Query was written much later, and it is not impossible that Newton is extending or developing an original position rather than just articulating it. See also McGuire, 1970b.)

Thus we see that the "cause" of the action is "the conspiring nature of both" bodies. For the "conspiring" to occur, the bodies must share a "nature" (Newton, 1969: 39). To sum up: the cause consists in the "nature" or "disposition" of two bodies (or a twofold cause, because involving two bodies), but it is one interaction or "nature."[7] What are caused are one interaction and two "actions upon bodies"; there are two impressed forces. As Howard Stein explains (while using somewhat anachronistic language), "exactly those bodies that are susceptible to the action of a given interaction-field are also the sources of the field" (Stein, 2002: 288).

Newton is offering his readers a radical new idea, one that he feared would encounter a lot of prejudice: a speculative hypothesis about the nature of matter. First, Newton emphasizes a single "action" between two bodies. (He uses the repeated *actio* in Latin.) This is, therefore, a very clear description of action at a distance; applying the third law of motion is not merely a mathematical statement, but action at a distance really takes place in nature.[8] In the *Treatise*, Newton explains that this is due to a shared quality of two bits of matter. Here Newton offers a hypothesis about the physical cause of the sort he came to reject firmly in the "General Scholium." Somewhat paradoxically, according to Newton the body with a "passive" disposition to be attracted is

---

and: "*Ad hunc modum concipe simplicem exerceri enter binos Planetas ab utriusque conspirante natura oriundam operationem*" (Newton, 1728: 26).

[7] For my argument below it is useful that Newton thinks of the interaction itself as a "nature." For context on Newton's "Platonic" distinction between "being a nature" and "having a nature," see McGuire, 2007.

[8] Janiak would agree with me that this reflects "Newton's contention in the *Principia*'s 'General Scholium' that *gravitas revera existat.*" But as George Smith first pointed out to me, Janiak's translation of that phrase is a bit misleading because the whole sentence is more plausibly read as a programmatic than an indicative statement.

part of the cause of the gravitational force. Instead of untangling this knot, in the *Principia* Newton focuses on defining measures of the gravitational force. In the *Treatise* Newton is silent, however, on whether or not this attraction is mediated through a medium.

So we should distinguish among (a) the force of gravity as a real cause (which is calculated as the product of the masses over the distance squared); (b) the cause of gravity (which is at least, in part, the masses to be found in each body); (c) "the reason for these [particular] properties of gravity" ("General Scholium," *Principia*, quoted in his own translation by Janiak, 2007: 129); and (d) the medium, if any, through which it is transmitted.[9]

So, the view presented here by Newton takes a complete stance on "a" and a partial stance on "b," but is silent on "c." In the *Treatise*, Newton is entirely silent on "d," the invisible medium, to explain in what way momentum could be exchanged between two bodies. Given that he uses the language of "action" and is completely silent on the possibility of a medium of transmission, the natural reading of this passage is (i) Newton's endorsement of action at a distance with (ii) the start of an explanation of the cause of gravity in terms of some of the qualities of matter. But instead of being distracted by the incomplete nature of explanation, Newton oriented mathematical natural philosophy toward a sophisticated program of piecemeal problem solving through measurement.

Janiak (2007: 137) quotes Leibniz's last letter to Clarke: "For it is a strange fiction . . . to make all matter gravitate, and that toward all other matter, as if all bodies equally attract all other bodies according to their masses and distances, and this by an attraction properly so called . . ., which is not derived from an occult impulse of bodies, whereas the gravity of sensible bodies toward the center of the earth ought to be produced by the motion of some fluid." I agree with Leibniz (*contra* Janiak, 2007: 141) that Newton can be read as endorsing the "strange fiction," and I reject Janiak's apparent willingness to accept Leibniz's insistence (against the fiction) that a motion of some fluid must be the "cause" of gravity.

Thus while Janiak criticizes Leibniz's argument against Newton, throughout his article, Janiak conflates (i) treating of the mechanism or medium by which gravity is transmitted with (ii) treating of the cause of the gravitational qualities of matter (Janiak, 2007: 136); while outside the

---

[9] In the body of the text, I distinguish between talking of "qualities" of matter and "properties" of gravity. I believe Newton and Locke use "quality" to pick out a property that can be causally efficacious on minds, a subset of "properties" more generally. I thank Lex Newman for sharing his unpublished ms.

Principia, Newton can sometimes be read this way, it appears that Leibniz is the source of Janiak's conflation.

Before I note some obvious objections to my *Treatise*-inspired interpretation of Newton, note three implications of the view I attribute to Newton. First, it treats gravity (understood as a causally real force) as itself "produced" by a relational quality of matter. It is only because of matter's special relationship to other parts of matter that it gravitates. That is, as Chris Smeenk has noted, we "need a pair of bodies for both the 'active' and 'passive' aspects of the disposition to gravitate to be in play; it is only with the pair that the 'active' disposition to produce force combined with the 'passive' disposition to respond causes an interaction that yields actual motions" (Smeenk, personal correspondence, May 23, 2008).

Second, gravity is, therefore, not an intrinsic quality of a single bit of matter. The relation only holds between bits of matter, which are said to share a "nature"—that is, a "disposition" to gravitate when conjoined.[10] So on my interpretation of Newton gravity is intrinsic to the relation, a "nature," but not to matter itself.

For those disinclined a priori to believe that Newton would hold such a position, it might help to note that this dispositional, relational account is analogous to a claim made in terms of capacities in De Gravitatione: "for when the accidents of bodies have been rejected, there remains not extension alone, as [Descartes] supposed, but also the capacities by which they can stimulate perceptions in the mind" (Newton, 2004: 35). A lone particle in the universe will not stimulate a perception until there is a mind present. (Of course, the official argument goes in the other direction: the description of corporeal nature is "deduced ... from our faculty of moving our bodies!" [Newton, 2004: 30]).

Third, because gravity is not an intrinsic quality of a particle of matter, we can grasp why for Newton gravity is not essential to matter. John Henry has argued that this is due to Newton's commitment to God superadding gravity to matter (see Henry, 1999; 2007. See also Downing, 2009: 364ff.). My view cannot rule out this possibility, and Newton was probably eager to encourage it among certain of his more orthodox interlocutors, such as Bentley. However, I read Newton as claiming in the *Treatise* that gravity is in matter not by, say, God's superaddition, but rather generated by (for lack of a better

---

[10] In order to avoid confusion: Newton's dispositional language is used here not to advance an instrumentalist reading (e.g., McMullin, 2001 discussed in Janiak, 2007: 135).

term) the interaction between at minimum two bits of matter due to their shared nature—that is, mass.

Now, as Ori Belkind has reminded me, the shared mass of the two particles is reducible to the distinct masses that each object possesses separately, and this may be thought to make gravity an inherent property; or, given that mass is not a relational quality, one can have the cause of gravity defined nonrelationally, and gravity itself defined relationally. (Some of this will be clearer below from my discussion of Rule III and the difference between universal and essential properties of matter.)

But it is only in virtue of the masses entering into an "interaction" that they "produce" gravity. Thus, even after creation, a lonely part-less particle of matter in the universe would not be said to gravitate. The disposition is only actualized when the universe contains at least two bodies (or a body with parts). So first, when it comes to gravity, we are dealing with a relational and inessential property of matter. Second, even while we can think of "part" of the "cause" of gravity, mass, as intrinsic to matter, it only "is" a cause when related to other masses.[11]

My reading has seven nice features associated with it. First, while it is a species of speculative metaphysics that Newton increasingly came to deplore, it provides an account of part of the physical cause of gravity in an ontologically sparse way, in accord with Newton's first rule of reasoning: "No more causes of natural things should be admitted than are both true and sufficient to explain their phenomena" (Newton, 1999: 794).

No new entities are introduced. Of course, the downside is that surprising and strange qualities are attributed to known entities. Thus, second, it accords with the ontological priority of matter over the laws of nature in Newton's late cosmological query that: "it may be also allowed that God is able to create particles of matter of several sizes and figures, and in several proportions to space, and perhaps of different densities and forces, and *thereby* to vary the laws of nature, and make worlds of several sorts in several

---

[11] One might worry that Newton's second letter to Bentley might cause a problem for my reading. Newton writes, "You sometimes speak of gravity as essential and inherent to matter. Pray do not ascribe that notion to me; for the cause of gravity is what I do not pretend to know, and therefore would take more time to consider of it" (Newton, 2004: 100). While this is evidence for my reading that Newton denies that gravity is essential and inherent to matter, it also appears to claim that Newton knows nothing about the cause of gravity. But Newton's "therefore would take more time to consider of it" suggests, at least, perhaps, he thought the subject was within his apprehension and worth inquiring about further (i.e., that the cause might be discovered). By my lights Newton should have said he does not know the full cause so either Newton was not careful or there is a genuine shift from the *Treatise* (perhaps due to considerations of "audience"). I thank Sarah Brouillette for discussion.

parts of the universe." (Query 31, Newton, 1979: 403–404; emphasis added—see chapter 6 for analysis.) It is the relationships among the densities and forces of matter to space that accounts for the varying laws and worlds.

Third, related to these, is an additional attractive feature: it is theologically neutral. Newton leaves room for a possible role for God (e.g., as the medium, or as cause of the world), but he is not required to commit to it. Newton writes in a different context in *Opticks*, Query 31, that it only seems probable to [Newton] that God in the beginning formed matter in solid, massy, hard, impenetrable, moveable particles, of such sizes and figures, and with such other properties, and in such proportion to space (Newton, 1979: 400).

Fourth, this probabilistic and empirical approach to matters related to our knowledge of God's attributes, which are treated "from phenomena," fits the overall character of the *Treatise*, where God shows up only once (and then very briefly) in the antecedent of a conditional statement about the placement of the planetary orbits (Newton, 1969: 33; in the published version of the first edition of Book 3 of the *Principia*, Newton firms up the claim in Proposition 8, Corollary 5 [Newton, 1999: 814]).

As an aside, it even fits the more theologically ambitious picture of De Gravitatione, where Newton's treatment of the idea of space is compatible with the claim that it is merely (a) "as it were an emanative effect of God" (Newton, 2004: 21) and (b) "an emanative effect of the first existing being" (Newton, 2004: 25) as well as (c) "the emanative effect of an eternal and immutable being" (Newton, 2004: 26); while the reader is invited to think of God as the first cause, all three versions are compatible with a theistic and pantheistic interpretation (see chapter 5). Regardless of Newton's private commitments, in public he is nearly always careful not to overstep the evidence.

Fifth, gravity is not essential to matter because not only can we conceive that God would initially create matter without gravity, but also—and more importantly, Newton writes in his comment on the third Rule of Reasoning: "I am by no means affirming that gravity is essential to bodies . . . Gravity is diminished as bodies recede from the earth." (Newton, 1999: 796; this was added to the third edition of the *Principia*.) That is, it is an empirical discovery that the strength of gravity can vary with distance. The value of gravitational mass does not vary; what varies is the strength of the interaction. A way to make sense of Newton's denial of having claimed that gravity is essential to bodies is to argue that gravity follows from a relation.

Against Janiak I see no textual evidence that Newton ever equated "attraction at a distance" with the claim that gravity is "essential" to bodies (Janiak, 2007: 128, n.5).[12] There is no evidence that Newton equated gravity being "essential" to matter with it being "inherent" in matter. To be an "essential" quality means being a quality that is required for (or a necessary condition of) the existence of matter, while gravity being "inherent" says something about its location (Newton, 1999: 404). McGuire puts this point nicely: "For Newton, a body can be a body without acting gravitationally" (personal correspondence, May 10, 2008). Of course were one to determine that gravity is an essential quality of body, one would also be inclined to claim it inheres in it. But it is true that Newton's third rule also asserts that while gravity is not essential to bodies being bodies—in the sense that it is not a necessary condition for their existence (bodies could have been arranged otherwise)—it is, until we find contrary phenomenon, a universal, empirical fact that gravity is not separable from pairs/systems of bodies (Miller, 2009).

In order to avoid confusion, I am not claiming that Newton argues from the universal fact of gravity to its being a relational quality of matter. These are distinct points. It also means that if I were to wish to argue directly against the superaddition thesis defended by Henry, I would have to challenge his privileging of the exchange with Bentley and the "General Scholium" over other texts. But that is not my aim here. In this chapter, all I wish to show is that Newton's conception while writing the first draft of the *Principia* is theologically more austere than scholars tend to recognize. This conjoined with the fact that my reading fits with a whole range of Newtonian commitments and the fact that Newton made sure that the readers of the *Principia* were familiar with the existence of the *Treatise*, offer some grounds for thinking that I am offering a plausible alternative to Henry's approach.

Sixth, Newton is committed to the position that the various phenomena caused by gravity are such that mass and distance are the only salient variables in the causal chain that involves them. We express this precisely through the law of universal gravitation, asserting that gravity is as the masses of the objects in question and is inversely proportional to the square of the distance between them. (Janiak, 2007: 142) Janiak is correct to emphasize that Newton's mathematical account places constraints on what a

---

[12] Janiak and I agree in rejecting the so-called Cotes/Kant reading—gravity is an essential quality of matter—adopted in the Editor's preface of the second edition of the *Principia* (Janiak, 2007: 143, n.41).

physical account should look like because our goal is to supplement rather than replace the mathematical account (Janiak, 2007: 136). But Janiak ignores that Newton is also in a position to distinguish conceptually between investigating the material cause of gravity and the medium that facilitates the interaction. We can learn things about the nature of matter, mass, without learning anything about the medium. In fact, given that post *Principia* the medium must have negligible mass, one should say that, if it exists at all, it has an entirely different "nature" than matter.

Seventh, my approach avoids attributing to Newton the "absurd" Epicurean position in which passive matter can act at a distance; this is clearly rejected in the fourth letter to Bentley (Newton, 2004: 102).[13] But it does not follow from this, as Janiak contends, that Newton rules out action at a distance *tout court*. For Newton's position in the letter to Bentley permits us to understand that in the right circumstances matter can be viewed as "active." (See also Joy [2006] for the importance of Newton's distinction between active and passive principles.) This is, in fact, what Newton indicates in Query 31, where he contrasts the "passive principle by which the bodies persist in their motion or rest, receive motion in proportion to the force impressing it, and resist as much as they are resisted" with "active principles, such as are the cause of gravity, by which planets and comets keep their motions in their orbits, and bodies acquire great motion in falling; and the cause of fermentation," and so on. Shortly thereafter, he lists "gravity, and that which causes fermentation, and the cohesion of bodies" among the "active principles" (Newton, 1979: 400–401). This allows, as he explicitly says in his letter to Bentley, that the attractive "agent be material." I should emphasize that in my position Newton neither asserts that matter is altogether active nor passive.[14] It depends on the way we are conceiving things.

---

[13] Henry 1999 and 2007 make this point forcefully. In correspondence (May 11, 2008), Katherine Brading offered the following lovely observation: "If you put a lone Epicurean atom into the void, it will 'know' which way is 'down' and start moving in that direction. If you put a Newtonian [part-less] body into the void it won't 'know' that it's a gravitational body, or 'know' which way to move gravitationally, until you put in a second [part-less] body, and thereby bring the gravitational relation into being."

[14] While this may sound strange, it is by no means unique in Newton. See, for example, Newton's treatment of the "inherent force" of inertia. Newton claims that this force can sometimes be viewed "passively": "Inherent force of matter is the power of resisting"; but sometimes it is more "active": "a body exerts this force . . . during a change of its state, caused by another force impressed upon it" (quoted from the third Definition, Newton 1999: 404). This is why Bertoloni Meli, 2006a introduces the language of "potential" or "latent" force to treat the "passive" mode. He captures insightfully the "twofold perspective" (Bertoloni Meli, 2006a: 325) with which Newton analyzes such matters. (See also the postscript to chapter 2.)

Incidentally, because the laws of motion are also said to be "passive" (and in virtue of the ontological priority of matter over laws, noted in my second point above), I reject as un-Newtonian, Fatio's claim speculation that *"that Gravity had its Foundation only in the arbitrary will of God"* (quoted with emphasis in Dobbs, 1991: 189). If God is involved in accounting for action at a distance, it will be in the way he created matter.

Thus, I reject Janiak's claim "Newton considered any non-local action to be simply 'inconceivable' " (Janiak, 2007: 144) on two grounds: Janiak misidentifies what is "inconceivable" according to Newton in the letter to Bentley; and he relies on a distinction between "local" and "distant" action (Janiak, 2007: 145) that is not to be found in Newton (see chapter 2 and its postscript).

1.3 On the Conceivability of Local and Distant Action

Because my reading of the letter to Bentley conflicts with Janiak's and this ramifies through our approaches to Newton, I quote Janiak (2007) at length before I take up the details of his argument. The passage in question is from the letter to Bentley written in 1693, six years after the *Principia* first appeared:

> It is inconceivable that inanimate brute matter should, without the mediation of something else, which is not material, operate upon and affect other matter without mutual contact . . . That gravity should be innate, inherent, and essential to matter, so that one body may act upon another at a distance through a vacuum, without the mediation of anything else, by and through which their action and force may be conveyed from one to another, is to me so great an absurdity, that I believe no man who has in philosophical matters a competent faculty of thinking can ever fall into it. (Newton, 2004: 102)

After quoting the passage, Janiak claims that Newton, thus, "forcefully" rejects "the very idea of action at a distance," (Janiak, 2007: 128). In using this passage to motivate his reading of both Newton's rejection of distant action between parts of matter as well as Newton's search for properties of a medium, Janiak finds himself in excellent company with Maxwell, who also uses it to carefully distinguish between Cotes' attribution of "direct" distant action and Newton's position.[15]

---

[15] See Maxwell, 1890, 2 [Vol. 2?]: 316 & 487. Maxwell appears to be influenced by Colin MacLaurin's (1748) account. We can understand Smith's "History of Astronomy" (Smith, 1982: 29–93) as an attempted correction to MacLaurin. For more on these matters, see van Lunteren, 1991.

But Janiak (and even Maxwell!) quotes the passage selectively. Janiak leaves out the crucial point of the passage—that is, Newton's denial that he is Epicurean. (John Henry [1999, 2007] has made this point forcefully in his papers.) The full passage reads:

> It is inconceivable that inanimate brute matter should, without the mediation of something else, which is not material, operate upon and affect other matter without mutual contact, as it must be if gravitation in the sense of Epicurus, be essential and inherent in it. And this is one reason why I desired you would not ascribe innate gravity to me. That gravity should be innate, inherent, and essential to matter, so that one body may act upon another at a distance through a vacuum without the mediation of anything else, by and through which their action and force may be conveyed from one to another, is to me so great an absurdity, that I believe no man who has in philosophical matters a competent faculty of thinking can ever fall into it. Gravity must be caused by an agent acting constantly according to certain laws; but whether this agent be material or immaterial, I have left to the consideration of my reader. (Newton, 2004: 102–103)

By contrast, my preferred reading of the letter to Bentley is that it is "inconceivable" that "inanimate brute" matter can produce action at a distance (especially) if we conceive gravity along Epicurean lines—that is, as innate, essential, and inherent to matter. But this entirely allows other conceptions of matter with more "active" properties (and other conceptions of gravity). The last sentence of the letter to Bentley quoted previously qualifies the reading of absolutist-denial of action-at-a-distance offered by Maxwell and Janiak: Newton says that he will leave it to the reader to decide if gravity is "caused" by a "material or immaterial" "agent"! This means that Newton does not rule out the existence of (properly reconceived) matter as an active agent or cause of gravity. It would, of course, be a contradiction in terms for "passive" matter to be an "agent"; but Newton never claims in his own voice that matter must always be passive (see also chapters 2 and 8). Newton and later Cotes did consistently deny that "mere mechanical" causes or "necessity of nature" (Newton, 1999: 397) or "blind fate" (Newton, 2004: 138) can explain the universe. He wants to avoid being read as an Epicurean or Spinozist. It is surprising that Janiak misses this concern. For, in the fifth reply to Leibniz, which plays a crucial role in Janiak's argument, Clarke forcefully rejects Leibniz's attempts to tag Newton as an Epicurean (Sections 128–130 LC: 96).

Of course, everything I say here can be embraced by somebody defending the superaddition thesis.

Unfortunately, Janiak uses his reading of the letter to Bentley to attribute to Newton the view that action at a distance is inconceivable and, thus, that Newton relies on a (tacit) distinction between (inconceivable) distant and (conceivable) local action (e.g., "all action between material bodies must be local—on pain of there being an 'inconceivable' distant action," [Janiak, 2007: 143 and 141]). I find no evidence in Newton for such a distinction between local and distant action (see also chapter 2 and its postscript). To be clear, in his analysis of "local" action, Janiak (2007: 143) is careful to distinguish Leibniz's restrictive position, which only permits surface action (any causal interaction that involves the surfaces of two or more bodies), from a less stringent version (gravity may involve some kind of medium that does not act on the surface of bodies, but somehow penetrates them) that Janiak attributes to Newton.

In his paper Janiak usefully cites Newton's Query 28 as rejecting the Leibnizian hypothesis (Janiak, 2007: 143). Janiak also appears to cite Newton's Query 21 to illustrate his claim that Newton is tacitly relying on a local/distant action distinction. But unfortunately, most such claims by Janiak presuppose his further claim that Newton thinks distant action is inconceivable.[16] Janiak ignores very clear evidence to the contrary from the *Principia*. For in the Scholium to Proposition 69 of Book I of the *Principia*, Newton lists as the very first of four possible, physical explanations of "attraction" that it may be "a result of the action of the [distant!] bodies either drawn toward one another" (Newton, 1999: 588). If Newton had thought it inconceivable it would be peculiar for him to leave it in the *Principia* after his exchanges with Bentley.

---

[16] One might also be tempted to point to a passage from Newton's anonymously published "Account of the *Commercium Epistolicum*," where Newton rejects Leibniz's view of God as "an intelligence above the bounds of the world; whence it seems to follow that he cannot do anything within the bounds of the world, unless by an incredible miracle" (Newton, 2004: 125). Now it is true that Newton has just affirmed that God is omnipresent. So, a natural reading of the "incredible miracle" that Newton attributes to Leibniz might be that God acts at a distance (for an omnipresent God can always—in Janiak's terminology—act locally). The passage would then be very ironic: Leibniz accuses Newton of being committed to action at a distance, and Newton turns the table on Leibniz. But this misunderstands Newton's point here; if (Leibniz's) God is above the bounds of the world, this means he is outside of space and time altogether. So, it would not be appropriate at all to say that God is acting at a distance; God would literally be acting from nowhere—that is, an incredible miracle. In De Gravitatione Newton had claimed "no being exists or can exist which is not related to space in some way" (Newton, 2004: 25). Thus, in his response to Leibniz, Newton is echoing the doctrine of the sixth chapter of *Spinoza's Theological-Political Treatise*, where in his discussion of miracles Spinoza rejects the very intelligibility of placing God above the bounds of the world.

Equally unpromising is the way that, from Newton's explicit denial of gravity being essential to matter, Janiak moves to attributing to Newton a denial of action at a distance. As noted above, to deny that gravity is essential to matter is just to deny that gravity is a necessary condition for the existence of a lone, part-less particle, or one of the primary qualities of matter. But I have argued against such a view: Newton is committed to gravity being a relational quality.

My position does come into conflict with an alternative, influential reading of Newton.[17] Howard Stein's "field" interpretation of gravity was initially and influentially offered as an interpretation of Newton's example in the treatment of Definition 8 of the *Principia*, where Newton writes that,

> accelerative force [may be referred to], the place of the body as a certain efficacy diffused from the center through each of the surrounding places in order to move the bodies that are in those places; and the absolute force [may be referred], to the center as having some cause without which the motive forces are not propagated through the surrounding regions, whether this cause is some central body (such as the lodestone in the center of a magnetic force or the earth in the center of a force that produces gravity) or whether it is some other cause which is not apparent. (Newton, 1999: 407)

Note that Newton distinguishes the cause of the force and the force in turn from gravity, offering independent support for my claim above not to conflate such matters. Given that Stein was the original source of my treatment of the *Treatise* passage as authoritative, I conclude by briefly distinguishing my view from Stein's.

Stein treats Definition 8 as claiming that a single particle generates a force field in the places around it (see Stein [1970]; Belkind [2007] inspired my questioning of Stein's use of Definition 8). But this does injustice to the fact that according to Newton it is the "shared nature" that generates the inverse square force. So, on my view it is only a pair of bodies that generates anything. My disagreement with Stein turns on how to interpret Newton's empirical approach, especially in terms of (i) the exact wording of the explanation of Definition 8, (ii) the formulation in the third Rule of Reasoning,

---

[17] The discussion below has benefitted from private correspondence with Howard Stein, although the reader should be aware that he objects to my characterization of his position.

and (iii) perhaps one's sense of Newton's aesthetics. Let me explain these three differences, and offer responses to Stein's arguments.

In arguing for his field interpretation, Stein is committed to the claim that bodies generate a gravity field around them even in places where there are no other bodies. My problem with Stein's reading is that the use of "field" can suggest to some readers an ontology in conflict with the interpretation that I have been developing. Against Stein, I argue, first, that in Definition 8, Newton does not claim that there would be an accelerative force or an accelerative measure of a gravitational force in the absence of a second body; rather, the definition of accelerative forces is given in terms of the disposition of a central body to "move the bodies that are in" the surrounding places. Thus the example following Definition 8 is quite clear that we are dealing with an interaction (or shared action). The example provides, as Ori Belkind writes, "evidence for the fact that Newton distinguished between the cause of gravity (the masses of bodies) and the force of gravity [in] the distinction between the absolute, accelerative and motive measures of the force. The absolute measure of the force is associated with the cause of the force, located inherently in the body at the center of attraction. The accelerative and motive measures can be relational" (Belkind, private correspondence, May 4, 2008).* Moreover, the whole example of Definition 8 is about weight; that is, Newton is giving a treatment of forces not in terms of their (counterfactual) impact on empty places but on places filled with matter. This fits nicely with my emphasis above on the ontological priority of matter.

Second, on my reading of the third Rule of Reasoning, we can infer universally that pairs or systems of bodies, or shared actions, generate forces where bodies are, but no more. (See Miller, 2009) The earliest draft of Rule 3 reads: ""The laws of all bodies in which experiments can be made are the laws of bodies universally," quoted in McGuire (1970a: 15), We have no empirical evidence whatsoever for how to think about a lone part-less particle. In fact, Rule 3 and Newton's explication of it is articulated in terms of the plural "bodies" and their plural "parts." (Recall: "Those qualities of bodies that cannot be intended and remitted and that belong to all bodies on which experiments can be made should be taken as qualities of all bodies universally" [Newton, 1999: 795]) The only lone body mentioned in the long

---

* Because I did not comment on it, it is natural for readers of the original publication to assume that I agreed with Belkind that masses are the full causes of gravity. Please see my response to Parker in the postscript to chapter 2 for my considered view.

discussion accompanying Rule 3 is a hypothetical disconfirming divisible body, but this is unrelated to the universal nature of gravity and is thus evidence for my approach.

Finally, my whole approach makes sense of the fact that Newton's earliest and most able readers, including Leibniz, found even the cleansed view of the *Principia* unabashedly occult. A reading of Newton that makes him out to appear plausibly as a certain kind of innovative scholastic should, thus, have something to be said for it, especially because we are also trying to explain why he would have suppressed it in light of expected prejudices.

It is to be admitted that in my view of Newton the infinite universe is populated by infinite number of pairs or systems of interactions within and among bodies generating infinite numbers of attractions. But that is no stranger than the mere fact of universal attraction. Newton is leading us to this view when he writes in his account of the third Rule of Reasoning,

> if it is universally established by experiments and astronomical observations that all bodies on or near the earth are heavy toward the earth, and do so in proportion to the quantity of matter in each body, and that the moon is heavy toward the earth in proportion to the quantity of its matter, and that our sea in turn is heavy towards the moon, and that all planets are heavy toward one another, and that there is a similar heaviness of comets toward the sun, it will have to be concluded by this third rule that all bodies gravitate toward one another. (Newton, 1999: 796)

So, on balance, my position should be favored over Stein's because it avoids anachronism and does justice to the text of Newton and the expected reactions to it by his contemporaries. It does saddle Newton with a potentially unbalanced mixture of speculative hypothesis and strict verificationism, and this may be thought to count against the view. (But this offers no solace to Janiak, because he involves Newton in more substantial and less empirically grounded speculative hypotheses.)

1.4 Conclusion

Let me end with two disconnected observations. In my treatment I have said nothing about a very important and much neglected passage (Book 1, Section 11, Scholium): "finally, it will be possible to argue more securely concerning the physical species, physical causes, and physical proportions of these forces" (Newton, 1999: 589). By talking about the conspiring nature of matter, my treatment of the *Treatise* has at least started an interpretation of what Newton thought the physical causes were and what species they belonged to. But this is only a tentative start because I have said nothing about Newton's concept of a "natural power."

Finally, I am aware that my somewhat Talmudic reading of Newton is not sexy; other commentators will note three omissions: I neither connect my account of Newton's views to other streams of intellectual thought; nor do I provide Newton with a view that is easily assimilated in our contemporary debates; nor, finally, is my Newton a thoroughgoing theist or Platonizing mystic. But while unfashionable, my Newton is a refined and subtle metaphysician.[18]

[18] I have received detailed comments on earlier drafts of (parts) of this chapter by Ori Belkind, Zvi Biener, Sarah Brouillette, John Henry, David M. Miller, Lex Newman, Chris Smeenk, and Howard Stein. I am also grateful to Ted McGuire for helpful suggestions, and to audiences in Dublin, Oxford, and Gent, including Dana Jalobeanu, Maarten Van Dyck, Karin Verelst, and Carla Rita Palmerino, to which I presented on an earlier drafts of some of this material. Moreover, Katherine Brading helped me formulate my thesis and gave penetrating comments on the whole of the chapter. I am also grateful to Bill Harper, who, despite misgivings about the position, encouraged me to pursue this line of thought after an extensive discussion in an airport lounge. Special thanks are due to Dana Jalobeanu and Peter Anstey, who were the editors of the volume in which it first appeared, for their many insightful comments. I also thank the anonymous referees of *Journal of the History of Philosophy*, who provided helpful comments on a rejected paper that includes substantial material discussed here. Finally, special thanks are due to Andrew Janiak, who not only gave detailed comments on this chapter, but thanks me (among others) in the final footnote to the paper I criticize here. Ever since Michael Friedman chose to deflect one of my questions to Janiak at a symposium at New York University (November, 2006), my views have been developed in dialogue and correspondence with Janiak.

2

Newton's Substance Monism, Distant Action, and the Nature of Newton's Empiricism

2.1 Introduction

Hylarie Kochiras (2009) is to be commended for taking a fresh look at many well-known Newtonian passages[1]. In doing so she diagnoses the following two problems for Newton: (1) the "causal question about gravity" turns "into an insoluble problem about apportioning active powers," (2) "More seriously" Newton has no "means of individuating substances," that is, "Newton's Substance Counting Problem" (Kochiras, 2009: 267). The final sentence of her carefully argued paper hints at a further problem: (3) Newton's Substance Counting Problem is said to threaten "even to undermine his concept of body" (Kochiras, 2009: 279). Kochiras appears to think the second and third problems are connected because if Newton cannot individuate substances he has no way of individuating bodies. Kochiras' argument turns on three "principles" she discerns in Newton: (1) the "very broad reach of his empiricism" (Kochiras, 2009: 267), (2) a "rationalist feature," the principle "that matter cannot act where it is not," and (3) "substances of different kinds might simultaneously occupy the very same region of space" (Kochiras, 2009: 267, 277).

I argue that Kochiras mistakenly attributes the second and third principles to Newton. Moreover, the first principle can only be articulated in Kochiras' manner if one ignores the development of Newton's views. Nevertheless, I grant that Kochiras has identified genuine tensions in Newton's system, but I argue she fails to appreciate how Newton handles these. In what follows, I tackle Kochiras' analysis and deployment of these principles in reverse

---

[1] This chapter originally appeared as Eric Schliesser, "Newton's substance monism, distant action, and the nature of Newton's empiricism, a discussion of H. Kochiras, "Gravity and Newton's substance counting problem." *Studies in History and Philosophy of Science Part A* 42.1 (2011): 160–166.

Newton's Metaphysics. Eric Schliesser, Oxford University Press. © Oxford University Press 2021.
DOI: 10.1093/oso/9780197567692.003.0003

order. In doing so, I dissolve the three problems diagnosed by Kochiras in Newton's metaphysics. In what follows I generally accept Kochiras' terminology for the sake of argument and brevity, except that when I am describing my counterarguments or proposals I substitute "speculative" where she had used "rationalist."

2.2 The Substance Counting Problem and Collocation

The substance counting problem is generated because for Newton all beings need to be spatially located. Newton identifies four such beings: God, minds, spirits, bodies. Beings may be material, immaterial, and, perhaps, both. Kochiras offers considerable evidence from Newton's manuscripts that Newton accepts the "Scholastic Commonplace" that two material bodies cannot occupy the same place at the same time (267; ironically in contemporary metaphysics this commonplace is known as "Locke's Principle"—see Doepke, 1986; Simons, 1985). Newton also indicates that immaterial beings may co-occupy the same place as material beings. For example, in the "General Scholium" to the Principia Newton argues that God is omnipresent; he "will not be never or nowhere" (Newton, 2004: 91). Against the Cartesians, Newton insists in De Gravitione, that "created minds are somewhere" (Newton, 2004: 23), while bodies are always at a place. Clearly, bodies and minds can co-occupy the same places all of which are also occupied by God. So, it appears there is ample evidence for Kochiras' third principle.

Nevertheless, throughout her analysis Kochiras conflates Newton's treatment of beings with talk of "substances." In doing so Kochiras fails to appreciate fully Newton's conceptual innovations. To be clear Kochiras recognizes that Newton "eliminates" the "Aristotelian Scholastic notion of prime matter" (Kochiras, 2009: 269). She also recognizes Newton's Lockean assertion of epistemic ignorance about the inner substance of things in the Principia's "General Scholium" (Kochiras, 2009: 269–270). But she fails to recognize the significance for Newton that one can have being (or reality) without being either a substance or a property (or attribute) of a substance.[2] In De Gravitatione Newton made this point clear in his discussion of extension (or

[2] Ernst Cassirer seems to have been the first to call attention to Newton's role in the demise of substance/property metaphysics—see Cassirer, 1951: 61–64; Cassirer's argument relies on the reception of Newton by Dutch Newtonians. (Cassirer was, of course, unaware of many of Newton's manuscripts.)

space): "it is not substance; on the one hand, because it is not absolute in it-self . . . on the other hand, because it is not among the proper affections that denote substance, namely actions, such as thoughts in the mind and motions in bodies" (Newton, 2004: 21—see also chapter 5).

Extension is neither a substance nor can it be the kind of thing that can be said to 'belong to' a substance. Moreover, as Kochiras recognizes (Kochiras, 2009: 269) extension takes the place of the 'eliminated' prime matter; it contains body (Newton, 2004: 29). In De Gravitatione, the ontological dif-ference between space and body is one of different degrees of being; bodies have more reality than space "whatever has more reality in one space than in another space belongs to body rather than to space," (Newton, 2004: 27).

One might be tempted to argue on Kochiras' behalf that while extension is removed from the ontology of substance-property/quality framework, surely bodies are not. After all, Newton appears to be claiming that we are "able to" call bodies "substances" (Newton, 2004: 29). So, if Kochiras conflates beings and substances, so does Newton! But in context, Newton is not asserting that bodies are substances. Rather, he is claiming that his thought experi-ment about "a certain kind of being similar in every way to bodies" (Newton, 2004: 27; the thought experiment ends on p. 33) permits the following in-ference: what can be said legitimately about bodies can be said legitimately about his imagined beings.

Even if Kochiras were to grant my reading of De Gravitatione (Newton, 2004: 29), she might still be tempted to claim that Newton asserts that bodies are substances when he writes about bodies as "created substance" and that "substantial reality is to be ascribed to" them (Newton, 2004: 32). But in context Newton is making three subtly different points: first, he rejects the "atheists" who presuppose that bodies have a "complete, absolute, and independent reality in themselves." Second, Newton affirms the position that bodies have a "degree of reality" that "is of an intermediate nature be-tween God and accident" (Newton, 2004: 32). Now Newton asserts the doc-trine of different degrees of reality to help explain (as a kind of error-theory) the Scholastic "prejudice" in applying "the same word, substance . . . univo-cally . . . to God and his creatures."[3] Thus, third, bodies have some substantial reality but are not themselves substances. The third point makes sense in light of the passage quoted before (Newton, 2004: 21); for Newton "a substance is

---

[3] This is one of the few places in De Gravitatione, where Newton in in agreement with Descartes (*Principles of Philosophy* 1.51).

absolute in itself"—that is, it has full reality, and it is the kind of thing that can be the cause (or source) of actions (i.e., agent causation). The effects of these actions are thoughts in the mind and motions in bodies.

To be clear, on my reading according to Newton one is only a substance if one is both self-sufficient and the cause of actions. So because bodies are not absolute in themselves (they presuppose space and need to be created) they are not the kind of beings that are properly or strictly speaking called "substances." (Newton does allow a different, innocent use of "substance" to mean what we would call the [chemical] stuff or composition of matter—see Newton, 2004: 34.)[4]

The situation is even more complex when it comes to "created minds," which are said to be "of far more noble nature than body, so that perhaps it may eminently contain [body] in itself," (Newton, 2004: 30). Newton is eager to show that "the analogy between the divine faculties and our own may be shown to be greater than has formerly been perceived by philosophers" (2004: 30). This analogy and what we may call, "the eminent containment thesis"[5] are all suggestive evidence that, according to Newton, minds have more reality than bodies; perhaps minds are substances, especially because there is ample evidence that for Newton created minds have ideas (e.g., Newton, 2004: 21–22) and volitions (e.g., Newton, 2004: 27) and, thus, can be the source of actions. Even so, in virtue of being created, finite minds are dependent beings and so not substance.

Thus for Newton there is strictly speaking only one genuine substance, God. (For more on the monism of Newton and Spinoza, see chapter 5.) Extension has many of God's attributes (eternal, indivisible, etc.), but crucially lacks agency. Bodies and created minds are dependent on God (and space) and, thus, not absolute in themselves. This is, in fact, the main point of a passage in the Principia's "General Scholium" that receives much attention from Kochiras (as evidence for the second "rationalist" principle): "In Him all things are contained and move." All entities are contained by and, thus, dependent on the sole substance, God.

To conclude: the Substance Counting Problem is a pseudo-problem because there is only one substance. It rests on a mistaken conflation between

---

[4] I am ignoring a further complication: We might have to distinguish between being what one may call conceptually dependent on God (e.g., God as the emanative cause of space) and being causally dependent on God (e.g., God as the creator of body). It is unclear which dependency is required to rule out being a substance. I thank Ori Belkind for pressing this point.

[5] I borrow the term from Gorham, 2011a.

beings and substances and it ignores the doctrine of different degrees of reality, which accounts for Newton's use of "substantiality" when discussing bodies. Finally, Newton's Lockean epistemic insistence that we are ignorant of "inner substances" of bodies or the "idea of substance of God" is, of course, quite compatible with the more radical ontological claim that there are no dependent substances at all.

Even if the letter of the Substance Counting Problem is deflated, one might well think that Kochiras has adequately identified another, related problem: can Newton determine how many entities can occupy a place? (This sounds fearfully close to the question how many angels can occupy the tip of pin.) And, related to this, can Newton individuate bodies? I address these questions in turn.

2.2.1 The Substance Counting Problem and the Causes of Gravity

Kochiras is correct that for Newton two bodies cannot occupy the same region of space at the same time (Kochiras, 2009: 269–270). As Kochiras notes this is an empirical result for Newton. Moreover, one of her crucial insights is that according to Newton "secondary causes" and not God are the source of gravitational forces (Kochiras, 2009: 270–272). God "does not act on [all things] nor they on him" ("General Scholium," Newton, 1999: 941). To be precise God's activity plays only the two following explicit roles in Newton's thought: (1) God's ubiquity is required for the act of creation (Newton, 2004: 26) as Newton's God does not create the universe from outside of space and time, but from within,[6] and (2) Newton's benevolent God occasionally (and miraculously) reforms the machinery of the universe (Kochiras, 2009: 271; Newton, 1730: 378).* So even if God is omnipresent, his activity is limited.

But even if God is removed from the scene, Kochiras would claim that according to Newton minds and spirits, on the one hand, and bodies, on the other hand, can occupy the same space at the same time. There is no doubt

---

[6] On the nature and extent, if any, of Newton's voluntarism, see the exchange between Harrison (2004) and Henry (2009).

* I now would argue that Newton only offers the suggestion of God's "reformation" of the solar "system," but need not be taken to endorse it. So, the activity may be more limited than I argued in the original version of this chapter.

that in De Gravitatione the "union" of the "distinct" beings, mind and body, is left rather mysterious (Newton, 2004: 31), but body is dependent on and eminently contained by (nobler) mind.[7] Presumably this means that against Descartes, for Newton mind and body occupy the same place, when unified. So, there is no doubt that Kochiras could redescribe the substance counting problem in terms of the co-occupation of a space at a given time by material and immaterial beings. She might wish to call this the 'entity counting problem.'

On Kochiras' reading of Newton "he draws no sharp distinction between physics and metaphysics," (Kochiras, 2009: 270). For Kochiras this appears to imply that Newton's metaphysical commitments threaten to undermine his natural philosophy. But four considerations suggest the situation is nowhere near as dire as suggested by Kochiras.

First the substance counting problem only becomes problematic if one wants to identify a particular (created) entity at a particular point of space and time as the cause of gravity. Kochiras argues that if we are unable to distinguish material and immaterial entities then we can never hope to establish the cause of gravity. But Kochiras never shows that Newton is in no position to distinguish material and immaterial entities. For immaterial entities will have near zero mass and as such are very different from material beings. This means, of course, that in practice it will be very difficult to find a measure or quantity for immaterial beings; their being must be inferred from their effects.[8] But being difficult to detect is not the same as being unable to distinguish.

Second from the point of view of Newton's "rational mechanics," which does not concern itself with miracles, but only aims to establish "the motions that result from any forces whatever and of the forces that are required for any motions whatever," ("Author's Preface," Principia, Newton, 1999: 382), discovering the causes of these forces is not a fundamental aim. So, even if Kochiras were correct, she would be calling attention to a problem external to Newton's rational mechanics. Newton repeatedly declared that he was

---

[7] At first glance, in the "General Scholium," the same dependence is asserted: "Every sentient soul, at different times and in different organs of senses and motions, is the same indivisible person" (Newton, 1999: 941). Souls are the source of continuity of personhood. Yet, a few lines down Newton implies a more reciprocal relationship: "Every man, insofar as he is a thing that has senses, is one and the same man throughout his lifetime in each and every organ of his senses."

[8] In De Gravitatione Newton allows that even God can have a quantity of spatial and temporal existence (Newton, 2004: 25–26), so my proposal is strictly Newtonian (chapter 5). In the "General Scholium" God is inferred from his effects (i.e., the argument from design discussed in chapter 8).

unable to assign the cause of gravity; given the aims Newton articulated for his science in the first edition of the *Principia* (before all the controversy over action at a distance) this is a completely legitimate move.

Of course, within the *Principia* Newton also admits an interest in the "physical causes and sites of force" (Newton, 1999: 407). It may be thought dogmatic that at least in the body of the *Principia* (but not elsewhere), Newton does not allow for immaterial causes of the forces. So, Kochiras' position is really a complaint about what may be thought of as Newton's methodological stance that rules out consideration of immaterial causes of forces in the body of the *Principia*.

But within the *Principia* Newton's focus on physical causes has some empirical support. In discussing these matters we should distinguish among (1) the force of gravity as a real cause (which is calculated as the product of the masses over the distance squared); (2) the cause of gravity; (3) "the reason for these properties of gravity"; and (4) the medium, if any, through which it is transmitted (see chapter 1). Much discussion about Newton conflates (2) or (3) with (4), perhaps prompted by Newton's treatment of the ether. Of course, if one were to be able to establish empirically an ether, one might also wish to use its known properties to try to explain the cause of gravity and its properties. But this is not required.

For in the third Rule of Reasoning Newton offers, as Ori Belkind emphasized, "a criterion for universalizing properties. Qualities are universalizable if they are found through experience, and if they are reducible in each case to the ultimate ingredients of matter" (private communication, January 5, 2010). The third rule affirms that gravity is a real cause and a quality of matter, even though Newton is ignorant of the cause of the quality. One can be ignorant of the cause of gravity because according to Newton to be a universal property does not require it to be an essential (or in modern terminology, intrinsic) property (see chapter 1). So the methodological (dogmatic) point is affirmed by this addition to the second edition of the *Principia*. I return to this point in the next section in this chapter.

Finally, one further reason why Newton might keep (2) and (3) distinct is that in some of his works he may have been committed to action at a distance! Of course, in the context of my criticism of Kochiras this is question begging. So, another reason why Newton would keep 2 and 3 distinct is that when Newton was drafting the first edition of the *Principia* he was committed to the claim that at least part of the cause of gravity is a contingent, dispositional, and relational quality of matter. In chapter 1, I argued that Newton's posthumously published *Treatise*, the existence of which Newton advertised in all editions of the "preface" to Book III of the *Principia*, offers evidence for rich details of this view (recall chapter 1). Kochiras accepts that the *Treatise* argues for the claim that gravity is "a relational quality" (Kochiras, 2009: 272, n. 43). She disagrees with my claim that the Treatise endorses action at a distance, but for my present argument we can ignore our differences on that score.[9]

2.3 Individuation of Bodies

In the last line of her paper Kochiras suggests that the substance-counting problem threatens to undermine Newton's concept of body (Kochiras, 2009: 279). She offers no argument for this claim. Presumably she worries that if material and immaterial entities can co-occupy places simultaneously one might not merely mistakenly attribute some properties to bodies when one should, perhaps, attribute these to immaterial entities, but once this possibility is raised, one might leave body without any (stable set of) properties.[10]

Kochiras' worry gains strength from the observation that despite toying with the idea in the *Principia* Newton never explicitly defines body.[11] Of course, one might read a line in the third rule of Reasoning as an explanation why such a definition is not forthcoming: "the qualities of bodies can be known only through experiments; and therefore qualities that square with experiments universally are to be regarded as universal qualities." (Newton, 1999: 795) This is presumably an example of what Kochiras calls "the very broad reach of Newton's empiricism" (Kochiras, 2009: 267). Some philosophers have lauded Newton for allowing an open-ended conception of body (see Harper, 2012). Katherine Brading is surely correct to see in the three Newtonian laws of motion (all three refer to bodies) a kind of law-constitutive conception of body (Brading, 2012). These laws provide the revisable conception of body that is required for Newton's rational mechanics.

[9] In order to avoid confusion: I distinguish among: (i) Newton's acceptance of action at a distance; (ii) Newton's insistence that this is the product of an accidental, relational quality of matter; and (iii) this quality is an active and a passive disposition. I argue for ii and iii in order to explain why Newton might have distinguished between the cause of gravity and the medium, if any, through which it is transmitted.

[10] To argue that immaterial beings occupy space (as More and Newton did) was to open oneself up to ridicule. So, presumably few of Newton's readers would have worried about confusing material and immaterial beings and their properties.

[11] Before the third edition of the *Principia*, Newton thought of defining body in terms of resistance (Biener and Smeenk, 2014).

Now it's true that no experiment on bodies can, in principle, rule out that the revealed properties are not really properties of immaterial entities. This presumably helps explain why Newton is so cautious in attributing any essential (i.e., intrinsic) qualities to bodies. So Kochiras' claims might be thought un-refuted. Yet, the fourth rule of reasoning (added to the third addition of the *Principia*) offers a convincing methodological stance: until one has very compelling empirical evidence—that is, one has established a "phenomenon" (a stable, expert established generalization)—one should not worry about competing hypotheses. So, if experiments on bodies reveal certain properties, one should stick to interpreting these in terms of bodies (and forces). It is Newton's great insight that while one should certainly be open to speculation on the causes of gravity, one can continue with rational mechanics without knowing such a cause. The subsequent history of science has amply vindicated this hope (Smith, 2014).

2.4 Passive Matter and Local Action

Kochiras claims that Newton is committed to two "rationalist" features: (i) that all matter is "passive" (Kochiras, 2009: 275); and (ii) "that matter cannot act where it is not" (Kochiras, 2009: 268)—or what she also calls "the principle of local causation" (Kochiras, 2009: 277). These two features combined are crucial for Kochiras' claim that Newton rules out material action at a distance. Kochiras' evidence for both features is thin.

In fact, in support of the first principle Kochiras only cites a single unpublished manuscript sentence, "Matter is a passive principle & cannot move itself."[12] Yet, the published version of the *Opticks*, is agnostic on this matter. In Query 31 Newton contrasts the

passive principle by which the bodies persist in their motion or rest, receive motion in proportion to the force impressing it, and resist as much as they are resisted [with] active principles, such as are the cause of gravity, by which planets and comets keep their motions in their orbits, and bodies acquire great motion in falling; and the cause of fermentation." (Newton, 1952: 400–401)

---

[12] Kochiras gives the following reference: "ULC MS Add. 3970, fol. 619r, written in English, cited in McGuire (1968): 171, and identified by him as draft variants of the 1706 Optice" (Kochiras, 2009: 274, n.55).

Shortly thereafter, he lists "gravity, and that which causes fermentation, and the cohesion of bodies" among the "active principles." (Newton, 1952: 400–401) This is compatible with, as he explicitly says in his Letter to Bentley, that the attractive "agent be material" (Newton, 2004, 103). If attractive agents can be material then for Newton matter need not always be passive. Given the importance of this fourth Letter to Bentley as evidence for Kochiras' claim that Newton holds the principle of local causation, she is in no position to deny that matter can be active for Newton. (Recall that in the preface to the *Principia*, Cotes goes even further than I and claims that Newton has shown that gravity is primary quality of matter [Newton, 1999: 391–392].)

Of course, in a draft letter to Roger Cotes Newton does deny the "principle of self-motion" to bodies (Newton, 2004: 121), which accords with his denial to Bentley that he is an Epicurean about gravity (Henry, 1999 and 2007; for excellent background on the range of options on active matter available to seventeenth-century natural philosophers, see Clericuzio, 2001). But one can deny that matter has a principle of self-motion and still allow that it can be active—dispositional relational qualities of matter fit this category.

2.5 Local Action

According to Kochiras Newton is committed to the "principle of local causation." Nowhere does Newton enunciate such a principle. Nevertheless, as Kochiras notes, this reading of Newton dates back to Clarke's claim in response to Leibniz: "Nothing can any more act, or be acted upon, where it is not present" (Kochiras, 2009: 275, n.68). Kochiras offers two sources of evidence to attribute such a principle to Newton: the fourth Letter to Bentley and the "General Scholium" of the *Principia*. She generously acknowledges Andrew Janiak as the source of this reading. While Kochiras occasionally withdraws from attributing the principle to Newton (Kochiras, 2009: 373, n. 50 and its appeal to Query 21), she appears to accept the main features of Janiak's reading.

Now, as Kochiras acknowledges the "Letter to Bentley" has elicited extremely different readings. Janiak, 2007 and Friedman, 2009 accept Maxwell's old interpretation of the passage and see in Newton a complete denial of action at a distance (Maxwell, 1890: 316, 487). Henry (1999), Henry (2007), and I insist that Newton is only denying the Epicurean conception of gravity as innate, essential, and inherent to matter (recall chapter 1; see also Ducheyne, 2009). Henry uses Bentley's reaction to interpret Newton as endorsing a superaddition thesis (i.e., God adds gravity to matter at creation) to Newton. By contrast I claim that in developing the view of the *Principia* Newton thinks of gravity as a relational and accidental quality of conjoined matter. Kochiras argues against Henry's claim by noting that superaddition sits uncomfortably with Newton's empiricism. But Henry could argue that it is extremely likely that Newton wished to encourage the superaddition thesis in his pious readers (and may well claim that revelation provides empirical support). Thus far she has offered no argument against my reading.

Janiak's and Kochiras' reading of the "General Scholium" is no less problematic (Janiak, 2008: 173–174). Recall that Newton writes: "God is one and the same God always everywhere. He is omnipresent not; only virtually, but also substantially; for action requires substance. In him all things are contained and move, but he does not act on them nor they on him." Before I turn to how these lines are taken as evidence for an assertion of the "principle of local causation," it is useful to point out the two main points of this passage: first, as has been noted above, it contains Newton's denial that secondary causes do no work (Kochiras, 2009: 270–272); this reading of Newton goes back to David Hume, who noted that "It was never the meaning of Sir ISAAC NEWTON to rob second causes of all force or energy; though some of his followers have endeavoured to establish that theory upon his authority" (footnote at the end of *Enquiry Concerning Human Understanding*, 7.1.25). Second, Newton is asserting that all beings are dependent on God. God's substantial presence is required for all things to exist.

As I argued above, it turns out that for Newton God is the only substance. Now echoing Janiak, Kochiras read these lines as (also) offering the following kind of argument: "Why does Newton think that, in order for God's power to be omnipresent, the substance, God himself, must be present everywhere in space? The answer, it appears, is that Newton believes that a substance must be present where it acts. . . . If his very concept of God is grounded in a principle of local causation, he will be extremely reluctant to allow powers of distant action to any created substance" (Kochiras, 2009: 275–276). The crucial lines reads in Latin: "*Omnipraesens est non per virtutem solam, sed etiam per substantiam: nam virus sine substantia subsistere non potest.*" ('Virus' should almost certainly be 'virtus.')[13] There is a standard distinction

---

[13] I thank Dan Garber for discussion.

in seventeenth-century Latin: acting *"per virtutem"* means acting through one's power, used for something (God, a king, etc.) that can act on something without actually being physically present there. To say that something acts *"per substantiam"* means that it is actually physically there. These are ordinarily used contrastively. What makes Newton's locution so unusual is that he claims that God is not only omnipresent virtually, but also substantially. Newton is, thus, committed to: (1) God is omnipresent virtually; (2) Force/agency must subsist in substance; (3) God is omnipresent substantially.

Now, Janiak and Kochiras assume that there is a hidden premise, what I call the Principle of Local Action (PLA), which connects the second and third commitment. I offer eight arguments against attributing the PLA to Newton: First, I am unfamiliar with a single place where Newton supports it unambiguously in his own voice. So, why suppress it here in the "General Scholium?"

Second, while many seventeenth-century 'new' philosophers rejected action at a distance as 'occult,' in seventeenth-century planetary astronomy people as diverse as Kepler and Hooke embraced variants of it.[14] So when writing the first edition of the *Principia*, Newton may have been familiar with several authors who denied the principle of local causation. Thus, it should be no surprise that in the Scholium to Proposition 69 of Book I of the *Principia*, Newton lists as the very first of four possible, physical explanations of "attraction" that it may be "a result of the action of the [distant!—ES] bodies either drawn toward one another." Moreover, while Kochiras cites Clarke one can equally point to Cotes and—increasingly from Maupertuis onward—most eighteenth-century readers of Newton that were not embarrassed about action at a distance being explained by qualities of matter (see Maglo, 2003).

Third, how would Newton define the difference between local and distant action? Given that, as Janiak and Kochiras note (Kochiras, 2009: 268, 275), Newton rejects the requirement that all action is through contact he has no resources to draw such a distinction. Any distance cut-off point would be arbitrary.

This is related to a fourth point, Newton allows short range action at a distance to explain electricity and repulsion (Cohen in Newton, 1999: 61, drawing on *Opticks* Query 31); this would even be present in the ethereal medium he sometimes posits to account for the transmission of gravity. Kochiras

---

[14] Galileo famously called Kepler's theory of the tides "occult . . . and childish" (Kepler, 2003: 69). I thank Rhonda Martens for calling my attention to it. See also Purrington, 2009, especially 176, where Hooke is also quoted as distinguishing between the medium of gravity and the properties of matter.

appears to think that local action must occur at a given place (2009: 275), so on her account the Newtonian ether violates the PLA. It would be puzzling if Newton were committed to the principle of local action if he seriously entertained the ether to help explain the cause or transmission of gravity.

Fifth it would be very strange if Newton tacitly asserted the (speculative) PLA a few lines before denying that he frames hypotheses. We know that the "General Scholium" was crafted very carefully.

Sixth, the Janiak/Kochiras argument cannot really explain why Newton asserts God's virtual omnipresence in addition to his substantial omnipresence. Kochiras maintains that Newton includes it because it is widely accepted (see also Kochiras 270, n. 29).[15] Janiak/Kochiras assert it is only because of a missing premise that the inclusion of God's virtual omnipresence makes sense. Nevertheless, if one can offer an alternative reading in which there is no missing premise and this premise makes independent sense, then that reading should be preferred.

Seventh I offer an alternative reading: Newton makes no positive claims about the nature of God's actions in these lines. Rather he insists (negatively) that God "does not act on [things] nor they on him." So, it is completely unnecessary to include a missing premise about the PLA. On my reading Newton is agnostic about the way substances bring about effects. For that requires claims about how agents marshal or control the natural powers by the use of forces. Newton just spent the whole *Principia* teaching his readers how hard it is to establish such claims; it would be amazing if he would introduce metaphysical doctrine so carelessly (on what is very dangerous theological ground).

Eight, if we read the sentence in context, that is, as Newton permitting secondary causes and as asserting that God is the only substance, then Newton's phrasing makes perfect sense: God is present virtually in that he is not itself the (proximate) cause of all action in nature; God is present substantially in that all forces must subsist in some substance. Newton is claiming that God's power is latent everywhere in nature. (Recall his criticism against Leibniz for placing God outside of nature.) In the terminology of the De Gravitatione forces get their being from God. So in Newton there is not even an implicit argument that goes from stating that God can act only locally to the denial of action at a distance.

---

[15] In Janiak (2010: 664 and note 10), Janiak also cites More's letter to Descartes, December 1648, in AT 5: 238–239, as evidence for this.

2.6 Is Newton an Extreme Empiricist?

I argued that Newton does not endorse the PLA or passivity of matter. But to state this is not to endorse unqualifiedly the Stein/Di Salle reading in which Newton is a kind of extreme empiricist. Rather, I claim that Newton's empiricism is itself a hard-won achievement. In particular, prior to the second edition of the *Principia*, Newton has at least three nontrivial speculative principles.

First, the original third hypothesis reads: "Every body can be transformed into a body of any other kind and successively take on all the intermediate degrees of qualities" (Newton, 1999: 198). This (alchemical) transformation thesis is a very broad assertion of the homogeneity of matter.[16] The empirical failure to turn lead into gold does not favor it. Newton wisely dropped it in the second edition.[17] Mass as a quantity or measure also presupposes the homogeneity of matter, but does not require the transformation thesis (McGuire, 1970).

Second, Zvi Biener and Chris Smeenk (2013) have called attention to the importance in Newton's development of his 'geometric' conception of matter ("quantity of matter is measured by the volume a body impenetrably fills"). The geometric (as opposed to dynamic) conception of matter is most clearly present in *Principia*, Book 3, Proposition 6, Corollary 3. As my former student Wouter Valentin (2009) has demonstrated, this conception itself rests on the counterfactual (and, thus, speculative) assumption that when there would be no space between matter, all matter would be of the same density. This presupposes yet another homogeneity of matter thesis. In particular, as Biener and Smeenk also note, it is probably a consequence of Newton's atomism (well known from the *Opticks*), which is clearly evident in one of Newton's responses to Cotes' criticism of the corollary: "A body is condensed by the contraction of the pores in it, and when it has no more pores (because of the impenetrability of matter) it can be condensed no more." (quoted in Biener and Smeenk, 2012: 133). Newton's atomism plays a nontrivial role in Newton's matter theory. As Biener and Smeenk show Newton was slow to rid his matter theory of speculative elements.

---

[16] In the preface to the second edition of the *Principia*, Cotes ridicules the mechanists for holding to the homogeneity of matter (Newton, 1999: 385). For the alchemical nature of the third hypothesis, see I. B. Cohen (Newton, 1999: 62).

[17] But in handwritten notes to Book III, Proposition 6, corollary 2 (the one that asserts that gravity is a universal quality of matter and varies in proportion to quantity of matter) Newton was apparently tempted to appeal to the hypothesis (Newton, 1999: 809).

Third, in the De Gravitatione Newton confidently asserted that he possessed the "exceptionally clear idea of extension" (Newton, 2004: 22, 27) and he derived all kinds of nontrivial properties of space from it. In particular, that "there are everywhere all kinds of figures . . . even though they are not disclosed to sight. For the delineation of any material figure is not a new production of that figure with respect to space but only a corporeal representation of it," (Newton, 2004: 22; for a very important contextual reading of this material and Newton's views on omnipresence more generally, see McGuire and Slowik [2013]). Howard Stein and Robert di Salle have both used De Gravitatione to support their readings of space. An appeal to De Gravitatione undermines claims about the consistency of Newton's extreme empiricism.

This short list is not meant to be exhaustive nor have I worked out all the necessary details. Rather these examples are meant to suggest that we can find traces of Newton's speculative commitments in the core of his physical theory. This is not to deny that after all of Huygens', Leibniz's, and Cotes' criticisms, by the time of the reworked second edition of the Principia, Newton had committed himself to ever more methodological empiricism; nor do I wish to deny that Newton's great achievement consists in part to turn metaphysical issues into questions that through theory-mediated measurement permit successively approximated answers (Smith, 2014). But only with the addition of the fourth rule of reasoning is Newton's empiricism finally in place in the third edition of the Principia. That is to say, by the third edition Newton has conceived of his enterprise in such a way that he can ignore the kind of metaphysical challenge Kochiras mounts on methodological grounds.[18]

[18] I thank John Henry, Ori Belkind, and Zvi Biener for detailed comments on an earlier version.

Postscript to Chapter 2

Despite grounding my views in a textual analysis of the *Treatise*, and an attempt to illuminate it in light of early criticism by Kochiras, my view has not generated consensus. In this postscript I engage with the most significant lines of criticism known to me in order to illuminate and improve upon my interpretation of Newton.

In Section I, I respond to Ducheyne's argument that I misunderstand the nature of the *Treatise*. My response will focus on how to understand Newton's contrast between a mathematical and physical point of view in the *Principia* and a similar contrast between a mathematical and real way in the Treatise.

In Sections II–III I engage with critics who have contested: (a) my insistence that Newton conceives of matter as having an active disposition; (b) my rejection of their claim that Newton accepts a principle of local causation. Some scholars seem to think that the rejection of (a) implies the rejection of (b). While I treat (a) and (b) separately, one aim I have is to show that these are conceptually distinct issues. In both sections, I revisit manuscript evidence that has been taken to refute my approach and offer new interpretations of it.

In Section IV, I engage with a series of arguments proffered by Adwait Parker. Alone among my critics, Parker engages me on with the details of my interpretation of the *Treatise*. Because we agree on the significance of the dual conception of matter, Parker's criticisms allow me, in response to correct and restate some of my views. I do so by exploring Newton's evolving account of *vis insita* in his metaphysics of nature.

I. Mathematical Way

Steffen Ducheyne suggests I have misread the nature of the Treatise. In an influential paper he argues that:[1]

---

[1] Some of the material in this section first appeared in Schliesser (2013).

Newton's Metaphysics. Eric Schliesser, Oxford University Press. © Oxford University Press 2021.
DOI: 10.1093/oso/9780197567692.003.0004

To ascribe to Newton what I have called robust action at a distance, is to miss out on significant features of his methodology, namely his desire in *De mundi systemate* to remain neutral with respect to defining "a species or mode of action, or a physical cause or reason [*modum actionis causamve aut rationem physicam*]." . . . Furthermore, there are clear indications that Newton thought that *De mundi systemate* contained no speculations or hypotheses on the cause of gravitational interaction. Newton wrote, for instance, that he wanted "to avoid all questions about the nature or quality of this force, which we would not understood [sic] to determine by any hypothesis." (Ducheyne, 2014: 690 and note 95 on same page)

Ducheyne's paper goes through the scholarly debate over the physical and metaphysical underpinnings of law of gravitation in the *Principia*. It is agreed by all parties to the debate that [A] in the *Principia* itself Newton intends to be agnostic about its underlying cause. (In this postscript I use '[A]' to refer to this position.) That is compatible with four other claims: (i) that Newton's early readers attributed all kinds of non-agnostic positions to Newton; (ii) that the *Principia* itself constrains any possible physics and metaphysics of the law of gravitation; (iii) that there is evidence of Newton's own views on the physics and metaphysics of the law of gravitation in his other writings that may either bear on (ii) or explain (iv) what Newton himself held about the physics and metaphysics of the law of gravitation, but did not print in the *Principia*.

In the first two chapters above, critically following Stein (2002), I claim (and this bears on (ii)-(iii)-(iv)) that Newton's discarded original draft of the system of the world (which ended up being rewritten as book 3 of the *Principia*), the *Treatise*, offers evidence for attributing to Newton the idea that, while drafting the *Principia* in the mid-1680s, he thought that gravity was the effect or manifestation of a twofold disposition, a nonintrinsic relational quality of matter.[2]

As I argue in chapters 1–2, on my view a body has two dispositions: a "passive" disposition to respond to impressed forces which is codified in the second law of motion, whereas an "active" disposition to produce gravitational force is treated as a distinct interaction codified in the third law of motion. And the "cause" of the action is "the conspiring nature of both" bodies. For the "conspiring" to occur, the bodies must share a "nature." In my argument, following Stein (2002), I relied on the following passage from Newton's *Treatise*:

---

[2] And so lurking in the *Principia* is a kind of matter theory (something Katherine Brading [2012, 2018 forthcoming] and Biener and Smeenk [2012] have argued on different, more solid grounds).

For all action is mutual, and makes the bodies mutually to approach one to the other, and therefore must be the same in both bodies. It is true that we may consider one body as attracting, another as attracted. But this distinction is more mathematical than natural. The attraction is really common of either to other, and therefore of the same kind in both.

And hence it is that the attractive force is found in both. The Sun attracts Jupiter and the other Planets. Jupiter attracts its satellites. And for the fame reason, the satellites act as well one upon another as upon Jupiter, and all the Planets mutually one upon another.

And though the mutual actions of two Planets may be distinguished and considered as two, by which each attracts the other; yet as those actions are intermediate, they don't make two, but one operation between two terms. Two bodies may be mutually attracted, each to the other, by the contraction of a cord interposed. There is a double cause of action, to wit, the disposition of both bodies, as well as a double action in so far as the action is considered as upon two bodies. But as betwixt two bodies it is but one single one. 'Tis not one action by which the Sun attracts Jupiter, and another by which Jupiter attracts the Sun. But it is one action by which the Sun and Jupiter mutually endeavour to approach each the other. (Newton, 1728: 38–39)[3]

---

[3] George Smith has generously provided me with a new translation from the manuscript. In the body of the text I have used the 1728 translation because this could have been known to eighteenth-century readers of Newton. Here is Smith's and Whitman's translation (with their editorial apparatus left in) from Newton forthcoming:

> For the action is mutual, and makes the bodies by a mutual endeavor [conatu mutuo] (by Law three) approach one another [ad invicem], and accordingly this [action] ought to be in confor-mity [conformis esse debet] with itself on both bodies [in corpore utroque]. One body can be considered as attracting [attrahens], and the other as attracted [attractum], but this distinc¬tion is more mathematical than natural. The attrac¬tion [attractio] is really [revera] both bodies on both, □and so of the same kind in both.
> And hence it is that the attractive [attactiva] force is found [reperiatur] in both. The Sun attracts [trahit] Jupiter and the other Planets, Jupiter attracts [trahit] its Satellites and similarly the Satellites act on one another [in se invicem] and on Jupiter, and all the Planets act on each other [in se mutuò].
> And although, in a pair of Planets, the actions on each can be distinguished and can be con¬sidered as paired actions by which each attracts [trahi] the other, they are not two but a simple operation between two termini. By the contraction of one rope insofar as between yet insofar as they are between [intermediae] [the two], they are not two, but a simple operation between two ter¬mi¬ni. Two bodies can be drawn [trahi] to each other by the contraction of a single rope between them. The cause of the action is two-fold, namely the disposition [dispositio] of each of the bodies; the action is likewise two-fold, insofar as it is upon two bodies: but the operation by which the Sun insofar as it is between two bodies it is simple and single. There is not, for example, one operation by which the Sun attracts [trahit] Jupiter and another operation by which Jupiter attracts the Sun, but a single operation by which the Sun and Jupiter endeavor to approach on another.

When confronted by my interpretation of this passage, Ducheyne argues that Newton's language is "rather loose, and without aiming at 'speculatively metaphysical' conclusions." (2014: 689–690) One line of evidence for this, according to Ducheyne, is the following passage that he (partially) quotes in an accompanying footnote:

> But our purpose is only to trace out the quantity and properties of this force from the phænomena, and to apply what we discover in some simple cases, as principles, by which, in a mathematical way, we may estimate the effects thereof in more involved cases. For it would be endless and impossible to bring every particular to direct and immediate observation.
>
> We said, *in a mathematical way*, to avoid all questions about the nature or quality of this force, which we would not be understood to determine by any hypothesis; as it is a force which is directed towards some center; and as it regards more particularly a body in that center, we call it circumsolar, circum-terrestrial, circum—jovial, and in like manner in respect of other central bodies. (Newton, 1728: 4–5; emphasis in original)

I agree with Ducheyne that this passage seems to be a problem for my interpretation. It seems that when Newton composed the *Treatise*, he held the agnostic, methodological stance of the *Principia*, that is [A]. In context, it is clear that "this force" refers to the force that keeps planets in orbits. And, in context, Newton discusses and rejects a number of hypotheses (solid orbs, vortices, principle of impulse or attraction, etc.) which he attributes to ancient and seventeenth-century thinkers.

On Ducheyne's reading of the passage, Newton's mathematical way avoids discussion of the ontology and metaphysics of forces—that is, it avoids speculation. However, if we go back to the first paragraph of the passage quoted from *Treatise* 38–39 that I quoted above, in light of Ducheyne's emphasis on the mathematical way, we notice something fascinating:

> For all action is mutual, and makes the bodies mutually to approach one to the other, and therefore must be the same in both bodies. It is true that we may consider one body as attracting, another as attracted. But this distinction is more mathematical than natural. The attraction is really common of either to other, and therefore of the same kind in both.

Here Newton explicitly offers a contrast between the "mathematical" and a "natural" way. And he treats the natural way as what "really" is the case. So, here Newton explicitly violates the injunction Ducheyne attributes to Newton. Newton does so in terminology that evokes the passage Ducheyne offers as evidence against the legitimacy of this happening.

One way to handle this is to claim that Newton is contradicting himself in the *Treatise*. (Nobody has suggested that, as far as I know.) Another, which is Ducheyne's approach, is, despite the distinctions invoked by Newton, to treat the passage on *Treatise* pp. 38–39 as "loose" talk. But this means that Ducheyne must explain away Newton's invocation of the very contrast that grounds Ducheyne's original interpretation—and Ducheyne has not done so. So Ducheyne's own argument is at best greatly underdeveloped.

Another way, which is the one I prefer, is that Newton means what he says and says so explicitly. But does that leave me attributing to Newton a singular violation of his own methodological principle [A]? I don't think so. Because I deny that Newton holds [A] in the *Treatise*.

For, in the *Treatise*, the point of Newton's 'mathematical way' is to use it as a means to infer features of the nature of reality, including that Copernicanism is to be adopted. This becomes clear if we look at what Newton does after he has explained the nature of centripetal forces. For example, he then goes on to "infer" that "[T]hat there are centripetal forces *actually* directed to the bodies of the Sun, of the Earth, and other Planets," (Newton, 1728: 10; emphasis added). In fact, Newton argues from "astronomical experiments" that it "follows by geometrical reasoning, that there are centripetal forces, *actually* directed (either accurately or without considerable errour) to the centers of the Earth, of Jupiter of Saturn, and of the Sun." (12; emphasis added. The nature of an astronomical experiment is of interest, but I leave it aside here.)[4]

So, it is misleading to use Newton's claims about the "mathematical way" as ruling out the possibility that Newton is making claims about the nature of reality. Rather, Newton treats the mathematical way as a means to infer all kinds of things about what actually or really is the case. That is to say, the

---

[4] In an edited Newton translation forthcoming, Smith and Hesni note an important feature of this: the *Treatise* "never so expressly identifies the centripetal forces in our planetary system with terrestrial gravity. Where Book 3 becomes an exposition of Newton's theory of gravity, *Liber Secundus* remains instead an exposition of his theory of celestial centripetal forces!"

mathematical way is a mechanism to make inferences from theory-mediated measurements and (astronomical) experiments to the nature of reality. (Smeenk, 2016). Of course, to say *that* is not to do full justice to the nature of Newton's inferential arguments.

My disagreement with Ducheyne over the mathematical way in the *Treatise* echoes an analogous disagreement with him (and Janiak, 2008: 58–65 and Janiak, 2013: 404) over the contrast between the mathematical and physical points of view in the *Principia*. In chapters 2–3 of his excellent (2011) book, Ducheyne divides the presentation of Newton's methodology into two distinct steps: (1) a phase of model construction and (2) a phase of model application cum theory formation cum theory testing, respectively. (Ducheyne, 2011: 56–57) Echoing Reichenbach, Ducheyne is not interested in exploring how Newton may have generated his results, only their justification. (Ducheyne, 2011: 64). Ducheyne's aim is to distinguish Newton's account from the more familiar hypothetico-deductivism of Newton's contemporaries and explore to what degree Newton's method was "more stringent and demanding" (Ducheyne, 2011: 62; see also Harper, 2011 and Smith, 2014).

Ducheyne identifies phase (1) with Book I of the *Principia* and phase (2) with Book III. Moreover, Ducheyne asserts that Book I is a "mathematical study" that (a) makes no claim about the "existence" of centripetal forces "in nature" and (b) abstracts "from the forces in the empirical world" (Ducheyne, 2011: 67). While in Book III, Newton "commits to centripetal forces" (67). This seems to track the distinction that Newton draws between mathematical treatment and a treatment in physics (e.g., in the scholium to section 11 of Book 1, quoted at Ducheyne, 2011: 66):

[C]onsidering in this treatise not the species of forces and their physical qualities but their quantities and mathematical proportions, as I have explained in the definitions.

Mathematics requires an investigation of those quantities of forces and their proportions that follow from any conditions that may be supposed. Then, coming down to physics, these proportions must be compared with the phenomena, so that it may be found out which conditions [or laws] of forces apply to each kind of attracting bodies. And then, finally, it will be possible to argue more securely concerning the physical species, physical causes, and physical proportions of these forces. (Newton, 1999: 588–589)

There are a number of problems with Ducheyne's approach. First, in Ducheyne's account, the status of Book II of the *Principia* is unclear. Initially, in Ducheyne's section 2.4.4, Book II is assimilated to the mathematical modeling side (see also 2011: 81 n. 148). But this is at odds with quite a bit of the contents of Book II. I offer some examples: (a) Ducheyne notes that the conclusion of Book II rules out the existence of Cartesian planetary vortices in the world (2011: 125). But the argument is, in part, very empirical: Section 9 of Book II argues that the vortex theory cannot satisfy both Kepler's second and third laws and is therefore incompatible with observed celestial motions (Smeenk & Schliesser, 2013). As Ducheyne recognizes, Newton formulates this argument as conditional on an explicit hypothesis that fluid friction leads to a resistance force proportional to the relative velocity (see section 3.3). (b) Section 8 of Book II offers physical models of and makes nontrivial experimentally established claims about the speed and nature of sound and light waves (Newton, 1999: 762–778; something Ducheyne does not discuss). In fact, Book II has a stunning number and variety of experiments (Smith, 2020). So, from the vantage point of Ducheyne's framework, Book II belongs more in Phase (2) then in Phase (1) (as Ducheyne seems to acknowledge at p. 108).

Moreover, upon closer inspection, even Book I causes problems for Ducheyne's framework. At the start of section 11 (of Book I), Newton acknowledges that the preceding results are all based on an unphysical assumption: they concern "bodies attracted toward an immovable center, such as, however, hardly exists in the natural world" (Newton, 1999: 561). In light of the third law, Newton then introduces more realistic assumptions into his model that (in propositions 58–61) apply to Keplerian motion (of Earth–Moon–Sun systems). Subsequent sections relax even more of the simplifying assumptions, and the models become increasingly less idealized. This suggests that Newton recognized that his mathematical models had *differing* relationships to reality (Janiak, 2008: 60). This is especially evident in the scholia that Newton included in Book I—these often deal with issues of traditional natural philosophy; for example, the last two scholia and the surrounding propositions in the final section (14) of Book I treat of Snell's law, the nature of lenses, and the speed of light (Newton, 1999: 625–629).

All these examples also raise a further worry, that Newton's contrast between the mathematical and the physical has not been properly understood. In particular, a number of commentators have a tendency to treat the mathematical models as somehow lacking in "empirical" or "referential" content (e.g., Ducheyne, 2011: 105–106).[5] There are, in fact, three problems with Ducheyne's claim: first, it sits uneasily with Newton's qualified empiricism about geometry, which is proudly announced in the Author's preface to *Principia*. That is, "geometry is founded on mechanical practice" in the sense that it turns to mechanics for the construction or generation of the objects used in geometrical reasoning, (Newton, 1999: 382) A geometric object such as a curve is understood in terms of how it can be generated by motion (Smeenk, 2016).

Second, it misunderstands Newton's account of abstraction; by 'abstraction' Newton does not mean 'un-empirical'; rather he means—in accord with his Scholastic and Platonic sources—something more akin to focusing on some isolated qualities found within the empirical world without getting distracted by other empirical qualities (Domski, 2012). Abstracted, mathematical models give a partial account of the world.

Third, Newton's laws and definitions, which enter into the models of Phase (1), are by Newton's lights "accepted by mathematicians and confirmed by experiments of many kinds" (Newton, 1999: 424, quoted in Ducheyne, 2011: 72). They are, in fact, a very creative assimilation of the major result of seventeenth-century mechanics. As Chris Smeenk and George Smith have taught me, one of the highlights of Book I of the *Principia*, Section 10, shows how to recover the Galileo-Huygens results on constrained motion in his framework in order to evaluate Huygens' measurements of surface gravity (Smeenk and Schliesser, 2013). So, the laws and definitions themselves *presuppose* a certain amount of empirical adequacy.

I offer three, jointly supporting diagnoses of these problems. First, Newton's contrast (Newton, 1999: 407–408) between the mathematical and physical is drawn *within* the empirical realm. Sometimes Ducheyne seems to recognize this by calling the models "physico-mathematical" (Section 2.4.4). By this Ducheyne means that "they have the potential of providing information about real physical forces once relevant empirical measurements are provided" (Ducheyne, 2011: 80). But, second, Ducheyne's way of framing this understates the extent to which models presuppose considerable commitment to what the nature of reality is (as is also indicated by Newton's treatment of loosening idealizing assumptions through Book I shows). Ducheyne is right to think that the models do not overdetermine which particular forces

---

[5] To be sure, this is not Janiak's official position because for him forces are quantities and "they can be measured by measuring other, obviously physical, quantities" (Janiak, 2008: 60). It is not entirely obvious why these other quantities are "obviously physical."

(i.e., causes) are to be found in the world, but they do presuppose a positive commitment to nontrivial elements of the world (mechanical principles, nature of space, time, motion, etc., as he recognizes at Ducheyne, 2011: 68–74).

So, I claim that in the *Principia* and in the *Treatise*, when Newton is describing a mathematical way (*Treatise*) or a mathematical point of view (Newton, 1999: 408), he is not emptying these claims from empirical content. Rather, he is describing the inference-licensing status of a particular claim (Smeenk, 2016).

So, to return to my disagreement with Ducheyne over the *Treatise*—that Newton does not treat the passage (Newton, 1728: 38–39) as "loose" talk is also evinced by his illustrative claim directly following it that the "action betwixt the load-stone and iron is single, and is considered as single by the philosophers." (40) In fact, it is no coincidence that Newton draws attention to magnetism because his treatment here (of the analogy between magnetic and gravitational action) evokes the wording of Gilbert's (1651) *De Mundo*. (Schliesser, 2015a: 13; for fuller details on the relationship between Gilbert and Newton see Parker, 2019.)

That is to say, Newton draws attention to the fact that the very conceptual apparatus he is deploying is one used by other philosophers (including the seventeenth-century expert on magnetism). That would be very odd if he were merely talking loosely. Of course, my position entails that Newton came to see that his way of putting things in the Treatise was likely to be thought a kind of speculative talk he was explicitly ruling out when he rejected the explanatory demands (the "hypotheses") of the mechanical philosophers to put all explanations in terms of the size, figure, and motion of colliding bodies.[6] And so in *Principia* he embraced [A].

II. Passivity of Matter

While I insist that on my interpretation of Newton, from a certain perspective matter has a "passive" disposition to respond to impressed forces which is codified in the second law of motion, most of my critics have concluded that on my interpretation Newton rejects the passivity of matter, which, say,

---

[6] In his commentary on the *Treatise*, George Smith notes, "Nowhere in any edition of the *Principia* is there any such attempt to argue that the gravitational interaction of two bodies consists of a single action, not a pair of actions on the respective bodies. In other words, no edition of the *Principia* contains even a remote counterpart of the Article."

a mechanical philosopher would accept. This is no surprise because I have emphasized the "active" disposition to produce gravitational force, which is treated as a distinct interaction codified in the Third Law of Motion. In this section I return to an unpublished manuscript which is thought to refute decisively my interpretation of Newton.

In particular, Steffen Ducheyne (2014: 691–692) draws on an unpublished manuscript sentence, "Matter is a passive principle & cannot move itself." He refers back to it twice more (Ducheyne, 2014: 692–693). The reason why it is important to Ducheyne's argument is that it seems to give him the smoking gun he needs. It's one of the few places where Newton explicitly says that matter is passive. In the initially published version of chapter 2, I had discussed this passage, when responding to a (2009) criticism by Hylarie Kochiras (who had criticized an early version of my argument now published as chapter 1) I wrote (now quoting chapter 2):

> Yet, the published version of the *Opticks*, is agnostic on this matter. In Query 31 he contrasts the "passive principle by which the bodies persist in their motion or rest, receive motion in proportion to the force impressing it, and resist as much as they are resisted" with "active principles, such as are the cause of gravity, by which planets and comets keep their motions in their orbits, and bodies acquire great motion in falling; and the cause of fermentation.".
>
> Shortly thereafter, he lists "gravity, and that which causes fermentation, and the cohesion of bodies" among the "active principles." (Newton, 1952: 400–401)

My response to Kochiras clearly did not convince Ducheyne (who cites Kochiras' exchange with me, so he is aware of it) nor Brown (2016: 41, n.4), who cites Ducheyne approvingly. So, normally I would now throw up my hands and leave it to others to figure out who gets the better of the debate. But it may be worth elaborating on two distinct new, additional arguments. The first argument would be to concede Ducheyne's and Kochiras' reading of the manuscript evidence, and fall back on the possibility that in this, as in other matters, Newton changed his mind and that after 1700 he was more inclined to accept the passivity of matter altogether. This response would be open to me because I am primarily committed to attributing to Newton the view sketched above to the mid-1680s (as he was drafting the *Principia*). In the *Principia* itself he is explicitly agnostic [A], after all—and this is agreed by everybody.

This (first) concessive response would allow me to take on board some of Ducheyne's arguments about the theological significance of embracing the passivity of matter; or at least recognizing that Newton actively distanced himself from being taken to embrace active matter after the initial reception of the *Principia*. These theological issues and the question of the nature of matter are related to the challenges to Newton of being taken as a kind of neo-Epicurean, as Leibniz, Kant, and Adam Smith hint (see chapter 3), or being taken to be a crypto-Spinozist, as Leibniz also suggests at the start of his correspondence with Clarke (Letter 2.8). In chapter 8, I discuss how responding to Toland pushed Clarke, who is often cited as an authority by my critics, into embracing the passivity of matter.

But as Ducheyne correctly notes in his paper, I have been greedy; I have also used other passages from different periods in Newton's life as converging evidence for my interpretation of his view in the mid-1680s. And so I have suggested that except for when Newton toys with ether theories, my reconstruction of his dispositional account of gravity is Newton's mature fallback position.

Another approach is not to be concessive. So, second, I simply deny that in the unpublished draft version of what was to become Query 31, Newton is presenting his own view about the passivity of matter. For I claim that when Newton writes that "Matter is a passive principle & cannot move itself," he is simply reporting or summarizing the Cartesian position on matter! This option is overlooked by my critics. Before I explain that let me quote a wider version of the passage (with a lot of editorial markings removed—Newton obsessively corrected his own texts):

Qu. 23. By what means do bodies act on one another at a distance. The ancient Philosophers who held Atoms & Vacuum attributed gravity to Atoms without telling us the means unless perhaps in figures: as by calling God Harmony & representing him & matter by the God Pan & his Pipe, or by calling the Sun the prison of Jupiter because he keeps the Planets in their orbs. Whence it seems to have been an ancient opinion that matter depends upon a Deity for its laws of motion as well as for its existence. The Cartesians make God the author of all motion & its as reasonable to make him the author of the laws of motion. Matter is a passive principle & cannot move it self. It continues in its state of moving or resting unless disturbed. It receives motion proportional to the force impressing it. And resists as much as it is resisted. These are passive laws & to affirm that there are no other is to speak against experience. For we find in our selves a power of moving our bodies by our thought Life & will are active Principles by which we move our bodies, & thence arise other laws of motion unknown to us.[7]

Newton is explicitly taking on the question of action at a distance. He first describes the innate gravity position of the ancient atomists (or Epicureans). He also calls attention to the possibility that they used esoteric and figurative speech to explain the mechanism behind their position. ("The God Pan & his Pipe" inspired the title of McGuire and Rattansi, 1966, one of the most famous papers in contemporary Newton scholarship.) Newton sums up his treatment of the ancients by suggesting they thought God was needed to account for the laws of motion and the existence of matter.

As an aside, something of this view of the Ancients survives in the "General Scholium" added to the second (1713) edition of the Principia, even against the evidence of the Ancients, who freely speculated about to what degree God/Gods are dispensable to the Epicurean natural philosophy. For example, in Cicero we find the remark, inspired by Posidonius on the *Nature of the Gods*, that "Epicurus does not believe in any gods, and that the statements which he made affirming the immortal gods were made to avert popular odium" (Cicero *De Natura Deorum* 1.123). Newton reads the Epicureans as natural philosophical theists of a certain sort. My view is that he is doing this in order to try out saying that the atypical Epicureanism his critics are attributing to him need not be "atheist" in character. (He may have even sincerely thought that this was a solid reading of the Epicureans.) That fits a lot of other material in this manuscript (which offers a robust assertion, inter alia, for the reality of final causes in experience).

After discussing the atomist position, Newton then moves on to describing the Cartesian position. Now on the reading advanced by Ducheyene and Kochiras the description of the Cartesian position is just one sentence: "The Cartesians make God the author of all motion." This would attribute to the Cartesians, in effect, the idea that Descartes is a kind of occasionalists such that God is the author of all motion.[8] On the reading advanced by Ducheyne and Kochiras, Newton then moves on to state his own view.

---

[7] Copied from the Newton project, http://www.newtonproject.ox.ac.uk/view/texts/normalized/NATP00125

[8] For a useful introduction to how some followers of Descartes attributed occasionalism to him, see Garber (1987).

But first, Newton usually does not treat the Cartesians as occasionalist. There is an equally respectable Cartesian interpretation in which occasionalism is not the proper Cartesian position (Hattab, 2000, 2007), but rather that God is responsible (or the author of) the laws of motion.[9] In fact, that's the next half of the sentence, that is, "it's as reasonable to make him the author of the laws of motion."

Second, it is more plausible, thus, that here Newton gives a summary of the Cartesian metaphysics of nature: "Matter is a passive principle & cannot move it self. It continues in its state of moving or resting unless disturbed." So, *pace* Ducheyne and Kochiras, the natural reading here is that what they take to be Newton's position is really just Newton giving a quick and dirty summary of the Cartesian natural philosophy that he rejects!

To be sure, I do not mean to deny that in this manuscript Newton does shift to explaining features of his own position—he regularly invokes the authority of experiments. And where exactly he does that shifting is a matter of judgment. But at the same time he also keeps shifting toward describing alternative views he rejects.

In addition, if we read the manuscript in the way proposed by Kochiras and Ducheyne then Newton draws a sharp contrast between entities with minds, who can be active ("thinking is an active principle"), and passive material entities that lack such power. In fact, Kochiras and Ducheyne end up treating Newton by implication as a property dualist; I return to this issue in the next section.

But while it is not impossible that Newton is a property dualist, Newton himself is much more cautious than that in the manuscript quoted. He makes it clear that all living things are active. And it is, for Newton in this very document, an open question whether (as the Stoics thought) all of nature is alive in that relevant sense: "We find in our selves a power of moving our bodies by our thoughts & see the same power in other living creatures but how this is done & by what laws we do not know. We cannot say that all Nature is not alive. not know her laws or powers any further then we gather them from Phænomena." Nobody has doubted that this is Newton's own position. And this suggests to me that Newton is much more agnostic about the nature of matter in this document than, say, Kochiras and Ducheyne have allowed. Because Newton allows that there may be as of yet undiscovered principles

---

[9] Hattab is responding to modern interpreters of Descartes that treat him as an occasionalist.

and laws that can explain the materiality and mindedness of living things. On my reading, these would also be laws and principles of matter.

What Newton does explicitly deny is that *vis inertia* can explain the generation of (new) motion: "vis inertiæ they continue in their state of moving or resting & receive motion proportional to the force impressing it & resiste as much as they are resisted, but they cannot move themselves; & without some other principle then [sic] the vis inertiæ there could be no motion in the world." Ducheyne thinks this also shows that for Newton matter is passive (Ducheyne 2014: 691). But that's not right. For even I assert—and this also follows naturally from the second law of motion—that *vis inertiæ* is a passive principle, but it does not constitute the whole of Newton's (implied) matter theory in the *Principia*.

III. Principle of Local Action

In chapter 2, and its original version, I granted that Clarke uses the "General Scholium" in support of attributing a PLA to Newton in his correspondence with Leibniz. I offered arguments against accepting Clarke's claim as an interpretation of Newton. (In chapter 8, I explore the circumstances that help explain Clarke's choices.) So, I was surprised to see Gregory Brown argue that Clarke's embrace of the PLA offers "presumptive evidence that" Newton adhered to the PLA (Brown, 2016: 42). And while not going that far, Andrew Janiak (2013) also repeatedly draws on Clarke's interpretation of Newton as evidence for Newton's acceptance of the PLA.

I have to admit that this reliance on Clarke surprised me. For, since the pioneering work of McGuire and Stein, I took it to be common ground among philosophical, contemporary Newton scholars that one should not read Clarke's claims in the Correspondence with Leibniz as Newton's self-understanding; and that Clarke's other writings, while instructive, do not provide us with an authoritative interpretation of Newton's philosophical commitments.

In particular, it is very odd that Cotes' interpretation of Newton is, if not neglected altogether, discounted. Brown (2016) makes no mention of Cotes at all, and Janiak (2013) does not engage with Cotes' interpretation of Newton's matter theory. For, in his editorial preface to the second edition of the *Principia*, after surveying the empirical evidence and finding that gravity was a universal quality of matter, Cotes went farther than Newton and claimed that "among the primary qualities of all bodies universally, either gravity will have a place, or extension, mobility, and impenetrability will not." (Newton, 1999: 392) And Cotes is quite explicit that gravity *is* a "property of all bodies" (Newton, 1999: 392) and, in context, he is naturally read that this property gives rise to action at a distance, or, at least the propagation of forces across vast distances toward other bodies. (Newton, 1999: 390)[10]

I do not claim that Cotes and Newton have the same position. But on my reading of Newton, Cotes, who is by all accounts a more subtle reader of Newton than Clarke, has a position closer to Newton's than Clarke's; and I do not see why we should treat Clarke's interpretation as more authoritative than Cotes'. We still need to make up our own minds.

Of course, be that as it may, if there were direct evidence that Newton embraced the PLA or explicitly denies the possibility of action at a distance, then we would have reason to treat Clarke as more authoritative than Cotes. Rather than rehearsing and repeating my interpretation of the famous letter to Bentley,[11] I discuss three other passages that have been brought to bear against my approach. The first and third passages may be thought to offer straightforward refutations of my views. In my response to the first passage I return to the question of Epicurean gravity. The second passage is more complicated and will lead me to discuss Newton's views on the metaphysics of mind-body interaction.

The first passage is quoted by Janiak (2013: 403):

For two planets separated from each other by a long empty [*vacui*] distance do not attract each other by any force of gravity or act on each other in any way except by the mediation of some active principle [*movente principio*] interceding between them by which the force is transmitted from one to the other. . . .[12]

Janiak then adds for good measure, "Unlike the Bentley correspondence, this is a direct expression of Newton's own understanding of gravity, one that is entirely independent of the questions about Newton's views of the essence of matter and his attempts to distance himself (perhaps) from what were then called 'Epicurean' conceptions of atoms in the void (cf. Henry, 2011). There

---

[10] To be clear, Cotes, too, thinks this leaves the cause of gravity itself unexplained. (392)

[11] Luckily I have the indefatigable John Henry (2019, 2020) on my side!

[12] In the accompanying Janiak cites it as follows: "unpublished manuscript, University Library Cambridge, Add. MS 3965.6, f.269; quoted in Casini (1984, 38)."

cannot be a clearer statement of his views: two planets do not act on each other in any way except through some mediating item." (Janiak, 2013: 403). I am, of course, pleased that Janiak grants that the Bentley exchange is not as straightforward as sometimes is suggested.

However, I am not convinced that this way of reading the passage is indeed a direct expression of Newton's own understanding of gravity. The broader passage reads as follows:

> The Epicureans making a distinction of the whole of nature into body and void, denied the existence of God, but very absurdly. For two planets separated from each other by a great expanse of void do not mutually attract each other by any force of gravity or act on each other in any way except by the mediation of some active principle that stands between them by means of which force is propagated from one to the other. [According to the opinion of the ancients, this medium was not corporeal since they held that all bodies by their very natures were heavy and that atoms themselves fall through empty space toward the earth by the eternal force of their nature without being pushed by other bodies.] Therefore the ancients who grasped the mystical philosophy as Thales and the Stoics more correctly taught that a certain infinite spirit pervades all space, and contains and vivifies the entire world; and this supreme spirit was their numen; according to the poet cited by the Apostle: In him we live and move and have our being. Hence the omnipresent God is recognized, and by the Jews is called 'place'. To the mystical philosophers, however, Pan was that supreme numen . . . By this symbol, the philosophers taught that matter is moved in that infinite spirit and by it is driven, not at random, but harmonically, or according to the harmonic proportions as I have just explained.[13]

As McGuire notes (1968: 169), the manuscript probably dates from the mid-1690s when Newton became eager to situate his own work in light of ancient Epicureanism (chapter 3 tries to explain why this would be so). Here the mediating active principle just is the Stoic world soul, which appears to be an electric spirit of some sort.

However, a very natural reading of the "General Scholium" of the *Principia* is that he denies that God is a world soul (Newton, 1999: 940; I quote and discuss the passage below). So, we are by no means required to accept this

---

[13] I have consulted Kochiras (2008: 109) and McGuire (1968: 169).

manuscript (MS 3965.6, f.269) as an authoritative expression of Newton's all-things-considered views.

In addition, and more controversially, it is not obvious to me that we are supposed to treat the key sentence—"For two planets separated from each other by a great expanse of void do not mutually attract each other by any force of gravity"—as really expressing Newton's own position. For, he could be expressing the Epicurean position here. For a natural way to read the Epicureans is that two planets separated from each other by a great expanse of void do *not* "mutually attract each other by any force of gravity."

For, Epicurean gravity is directed down toward a privileged place (Epicurus, "Letter to Herodotus," Section 61). And this is so in infinite worlds (Section 45). In a void, atoms, which make up bodies, do not interfere with each other unless they accidently deflect each other from their path (Letter to Herodotus, section 44). So, in an Epicurean cosmology, the atoms on a planet move down, barring deflection and the swerve, but planets or the atoms that compose them do not attract each other. This also helps explain why early modern critics of Newton such as Huygens were inclined to distinguish conceptually among terrestrial gravity, celestial gravity among the planets, and universal gravity.

So, if one is antecedently convinced that Newton rejects the possibility of action at a distance, and embraces the PLA, then indeed taken in isolation— "for two planets separated from each other by a great expanse of void do not mutually attract each other by any force of gravity or act on each other in any way except by the mediation of some active principle that stands between them by means of which force is propagated from one to the other"—this does seem like a smoking gun. But the sentence before it and the sentence after it discuss ancient theories. In fact, the following sentence—"bodies by their very natures were heavy and that atoms themselves fall through empty space toward the earth by the eternal force of their nature without being pushed by other bodies"—just conveys the Epicurean position.

So a more natural reading of the first three sentences of the quoted part of the manuscript is as a description of a debate between Epicurean atheists and ancient (mystical) critics. I am willing to believe that Newton agreed with the critics that the Epicurean position was absurd. Not just because the Epicureans could be taken to deny God's existence, but also because Newton himself has established that Jupiter and Saturn influence each other (Newton, 1999: 818). So, this passage does not support commitment to PLA.

Second, in my response to Kochiras, and in Janiak's reply to me, the following passage from the "General Scholium" plays a crucial role.

Every sentient soul, at different times and in different organs of senses and motions, is the same indivisible person. There are parts that are successive in duration and coexistent in space, but neither of these exist in the person of man or in his thinking principle, and much less in the thinking substance of God. Every man, insofar as he is a thing that has senses, is one and the same man throughout his lifetime in each and every organ of his senses. God is one and the same God always and everywhere. He is omnipresent not only virtually but also substantially, for active power [virtus] cannot subsist without substance. In him all things are contained and move, but he does not act on them nor they on him, experiences nothing from the motions of bodies; the bodies feel no resistance from God's omnipresence. (Newton, 1999: 941–942)

The first three sentences of the quoted paragraph were inserted into the third (and final) edition of the *Principia* (while the rest was added to the second edition). Janiak reiterates that in the quoted passage there is a hidden premise that "a substance cannot act where it is not substantially present" and this reinforces his argument that Newton (always) rejects unmediated action at a distance (Janiak, 2013: 399). My present strategy is not to go over his responses to my initial argument against this claim, but rather to offer a different interpretation of the passage altogether. My interpretation shows, if correct, we are not in a position to attribute the PLA to God according to Newton.

In context, in the last sentence of the previous paragraph, Newton has just asserted that God is eternal and omnipresent ("will not be never or nowhere," [Newton, 1999: 941]). In the paragraph following the one just quoted he claims that "the supreme God necessarily exists, and by the same necessity he is always and everywhere" (Newton, 1999: 942). That is to say, and as argued at length in chapters 3–5 & 7–8, Newton's God is immanent in the universe.

And, in fact, as "the same necessity" suggests, Newton's God's relationship to space and time is very tight, perhaps even an internal relation; rather than creating space and time, by an act of will, they exist when God exists in virtue of the same necessity. How to think of Newton's modal metaphysics here is no simple matter (see chapter 8). In De Gravitatione, Newton had tried to characterize this relationship in terms of an emanation doctrine, and there, too, it signals that space and time are not created (or so I argue in chapters 5–7).

It's crucial to Newton's general argument that space and time are neither nothing nor merely ideal or useful conventions. (As I explain in chapter 5, in De Gravitatione Newton emphasizes this. In the "General Scholium" it's treated as self-evident.) And it follows from this that there are structural or, as I put it in chapter 9, ordered features of the universe (space and time) that are as infinite and eternal as God is. My interest in this postscript is to characterize one of the relationships between sentience and this spatial-temporal structure.

A reading of the quoted passage (from Newton, 1999: 941–942) is that God is ensouled sentience. (I return to this below and qualify it.) And such ensouled sentience, of the right sort, is constitutive of a person. But unlike say, in Descartes' metaphysics, Newtonian ensouled sentience can be 'spread out' in space and time. Yet, like a Cartesian soul, such a Newtonian ensouled sentience does not, despite existing in space and time, itself have parts or is even divisible.

Now, an obvious way to make sense of the contours of Newton's position on ensouled sentience in the General Scholium is that he thinks sentience is a kind of emergent property of matter immanent in nature. And so Newton articulates a kind of property dualism.[14] This position works reasonably well for what he says about human persons, but, as I argue below, not for God.

In fact, Newton returns to the question of ensouled sentience in his fascinating closing paragraph of the "General Scholium":

A few things could now be added concerning a certain very subtle spirit pervading gross bodies and lying hidden in them; by its force and actions, the particles of bodies attract one another at very small distances and cohere when they become contiguous; and electrical [i.e., electrified] bodies act at greater distances, repelling as well as attracting neighboring corpuscles; and light is emitted, reflected, refracted, inflected, and heats bodies; and all sensation is excited, and the limbs of animals move at command of the will, namely, by the vibrations of this spirit being propagated through the solid fibers of the nerves from the external organs of the senses to the brain and from the brain into the muscles. But these things cannot be explained in a few words; furthermore, there is not a sufficient number of experiments to

---

[14] Gorham argues that in De Gravitatione all minds are "absolutely incorporeal and indivisible substances." And so that Newton is a kind of substance dualist (just not a Cartesian one). (Gorham, 2011b: 32) Crucially, on my interpretation of De Gravitatione mind is necessary for, and coextensive with, activities in the body and, simultaneously, body is intrinsically perceptible by minds (see chapter 5).

determine and demonstrate accurately the laws governing the actions of
this spirit. (Newton, 1999: 943–944)

As Dempsey (2006: 437–438) has argued, building on I. B. Cohen's interpretation of some draft manuscripts while Newton was developing a response
to Leibniz over the calculus controversy, Newton is referring to electrical
experiments done by Hauksbee. Here it looks like Newton is sketching how
sentience itself is ground in a powerful, "very subtle spirit" that pervades
bodies (see also McGuire, 1968: 175–180, 185–186). In particular, the presence of this electric spirit helps explain the exciting of sensation in beings
like us. Now 'spirit' is a fluid (in our age we use 'spirit' in that sense when we
describe liquor as 'spirits'). And for Newton spirit is, thus, a kind of subtle
matter. So, this seems converging evidence for the emergentism I attribute
to Newton. To be sure, I agree with Stein (2002) and Dempsey (2006) that
Newton viewed this as a budding research program because we simply know
too little about the nature of mind—so my claim is not meant to be doctri-
naire, although Newton is an adamant critic of some positions he rejects.

In particular, Dempsey calls attention to a passage, quoted by Cohen, "if this
spirit may receive impressions from light and convey them into the sensorium
& there act upon that substance which sees & thinks, that substance may mutually act upon this spirit for causing animal motions" (Newton, 1999: 282).
I agree with Dempsey that Newton thinks this spirit is a key element in mental
causation. But the most natural reading of the passage is that this subtle spirit
is a mechanism by which mental causation takes place. It is not obvious that
this spirit also helps constitute or ground the substance which sees and thinks.

Dempsey argues it is likely that Newton is "not here endorsing the independent existence of the mind, but rather is distinguishing it from the organs
of sense and the brain. What is more, one straightforward explanation of
the natural interaction of this spirit and the mind is precisely that the mind
and the spirit with which it interacts are quite similar in nature" (Dempsey,
2006: 438). And this fits very nicely with my own emergentist reading. But this
monistic interpretation of Newton's metaphysics of mind is rather speculative.

In chapter 5, I argue that Newton is a substance monist. (As I show in
chapter 8, this is, in fact, Clarke's position in 1704, which is what made him
a juicy target for Leibniz.) For Newton, a substance is a directed source of
activity, that is an agent.[15] (Janiak [2013] calls this "substantial causation,"

---

[15] I use "directed" to distinguish it from natural principles that are active, but have no self-directed
teleological orientation in their activity.

although Janiak would deny the monism.) And so on my reading God is the only truly active agent in Newton metaphysics.[16] The universal quantifier in the first sentence of the quoted paragraph seems to suggest that what Newton says about human minds must fit also God's mind: "Every sentient soul, at different times and in different organs of senses and motions, is the same indivisible person." And if this "every" includes God, this suggest that God is not just immanent in nature, but also material (e.g., "motions")!

But I have always hesitated about attributing to Newton an emergentist account of God's sentience because he writes shortly hereafter that God "totally lacks any body and corporeal shape" (Newton, 1999: 942). So, it is not obvious that God's sentience could be an emergent property of matter. For (a) God's sentience cannot be grounded in matter if God is incorporeal; and (b) in various places, Newton seems to suggest that conceptually and temporally God precedes matter.[17] God's sentience and personhood is something altogether mysterious.

In fact, a natural reading of the "General Scholium" is that Newton explicitly denies—and he adds a second denial to the third edition—that God is a world soul. The passage reads as follows:

> He rules all things, not as the world soul but as the lord of all. And because of his dominion he is called Lord God Pantokrator. [Newton adds a note: "that is, universal ruler."] For "god" is a relative word and has reference to servants, and godhood is the lordship of God, not over his own body as is supposed by those for whom God is the world soul, but over servants. (Newton, 1999: 940)

Drawing on early work by Dobbs, Rudolf De Smet and Karin Verelst have convincingly argued that here (and in a few other places of the General Scholium) Newton is echoing Lipsius' attack, inspired by Philo, on the old stoic notion of *anima mundi* (2001: 14–17). The effect of their reading is that God remains immanent to nature, but is not grounded in body.

---

[16] Newton uses the plural "substances" (Newton, 1999: 942) in order to deny, in Lockean fashion, that we have knowledge of their essences (and even less of God's essence). But, as I argue in chapter 4, when he uses "substances" he means it in the more innocent sense of a "material entity" not an "agent."

[17] In the "General Scholium" that is not explicit, but it is a kind of implication of his claims about the designed nature of the visible universe a claim echoed in the Queries of the *Opticks* ("such a wonderful Uniformity in the Planetary System must be allowed the Effect of Choice." [Newton, 1730: 378]); there are more passages that suggest it (see chapters 5–6).

While I agree with their claim that Newton is indebted to Lipsius, in rereading the passage it is not obvious that Newton rejects conceiving God as the world soul as such. Rather, he rejects analyzing God's rule in terms of the world soul because it does not capture the hierarchical nature of God's dominion. So, it is possible that Newton thinks God's sentience emerges from, say, the pervasive, subtle (electric) fluid/spirit which he discusses in the final paragraph of the *Principia* (1999: 943–944).

Even so, given that Newton is explicit that God "totally lacks any body and corporeal shape" (942), it seems unlikely that we're supposed to read Newton as claiming that God's sentience is grounded in, and emergent of a, subtle electric (etc.) spirit.[18] So, I deny it is correct to treat God's sentience in terms of emergent properties.

So, where does this leave us? First (at Newton, 1999: 941–942), Newton articulates an account of the relationship between sentience and body. For human sentience this is ground in a kind of emergent quality of body, most likely a subtle (electric) fluid.[19] Second, God is immanent in nature, and features of his existence can be derived from the study of nature ("to treat of God from phenomena is certainly a part of 'natural' philosophy" [Newton, 1999: 943]). Ordinarily, bodies are not hindered by his presence, and "he does not act on them nor they on him" (Newton, 1999: 941).[20] Third, I grant that it makes sense to ascribe the PLA to human agency in Newton. Our souls or minds are tied to our bodies. The denial of this by Descartes is criticized by Newton in a passage in De Gravitatione rightly singled out by Janiak (2013: 402).[21] The *Principia* adds to this that our volitions

---

[18] Interestingly, this spirit may be the same spirit that Newton mentions at the end of the *Principia* proper, where he discusses the role of comets in the cosmic economy, and which is a kind of lifegiving source: "most subtle and most excellent part of our air, and which is required for the life of all things" (Newton, 1999: 926 discussed in chapter 3). Cesare Pastorino (forthcoming) has written very helpfully on this.

[19] For the eighteenth-century significance of this see (Riskin 2002).

[20] Brown (2016: 43) argues that my position cannot account for Newton's providentialism. While it is true I downplay Newton's providentialism (see also chapter 5), I do claim that for Newton God could have created the material world otherwise and that God's creation exhibits his providence (see, especially chapters 6 and 9). However, I agree that on my view of Newton's natural philosophy God's special or particular providence is mysterious. But notice that even in his famous passage —"blind Fate could never make all the Planets move one and the same way in Orbs concentrick, some inconsiderable Irregularities excepted, which may have risen from the mutual Actions of Comets and Planets upon one another, and which will be apt to increase, till this System wants a Reformation" (Newton, 1730: 378)—Newton here does not actually say that God does reform the system. He leaves that inference to the reader. The same is true of the claim on p. 379, which I quote in the next footnote.

[21] The passage reads: "Nor is the distinction between mind and body in his [Descartes'] philosophy intelligible, unless at the same time we say that mind has no extension at all, and so is not substantially present in any extension, that is, exists nowhere; which seems the same as if we were to say that it does not exist, or at least renders its union with body thoroughly unintelligible and impossible"

are mediated, and perhaps constituted, by a certain kind of powerful, electric fluid.

But, fourth, in so far as God does interact with bodies and spirits in nature this is fundamentally mysterious, as mysterious as his substance. In fact, in a passage that echoes a skeptical trope in Locke (see chapter 9), Newton stresses the fact that we really have no epistemic access, not even analogical, to God's manner of acting,

> . . . he is all eye, all ear, all brain, all arm, all force of sensing, of understanding, and of acting, but in a way not at all human, in a way not at all corporeal, *in a way utterly unknown to us.* As a blind man has no idea of colors, so we have no idea of the ways in which the most wise God senses and understands all things. (Newton, 1999: 942; emphasis added)

So, when it comes to God's sentience and agency, Newton leaves us without any resources to make claims about the nature and manner of God's volitions (Levitin, 2016: 70). And this suggests, there are no grounds to attribute the PLA to Newton when Newton is considering God's agency. So while Janiak has a valid argument in which the PLA figures as a premise, *pace* Janiak (2013: 399) I deny that we can "apply" his "argument to God." And so we cannot infer the PLA from Newton's claims about God's omnipresence or agency and so cannot rely on this to make claims about what bodies can or cannot do. So, on my view of Newton's theology, we can know what God does and, thereby, learn about his "attributes" (Newton, 1999: 942) and design, but we cannot know God's substance nor how God does what s/he does (and so thereby obeys or violates the PLA).

The third passage is quoted, in part, by Brown (2016: 43). It is from the *Opticks*:

> Such a wonderful Uniformity in the Planetary System must be allowed the Effect of Choice. *And so must the Uniformity in the Bodies of Animals,* they having generally a right and a left fide shaped a like, and on either side of their Bodies two Legs behind, and either two Arms, or two Legs, or two Wings before upon their Shoulders, and between their Shoulders a Neck

(Newton, 2004: 31). It is clear Newton here is only focused on Descartes' views of mind-body union not God's relationship to the world. I claim it is a non sequitur to treat this as evidence for Newton's views about the nature of God's agency.

running down into a Back-bone, and a Head upon it; and in the Head two
Ears, two Eyes, a Nose, a Mouth, and a Tongue, alike situated. Also the first
Contrivance of those very artificial Parts of Animals, the Eyes, Ears, Brain,
Muscles, Heart, Lungs, Midriff, Glands, Larynx, Hands, Wings, swimming
Bladders, natural Spectacles, and other Organs of Sense and Motion; and
the Instinct of Brutes and Insects, *can be the effect of nothing else than the
Wisdom and Skill of a powerful ever-living Agent, who being in all Places,
is more able by his Will to move the Bodies within his boundless uniform
Sensorium, and thereby to form and reform the Parts of the Universe, than
we are by our Will to move the Parts of our own Bodies.* (Newton, 1730: 378–
379; the emphasized part is quoted by Brown.)

There is no doubt that if you are antecedently convinced that Newton
embraces the PLA, the italicized parts seems like conclusive evidence for this
view. An advocate of the PLA may well challenge me to answer why would
Newton otherwise emphasize that God is in "all places" and thereby be able
to "move the bodies?"

I grant that the passage is very suggestive of the PLA. But strictly speaking
Newton's claim *violates* the PLA. For, Newton's "more able" suggests that God's
omnipotent power and his intelligence is such that even if, counterfactually,
he had not been omnipresent, or his sensorium were not boundless (or uniform), he would still be *somewhat able* to move distant bodies at will. And
while I am the first to grant that this is a very peculiar claim, the point of the
larger passage is to emphasize God's design and intelligence ("Counsel of an
intelligent Agent," Newton, 1730: 378) not to suggest something about the
mechanism—which on my interpretation is mysterious anyway— by which
this intelligence is expressed or conveyed. And this point is, in fact, well
expressed by Newton's "more able."[22]

So, to wrap up: there are several passages that are often taken to express
something like the PLA. And this is especially tempting when one assumes,
on the authority of Clarke, that Newton is committed to the PLA. But upon
closer inspection it can often be doubted either that they express Newton's
own views or that Newton is really relying on the PLA. Along the way I have
argued that while Newton certainly thought we could learn about God's

[22] Zvi Biener has suggested another argument against the PLA (in personal correspondence): "that
God acts where he is isn't a limitation. Rather, God doesn't *need* to act at a distance, because he is everywhere. But that doesn't mean he can't. The structure of argument for PLA is usually: if God can't do
it, then nothing else can. But if God doesn't need to do it, that's a different story."

attributes through the study nature, he does not think we are in a position to say anything about the manner of God's agency in the world.

IV. The Metaphysics of Interactions

In section 2.2 of his dissertation, Adwait Parker (2019) challenges my reading of the *Treatise* in terms of Newton's conception of his measurement practice. This criticism raises issues that are of fundamental significance. Parker agrees with one of my key claims that "Newton neither asserts that matter is altogether active nor passive. It depends on the way we conceive things." For Parker this entails that discussions about the "seat" of gravity by Newton's early modern contemporaries (Parker focuses on Euler, Roberval, and Huygens) do "not capture Newton's position." (103) This much we agree on.

I also agree with his observation that the three measures of centripetal force have slightly different modalities. He wants to claim that the absolute measure is necessary; the accelerative measure is a potential; and the motive is actual. (Parker, 2019: 101) To the best of my knowledge Parker is the first to make these modal differences explicit.

Unfortunately, Parker and I disagree over the content of the absolute measure and so also on its modal status. On his reading this just pertains to the "body itself as the center of some efficacy no matter what other bodies there may be" (Parker, 2019: 101). This is the position I attributed to Stein and argue against in chapter 1.

In addition, I do agree with Parker that Newton often invites us to abstract away from the reciprocal contribution of the second body to consider alone the field generated by the first body (Parker, 2019: 106). It is this practice of drawing a line, as it were, around a closed system, and treating it in relative isolation, that is central to the theory-mediated measurement practices initiated by Galileo and Huygens, and revolutionized by Newton.[23]

Even so, I had argued that the actualization of the cause requires another body; the definition of accelerative forces is given in terms of the disposition of a central body to "move the bodies that are in" the surrounding places. And while Parker is meticulous and fair on most of my claims he simply

---

[23] Spinoza, who knew his Galileo and Huygens, is a fierce critic of this practice (see Schliesser, 2017).

seems to ignores the evidence and my argument for this (see also Harper, 2011: 93, 37–138).[24]

So if another body is needed to generate the joint action, or field, as I argue, the modal status of the absolute measure is not simply necessary without qualification. If Parker were right that the absolute measure is necessary then it is a bit mysterious why Newton would reject the claim that gravity is essential to body (as he does in the third Rule of Reasoning). Admittedly, the necessity at play here is merely, in modern jargon, physical necessity and perhaps Newton would have thought that to be an essential property requires a kind of logical or metaphysical necessity.

Now, Parker may respond that Newton does not do this because the measure is a magnitude and so distinct from body altogether. But Newton could have claimed that while gravity need not be a primary quality of matter, body necessarily generates a force-field (which comes close to Cotes' position). In fact, after reading a draft of this postscript, Parker explained that on his interpretation "the 'disposition' is necessary while its actualization is contingent" (Parker correspondence, October 10, 2020). This position is coherent and captures why it is so important on my reading that the actualization is an *interaction* between pairs of bodies. So, I am not far removed from Parker's approach.

But, if the disposition were not merely universal, but necessary to bodies, Newton could have said that and yet he seems to shy away from *that* very claim.[25] It is true that the position I attribute to Newton seems to imply something weird: that the presence of a quantity of matter capable of 'activating' the disposition and being activated in turn by it in surrounding regions affects the existence or degree of the activity in the first body. And so the efficacy, or field, just as a quantity, is not well-defined if there are no bodies in surrounding regions. But this weirdness just is a feature of my interpretation.

For Parker, by contrast, to say "that the disposition is necessary and its actualization is contingent . . . means [he is] committed to a kind of necessity also attaching to the accelerative measure. But this should be expected. The accelerative measure abstracts from the passive quantities of matter in the surrounding regions. It is given as a function of distance and the absolute

---

[24] Harper also discusses my interpretation carefully, but primarily in terms of the felt "tension" between Stein's acceleration field interpretation and my claim that gravity is constituted/produced by universal, pair-wise interactions. Harper's solution is to offer a developmental view.

[25] Parker claims that Newton does not and is unwilling to address the philosophical issue on what counts to be an essential attribute. (Parker, 2019: 101–102) That is true, but Newton does make a careful distinction between universal and essential qualities in the third Rule of Reasoning.

measure. So it is closer to the 'necessity' attaching to the absolute measure of force than the 'actuality' attaching to the motive measure of force (which, crucially, requires the passive quantity of matter)." (Parker, correspondence, 2020). So on this view, while accelerative gravity decreases, the body (active quantity of matter) remains the same.[26] Parker's is an interesting and co-herent position on the modality of these measures. I am unsure there is any additional text or implication I could cite to help you decide between us.

There is a more important disagreement: Parker objects to the idea that the natures or dispositions of bodies are "identical" to the masses of bodies. (Parker, 2019: 94) And he claims, by contrast, that "Newton's enterprise is not to give us the causes themselves, nor the reasons for those causes (or particular properties); rather, his is a way of setting up measurement practices for magnitudes proportional to the causes" (Parker, 2019: 94).[27]

Now, I *agree* with Parker that Newton proposed "measurement practices for magnitudes proportional to the causes." And, on my reading, this proportionality between the magnitudes we measure and the causes gives us some constraints on any possible explanation of those causes.

But I had *not* claimed that dispositions of bodies that give rise to pair-wise interactions are *identical* to the mass of those bodies.[28] I had claimed, rather that "the cause of gravity . . . is at least, in part, the masses to be found in each body." And I repeated the point by claiming that Newton had taken a partial stance on the cause of gravity. So, at best I asserted a partial identity. And the reason why I expressed myself more cautiously than Parker allows is that proportionality claims cannot be *merely* turned into identity claims.

So, perhaps to his surprise, this is why I agree with part of Parker's analysis of definition 8, that Newton considers a "single body to be the potential cause of a force . . . propagated outward in surrounding readings." (Parker, 2019: 96;

---

[26] For Parker, "the 'necessity' which attaches to the absolute measure is just the invariance of active quantity of matter." On his view Definition 7 "illustrates both the invariance of active quantity of matter and its relation to the necessity that attaches, on my reading, to the accelerative measure: Definition 7, which presents the accelerative measure, says that the force of gravity varies by distance, but which is everywhere the same at equal distances because it equally accelerates all falling bodies (heavy or light, great or small) . . . "[Parker quoting Newton], Here, the active quantity of matter of the central body is indicated to be invariant with respect to all possible passive quantities in surrounding regions" (Parker 2020, private correspondence).

[27] And so this means, as the previous footnote also reveals, that the ontology of what necessity is, as it were, attached to is going to be different.

[28] I had, however, quoted Belkind who could be construed to that effect. So I may well have caused confusion.

here Parker uses "potential"—that seems to undermine the necessity claim.)[29] And because I never argued that mass is identical to the cause, I see no reason to disagree with Parker's excellent analysis of definition 6 (on the absolute quantity of centripetal force), in which Newton distinguishes between the efficacy of the cause—a magnitude—and the cause itself (Parker, 2019: 95).

One reason, I think, why Parker attributes an identity claim to me is that I had claimed that the cause of gravity is, in part, a "quality of matter." And I had insisted that qualities are properties that can be causally efficacious on minds. So it seems like I would be committed to attributing to Newton the idea that mass is a quality of matter. (Parker, 2019: 97)[30] I realize now that I *should* have claimed that Newton's conception of mass is a feature shared by all bodies, or so I wish to suggest in the remainder of this postscript. And, if Parker is right, then on my reinterpretation of his position, this feature is an *abstract* quality of bodies.

However, because mass is a measure, a quantity, it is on Parker's account a kind of category error to think that mass is property of matter or part of a "start" of "an explanation" (Parker, 2019: 99). On his view what "will be explanatory of the action of gravity (the motions it produces) will be the forces as the causes of motions, and these forces will be proportional to the masses. They will . . . be proportional to the masses in a very specific way. The force towards a body is composed of the forces to its individual particles, and the quantity of matter of a body is the sum of the quantity of matter of its particles. We have, therefore, a more complicated relationship between mass and cause than one of identity" (Parker, 2019: 99–100). Since I had not claimed there was identity, I agree with most this.

Even so, there is a real disagreement here. On my view while it is true that mass is a measure, Newton does treat mass as a property of matter. My evidence for this is Newton's third Definition:

---

[29] In correspondence Parker agreed that this 'potential' was liable to confusion. For Parker "The cause is an entity which has contingent existence (which is what the 'sine qua' in Def 8 could be understood to refer to). But the absolute measure of centripetal force proportional to the cause, provided the cause exists, is not contingent—and specifically is not contingent on the existence of other bodies. Rather, it is necessary in the sense of being invariant. . . . This is indeed liable to confusion because Def 8 isn't only talking about absolute and accelerative quantities (which have the requisite 'necessity'), but is also talking about the motive, which involves the passive quantities in surrounding regions, which introduce the contingency your account wants to emphasize" (Parker, private correspondence, 2020).

[30] Because of this, and because I mention Locke's use of 'quality,' Parker also seems to attribute to me a naïve empiricist interpretation of Newton he tries to fit my use of quality into Locke's distinction among primary, secondary, and tertiary qualities (Parker, 2019: 97–99). His response is rhetorically effective, and no doubt fair; but mistaken because I had studiously avoided fitting Newton into that Lockean threefold distinction because I do not think it fits Newton's purposes!

Definition 3: Inherent force of matter, is the power of resisting [*Materiæ vis insita est potentia resistendi*] by which every body, as far as it is able, perseveres in its state either of resting or of moving uniformly straight forward.

This force is always proportional to the body and does not differ in any way from the inertia of the mass except in the manner in which it is conceived [*differt quicquam ab inertia Massæ, nisi in modo concipiendi*]. Because of the inertia of matter, every body is only with difficulty put out of its state either of resting or of moving. Consequently, inherent force may also be called by the very significant name of force of inertia. (Newton, 1999: 404)[31]

This *vis insita* or force of inertia is treated as a property of matter. And we are told this force is identical to the inertial mass except in the manner of conceiving. And while it is true that forces are the causes identified and tracked in Newton's mathematical natural philosophy, the really existing celestial and terrestrial bodies that are accelerated and impressed by such forces are not epiphenomenal bystanders. There are features of these bodies that generate forces of matter and these forces, in turn, are influenced by their relationships *toward* material bodies.

Now, we might say on Parker's behalf that the "body" mentioned in Definition 3 here is not matter but itself a quantity. For, in his comment on Definition 1—"Quantity of matter [*Quantitas Materiæ*] is a measure of matter that arises from its density and volume jointly"[32]—Newton had added, "I mean this quantity whenever I use the term 'body' or 'mass' in the following pages"[33] (Newton, 1999: 404). So, one may substitute "quantity of matter" (a measure) into the discussion of the third definition where it uses 'mass' and 'body.' This third definition would then read: "This [inherent] force is always proportional to the quantity of matter and does not differ in any way from the inertia of the quantity of matter except in the manner in which it is conceived."

---

[31] *Materiæ vis insita est potentia resistendi, qua corpus unumquodq; quantum in se est, perseverat in statu suo vel quiescendi vel movendi uniformiter in directum.*

Hæc semper proportionalis est suo corpori, neq; differt quicquam ab inertia Massæ, nisi in modo concipiendi. Per inertiam materiæ fit ut corpus omne de statu suo vel quiescendi vel movendi difficulter deturbetur. Unde etiam vis insita nomine significantissimo vis inertiæ dici possit.

[32] *Quantitas Materiæ est mensura ejusdem orta ex illius Densitate & Magnitudine conjunctim.*

[33] *Hanc autem quantitatem sub nomine corporis vel Massæ in sequentibus passim intelligo.*

Parker's position would then emphasize, as he does, that Newton offers us "coherent measurement practices" (Parker, 2019: 105; see also Smith, 2014). In this measurement practice forces and their measure are coupled together, but it is by no means obvious one can make inferences about the underlying particles of matter and how bits of inertial mass and bits of mass *simpliciter* hang together. Parker (2020) is a very fine analysis of the complications and subtlety of this latter problem. So, it is no surprise that Parker wishes to set aside my kind of approach as a "metaphysical reading" (Parker, 2020: 7, n. 28).

In fact, Parker's stance, a residue of a twentieth-century antimetaphysical strain in the philosophy of science, is a familiar one in Newton studies. I *illustrate* this claim in the process of partially *restating* the way I see the underlying debate between Parker's and my approach.

Rather than focusing on the metaphysics of action-at-a-distance or active matter, I focus on the nature of passive matter, or *vis insita*.[34] I do so by way of quoting a passage that is from one of the sequence of manuscripts that Newton wrote between Halley's famous visit to Cambridge and the *Principia*. It reads:

> *Def.* ~~11~~ 12. The ~~force of a body or inherent and innate to a body~~ The $_v$inherent, innate, and essential force [*vis insita, innata, et essentialis*]$_v$ of a body is the power by which it ~~endeavors to~~ perseveres in its state of rest or of moving uniformly in a straight line. It is proportional to the quantity of the body, and is in fact exercised proportionally to the change of ~~bearing~~ [*allatae*] state, $_v$and in so far as it is exercised it can be said to be the exercised force of the body endeavoring and struggling against [*conatus et reluctatio*] ~~Of this kind~~, of which one kind is the centrifugal force of rotating [bodies] [*gyrantium*]$_v$.—Isaac Newton "De Motu Corporum in mediis regulariter cedentibus," translated by George Smith. (Smith, 2020)

The first sentence of this passage has not been much discussed. For example, in his celebrated biography of Newton, Richard Westfall mentions the first sentence, in passing, as part of an effort to clarify the relation between inherent and impressed force (Westfall, 1983: 416). In his classic article on the subject of inherent and centrifugal forces, Domenico Bertoloni Meli quotes

---

[34] I do so, in part, to round out my version of Newtonian metaphysics, but also because Parker and I seem to read Definition 3 of the *Principia* in different fashion.

the passage (Bertoloni Meli, 2006a: 326). This is also true of another famous earlier article by Alan Gabbey (1971: 34). But the fact that Newton couples this *vis insita* with innateness and essence goes unremarked, as it does in Smith's article that has supplied the translation I used. I mention Westfall, Bertoloni Meli, Gabbey, and Smith as exemplars of a certain kind of reticence among leading Newton scholars to analyze some of Newton's metaphysics while he was developing the *Principia*. So Parker is not alone.[35]

Recently, George Smith, in a paper devoted to confirming Stein's conjecture "that Newton's conception of forces of nature as forces of interaction was actually developed by Newton *at the same time* that he was discovering the law of gravitation" (Smith quoting Stein) notes, while commenting on the disagreement between Stein and myself, "about metaphysics of Newtonian gravity," that "I have no objections to anyone's citing passages from Newton's writings in an effort to extract his 'metaphysical' thinking, for where else can one turn to? I question, however, how much weight can be put on Article 21 [that appeared posthumously]. As already noted, nothing remotely akin to it appears in any edition of the *Principia*, nor for that matter in any other publication or extant manuscript" (Smith, 2019: 170–171).

Before I get to the substantive point that nothing remotely akin to Article 21 of the *Treatise* appeared in *Principia*, I want to note that the implicature of Smith's comment puzzles me: on the first two pages of my paper—chapter 1— I had, in fact, asserted multiple times that I was attributing a view to Newton that he held when he was drafting/writing the *Principia*. And while it is true that I have used evidence from other periods of Newton's life, I also grant that Newton tried out different mechanisms.

Even so, while I am happy for anybody to get on with the business of analyzing measurement and evidence (see also Smeenk and Schliesser, 2013; Smith and Schliesser, forthcoming), metaphysical ephemera can be interesting, even illuminating not just of Newton's development (as Smith grants), but also of the conceptual trial and error and innovation that are the contents of and make the scientific revolution and philosophical development possible. So with that in mind let's return to the quoted passage from "*De Motu Corporum in mediis regulariter cedentibus.*"

---

[35] In correspondence, Parker called attention to Friedman 2020, which gives a *Kantian* account of the origin and very nature of this reticence, centered on how Newtonian methodological abstraction was understood after Kant. This also involves an interpretation of Stein that I do not fully endorse.

In the first sentence the power of a body by which it perseveres in its state of rest or of moving uniformly in a straight line is not treated as a property or quality, but as a *force* of *that* body. Descartes' first two laws of motion, for example, do no such thing (nor does Huygens suggest it anywhere).

So there are three questions worth asking: first, what is a force of a body? Second, what makes a force inherent, innate, and essential? Third, why, in the *Principia*, did Newton drop the claim that this force is "innate and essential" but left in that it is inherent? For (on this third question), when in the *Principia*, he returns to something very much like the idea of Definition 12, he writes that "Inherent force of matter is the power of resisting by which every body, so far as it is able, perseveres in its state either of resting or of moving uniformly straight forward." And in the very next sentence, Newton adds, "This force is always proportional to the body and does not differ in any way from the inertia of the mass except in the manner in which it is conceived." This is Definition 3 (of the *Principia*) already quoted in my response to Parker above. Of course, in comparison to the quoted passage from "De Motu Corporum in mediis regulariter cedentibus," in the *Principia* Newton adds the very striking claim, as we have seen, that *vis insita* is identical to the inertia of the mass, except (and this is no less striking) in the manner of conceiving them.

In the *Principia*, Newton treats forces as abstract mathematical quantities. (Smeenk, 2016; Smeenk and Schliesser, 2013) Parker and I seem to agree about this. Of course, it is not entirely obvious what to make of the claim that a force just is an abstract mathematical quantity. And it is understandable that Parker, when confronted with the claim that *vis insita* is identical to the inertia of the mass, just wishes to focus on the fact that two abstract mathematical quantities related to bodies are alike, even in a certain sense conceived to be identical.

Because it is common since the advent of relativity theory and the quantum revolution to present Newton's ontology as kind of commonsensical (with modest allowance for the strangeness of the possibility of action at a distance), it is sometimes forgotten or underestimated that even the building blocks of Newton's physics are, from an ontological perspective, quite unusual. That Newton was innovating helps explain that, as Zvi Biener and Chris Smeenk have shown in an eye-opening article, that there are tensions lurking here between a geometric conception of body and what they call dynamic conception of body (looking forward to Boscovitch) as a collection or bundle of forces. (Biener and Smeenk, 2012)

POSTSCRIPT TO CHAPTER 2   77

My hypotheses about why Newton dropped the very idea of inertia as "an innate and essential force" from the *Principia* is as follows: once one starts treating forces as abstract mathematical quantities that can be measured (mediated by theory) and that are *causes* of motions/accelerations, it is unclear what is gained by treating them as innate or essential or not.[36] That's just the kind of metaphysics George Smith finds irrelevant to the evidential practice. Fair enough.

But it is not *Newton's* own last thought, for he added the third rule of the study of natural philosophy to the second edition (Newton, 1999: 198). As it happens, in Rule 3 of the *Principia*, Newton explicitly *claims* that *vis insita* is "immutable" (Newton, 1999: 796). In addition, he is explicit that *vis insita* is a means to establish that movability and perseverance are universal properties in bodies given the third rule that encourages us to make an inductive leap from the finding of particular properties of bodies without exception to the commitment that these "be taken as qualities of all bodies universally" (Newton, 1999: 795). And Newton then suggests that there is a contrast between universal and essential qualities of matter. The third rule is a means toward establishing the former not the latter.

It is, in fact, quite natural to read Newton's third rule as claiming that while gravity is merely universal and not an essential quality of matter, in addition, and this is less frequently noted, *vis insita* is not just universal, but *also* an essential quality of matter as evinced by its immutability: "[B]y inherent force I mean only the force of inertia. This is immutable." (Newton, 1999: 796). And it is this claim that—*pace* Smith—echoes features of the discarded Definition 12. In addition to other aims (Belkind, 2017), the third rule is a contribution to the metaphysics of natural philosophy. But the metaphysics is treated as a kind of consequence of empirical work rather than a definition. (This is in the spirit of Brading, 2017.) So, while I grant there is no conclusive evidence that Newton always thought of mass as conceivable as a physical property of bodies (beyond it being an abstract quality of matter), there is thus very good evidence that Newton did think of *vis insita* as a property of matter.

At this point, I need to make a modest detour. The trio "inherent, innate, and essential" has a notorious afterlife because, as we have seen in chapters 1–2, in a famous letter to Bentley, of February 1693, Newton denies

---

[36] I do not mean to suggest measurement is absent in the discarded Definition 12; one can discern an interest in measurement behind the idea of "exercised proportionally."

"that gravity should be innate, inherent, and essential to matter."[37] What Newton has meant by his rejection that gravity is "inherent, innate, and essential" quality of matter has been subject of much scholarly controversy. And I do not wish to relitigate my position here. But because of that debate, it seems to be agreed that by an "essential quality" of matter Newton means it is a quality, or a *sine qua non*, required for the existence of matter. And so understood one can see why Newton would have been tempted in "De Motu Corporum in mediis regulariter cedentibus" to call *vis insita* essential, because it is *the* feature that is required for (to speak anachronistically) the physics he is pursuing. Without the power by which a body "perseveres in its state of rest or of moving uniformly in a straight line" it would not be the kind of entity Newton is studying and one cannot treat forces as the causes of change.[38]

Now, in the previous paragraph I moved from ontology to epistemology. That's partially because the "Rules for the Study of Natural Philosophy" in the *Principia* themselves seem to give the reader, as Zvi Biener and I argue in chapter 6, mostly an ontology apt for the epistemic attitude to be taken in the context of natural philosophical research. And this ontology is, therefore, more provisional than talk of "essential" qualities seem to suggest.

Andrew Janiak has suggested that to take a quality to be "innate" is to think of it as due to no other physical process, entity, or medium between material bodies (Newton, 2004: xxiv). So an innate quality is to be contrasted with a relational quality (which is generated from, say, a pair-wise interaction). To the best of my knowledge Newton calls no quality or power innate in the *Principia*.

That Newton dropped the claim that *vis insita* is an innate quality of body can then be readily explained. Between "De Motu Corporum in mediis regulariter cedentibus" and the *Principia*, Newton came to think of natural

---

[37] In the published (1693) version of Bentley's sermons, Bentley continued to tie Newton's theory to Epicurus, but limited the similarity to the claim "that the weight of all Bodies around the Earth is ever proportional to the quantity of their Matter" (Bentley, a famous classicist, adds a footnote to Lucretius). Bentley goes on to add that Newton and the Epicureans also agree on the existence of a vast vacuum of empty space in the universe. He then draws a distinction between two kinds of atheist cosmogonies (one he labels Epicurean, and another which is more crypto-Spinozist, of a sort later defended more explicitly by Toland [see chapter 8]). He then insists that both will be refuted on the assumption that "mutual gravitation or spontaneous attraction can neither be inherent and essential to Matter; nor even supervene to it, unless impressed and infused into it by a divine power." This superaddition of gravity thesis is defended by John Henry as indeed Newton's own view (in part on the basis of Newton's letter to Bentley).

[38] Here as elsewhere I am influenced by Brading (2012), which has a much more sophisticated version of this insight.

forces as forces of interaction. This is in fact the very point of Smith's (2019) recent article on Stein (2002). These forces are manifested in and measured by interactions (in part, as "struggle" against change, etc.). And in the *Principia*, Newton is explicit that these forces are both causes and effects of the interactions of bodies.

Famously, Descartes' metaphysics floundered on the mysterious interaction between bodies and souls. It is worth noting that in Newton there is an interaction between bodies and abstract mathematical quantities. If we focus on the measurement practice this is not mysterious at all and it generates, as Smith shows in his majestic "Closing the Loop" ever more stringent evidence (Smith, 2014). It is no surprise that Parker focuses on this. But if we step back, and ask ourselves how to conceive of the interaction between bodies and abstract mathematical quantities, one is left somewhat speechless.[39]

---

[39] This postscript has benefitted from generous comments by Adwait Parker and Zvi Biener.

3

On Reading Newton as an Epicurean

Kant, Spinozism, and the Changes to the *Principia*

3.1 Introduction

In this chapter I argue for three distinct, albeit mutually illuminating theses: first I explain why well informed eighteenth-century thinkers (e.g., the pre-Critical Immanuel Kant and Richard Bentley, who had a widely discussed correspondence with Newton), would have identified important aspects of Newton's natural philosophy with (a species of modern) Epicureanism[1]. Second, I explore how some significant changes to Newton's *Principia* between the first (1687) and second (1713) editions can be explained in terms of attempts to reframe the *Principia* so that the charge of "Epicureanism" can be deflected. To complicate matters, this thesis is compatible with a more thoroughgoing renewed and direct engagement with Epicurean ideas by Newton from the 1690s onward. For discussion, see below. In order to account for this I call attention to significant political and theological changes in the wake of the Glorious Revolution; as has been documented by others, Bentley plays a nontrivial role in these matters. Third, I argue that there is an argument in Kant's (1755) *Universal Natural History and Theory of the Heavens* that undermines a key claim in Newton's "General Scholium." I suggest that this argument reopens the door to "blind necessity"—that is, Spinozism.[2]

This chapter consists of three sections. First, I show that the *Principia* was explicitly associated with Epicureanism in various ways by the pre-Critical Kant and Bentley. This is prima facie puzzling because the "General Scholium" seems to argue straightforwardly against Epicureanism. To explain

---

[1] This chapter first appeared as Eric Schliesser. "On reading Newton as an Epicurean: Kant, Spinozism and the changes to the Principia." *Studies in History and Philosophy of Science Part A* 44.3 (2013): 416–428.

[2] In Leibniz's correspondence with Clarke, Leibniz explicitly raises the specter that Newton reduces to Spinozism (2.7, p. 29). All my references to it are by letter, paragraph number, and page number in Leibniz and Clarke, 1717.

Newton's Metaphysics. Eric Schliesser, Oxford University Press. © Oxford University Press 2021.
DOI: 10.1093/oso/9780197567692.003.0005

this puzzle, I show how the mathematical-physical core of the *Principia* could and was plausibly read as a tract in which Epicurean themes could be discerned (regardless of Newton's intentions), even if the "General Scholium" appears to disown that reading of the book. In particular, in addition to some substantial doctrines that are often associated with Epicureanism, I call attention to the fact that the *Principia*'s first edition starts (Halley's Ode) and ends (Newton's treatment of comets) with highly Epicurean themes. This section aims to contribute to a better understanding of the reception of Newton's views on gravity and action at a distance because Kant and Bentley both discern in Newton an Epicurean conception of gravity.

Second, I explain how many changes to the *Principia* resulted in a new framing of the book in the second (and third) edition(s). I focus on the "General Scholium," but my argument calls attention to a host of significant changes between the first and second editions. By focusing on the political and theological context of Bentley's correspondence with Newton, I also explain why Newton would have been motivated to alter the *Principia* in nontrivial ways. So this section means to contribute to a scholarly literature that recognizes the development of Newton's views.

Third, I argue that despite explicit efforts to distance himself from Epicureanism, Kant's UNH re-pens the door to what one might call neo-Epicureanism by undermining the "General Scholium"'s strongest argument against it. By comparing the "General Scholium" with an argument by Clarke from the *Demonstration*, I argue that the target of Newton's argument is Spinozism; so this chapter is also meant to contribute to research on Newton's and Kant's engagement with Spinozism. For, while one can see "Spinozism" and "Epicureanism" as radically different traditions, in the early modern period they sometimes get linked as defending "blind metaphysical necessity."[3] In particular, I call attention to nontrivial allusions to Spinozism in UNH.

3.2 The Ascription of Epicureanism to Newton's *Principia*

In this section, I show that the *Principia* was explicitly associated with Epicureanism in various ways. I focus on the pre-critical Kant and Bentley,

---

[3] For example, the entry on Spinoza in Diderot's *Encyclopedia* is framed by a contrast between the systems of Spinoza and Epicurus, but even that entry emphasizes that both reject Providence and embrace necessity (Diderot, 2007).

but they were not alone in this. For example, Adam Smith also links Newton to an Epicurean-atomist tradition in his piece "Of the External Senses."[4] Smith's essay and Kant's UNH were both written early in their careers before 1755; Smith's piece appeared only in the 1790s (posthumously). Kant's UNH seems to have appeared in the mid-1760s.[5] There is no evidence of mutual influence. But there may well have been common sources. For example, in his famous correspondence with Clarke, Leibniz links Newton's position to the Epicurean tradition (Leibniz to Clarke 2.1 [19–20]). Because Kant and Adam Smith had more than superficial knowledge of the contents of the *Principia* (on Kant and Newton, see Buchdahl [1992] and Friedman [1992]; for Adam Smith and Newton, see Schliesser [2005b] and Montes [2003]), the (shared) sources and reasons for their overlapping judgment are worth exploring.

3.2.1 Kant

Immanuel Kant's UNH is now primarily known for its pioneering contribution to speculative cosmology and cosmogony. In Kant scholarship it plays an important role in debates over Kant's so-called conversion to Newton (Watkins, 2013: 431). I focus on a narrower theme in it. At the start of the work, Kant raises the specter of Epicureanism.[6] I quote two passages before commenting on them:

[1] If the planetary structure, with all its order and beauty, is only an effect of the universal laws of motion in matter left to itself, if the blind mechanism of natural forces knows how to develop itself out of chaos in such a marvelous way and to reach such perfection on its own, then the proof of the primordial

---

[4] "[This] doctrine, which is as old as Leucippus, Democritus, and Epicurus, was in the last century revived by Gassendi, and has since been adopted by Newton and the far greater part of his followers" (Smith, 1982: 140).

[5] Michaela Massimi has called my attention to the Introduction by W. Ley to Kant's *Cosmogony* (1968) where he says on pp. viii and ix that "Years later—nobody knows the date—the printed books were released, probably in 1765 or 1766."

[6] I have relied on and quoted extensively (with minor modifications) from a translation (Kant, 2008) by Ian Johnston and compared it with the German version edited by Wilhelm Weischedel (Kant, 1977). I refer to the latter by page and then give parenthetical references to the Akademie edition. The translation has no page numbers associated with it; so in addition I always refer to the page number of (Kant, 1977) and in footnotes or parentheses to the page numbers in the Akademie edition.

Divine Author which we derive from a glance at the beauty of the cosmic structure is wholly discredited, nature is self-sufficient [*selbst genugsam*], the divine rule is unnecessary, Epicurus lives once again in the midst of Christendom, and an unholy worldly-wisdom [*Weltweisheit*] treads underfoot the faith which proffers a bright light to illuminate it. (UNH Preface, 222)

[2] But the defense of your system, it will be said, is at the same time a defense of the opinions of Epicurus, to which it has the closest similarity. I will not completely deny all agreement with him. Many people have become atheists through the apparent truth of such reasons which, with a more scrupulous consideration, could have convinced them as forcibly as possible of the certain existence of the Highest Being. . . . I will also not deny that the theory of Lucretius or of his predecessors (Epicurus, Leucippus, and Democritus) has much similarity to mine. Like those natural philosophers [*Weltweise*], I set out the first condition of nature as that state of the world consisting of a universal scattering of the primordial materials of all planetary bodies, or atoms, as they were called by these. Epicurus proposes a principle of heaviness which drives these elementary particles downwards, and this appears not very different from Newton's power of attraction, which I assume. He also assigned to these particles a certain deviation from the straight linear movement of their descent, although at the same time he had an absurd picture of the cause and consequences of this deviation. This deviation comes about to some extent with the alteration in the straight linear descent, a change which we derive from the force of repulsion of the particles. Finally, came the eddies, which arose from the confused movement of the atoms, a major part of the theories of Leucippus and Democritus. We will meet them also in our theory. But such a close affinity with a theory which was the true theory of atheism in ancient times does not lead mine to be grouped in the company of their errors. (UNH preface, 226)

In the first quote, Kant associates Epicureanism with the view of nature that is self-sufficient; one in which there is only "blind mechanism." In particular, Epicureanism is said to embrace the idea that the laws of motion can account for the origin of the present planetary structure from unordered ("chaos") beginnings.* In the second quote, Kant allows that the system he

---

* In the *Opticks*, Newton explicitly denies that this is possible (1730: 378); in the "General Scholium," Newton merely implies this (1999: 940 to be quoted below).

will put forward has a close resemblance to an ancient atomist-atheist tradi-tion in four important features: (i) the beginning of the universe consists of scattered atoms, (ii) these atoms have an innate downward motion, (iii) this motion deviates from a straight line, (iv) the world consists of vortices. Kant claims that the second resemblance has a close kinship to Newtonian attrac-tion. Moreover, Kant offers a Newtonian reinterpretation of the Epicurean causal analysis of the third feature.

One might object to my more general thesis—that Kant closely identifies his own Newtonian inspired cosmology with Epicurean themes—by saying that in the second quote Kant embraces a vortex theory; Newton criticizes the vortex theory in detail in Book II and the "General Scholium" of the *Principia* ("the hypothesis of vortices is beset with many difficulties" [Newton, 1999: 939]). But in Kant's argument the only role vortices play is as an im-portant *intermediary* step between the chaotic origins and the clustering of matter in stable orbits. In context Kant leaves no doubt that he accepts vast empty spaces beyond the solar system, and considerable empty space within it—both orthodox Newtonian positions; he is thus no Cartesian-Leibnizian vortex theorist. So of the four Epicurean features mentioned in the second quote, the middle two have Newtonian analogues; the last is at least compat-ible with Newtonianism (and may be thought of as a creative extension of Newtonian theory with Newtonian tools). This provides prima facie evidence for my general thesis that, at one stage, Kant explicitly associated significant features of the Newtonian natural philosophy with Epicurean commitments.

The first quote might be thought to generate another objection to my ge-neral thesis. For it can be interpreted as suggesting that if Epicureanism is accepted then prominent Newton arguments are refuted: the rejection of divine providence (that is, the universe reached its present "perfection" by itself) and the refutation of an argument for God's existence from the ap-pearance of cosmic beauty both can be said to target Newton's claims in the "General Scholium" ("this most elegant system of the sun, planets, and comets could have arisen without the design and dominion of an intelli-gent and powerful being" [Newton, 1999: 940]). The language with which the Epicurean position is described by Kant in the first quote [*"die blinde Mechanik der Naturkräfte"*] echoes the way Newton articulates the "blind metaphysical view" he opposes ("a God without dominion, providence, and final causes is nothing other than fate and nature" [Newton, 1999: 942]).

This objection is really an argument for my thesis. For, in both quotes from Kant's UNH, positions that can be ascribed to the body of the *Principia* (especially the first edition) are associated with Epicureanism, while the anti-Epicurean claims echo the "General Scholium" (and other editorial material to be discussed below) added to the second edition.[7] To put the thesis of this chapter starkly: without the "General Scholium" (added to the second edition) the *Principia* can—and was—naturally read as a neo-Epicurean tract. My argument does not require that all informed readers would have discerned Epicurean themes in the *Principia*. My argument below does account for the fact that despite the presence of the "General Scholium," people continued to attribute Epicurean themes to Newton.

There is a third quote from Kant's UNH that may seem even more problematic for the thesis developed in this chapter:

[3] Everything flows from it [the Godhead] according to unchanging laws which thus must display nothing other than appropriate [*welche darum lauter Geschicktes darstellen müssen*], because they are exclusively features of the wisest of all designs from which disorder is prohibited. The chance collisions [*ohngefäre Zusammenlauf*] of the atoms of Lucretius did not develop the world. Implanted forces and laws which have their source in the Wisest Intelligence were an unchanging origin of that order inevitably flowing out from nature, not by chance, but by necessity. (UNH Part 2, chapter 7, 358 (334))[8]

Kant contrasts (Lucretian) "chance" and "disorder" with law-governed "necessity" and "order." Kant claims that necessity and order are the product of divine design.* Kant's emphasis on the forces and laws' origin in "wisest intelligence" echoes Newton's "General Scholium." So, this fits my general thesis that the parts of Kant's UNH that are anti-Epicurean are indebted to the "General Scholium."

Nevertheless, there are three complications: first, the historical Lucretius *also* insists that "law presides over the whole creation"; he identifies law (*lex*) with necessity (*necessum*) (*De Rerum Natura*, Book V, 56–57). So, while for Kant there may be an opposition between the doctrines of chance and necessity, it is by no means obvious that others would have read Lucretius

---

[7] Strictly speaking, Kant does not attribute the doctrines to Newton. It is the point of this chapter to make clear that attributing them to Newton is not unreasonable.

[8] I have accepted the translation of the repeated *ohngefähre* as "chance"; Kant is clearly referring to the swerve of atoms here. (In context Kant does link the swerve with chance [*Zufalls*].) I thank Abe Stone, David Hyder, Thomas Sturm, and Stijn van Impe for discussion.

* Presumably *in virtue* of being law-governed (with God as lawgiver).

similarly. More important, Kant had already identified Epicureanism with law-governed "blind mechanism" (recall quote [1] above), so in this third quote, his treatment of Lucretius as defending "chance collision" as *contrasted with* law-governed necessity is at variance with himself and the normal way of treating these matters.[9] For example, in his exchange with Leibniz, Clarke explicitly equates "Epicurean chance" with "blind necessity."[10]

Second, in isolation the passage from Kant could be read as claiming that God implants order and necessity from *outside* of nature.[11] This, too, sounds like Newton's claims in the *Principia's* "General Scholium." For example, Newton writes: "And so that the systems of the fixed stars will not fall upon one another as a result of their gravity, he has placed them at immense distances from one another" (Newton, 1999: 940).

Yet as Schönfeld has persuasively argued, UNH as a whole seems to suggest that Kant's God is immanent in nature (2000: 107). Surprisingly enough this, too, echoes the "General Scholium" of the *Principia*, where "necessarily" existing God is said to be "omnipresent," "substantially" and "virtually." (Newton, 1999: 941) Newton's anti-Trinitarian God is, thus, spatially extended and also immanent in nature and closely identified with necessity (see chapters 5 and 8). In Kant and Newton the language of design/intelligence is compatible with an immanent God.

These two points relate to a third complication: in a letter to Bentley, Newton writes, "Gravity must be caused by an agent acting constantly according to certain laws; but whether this agent be material or immaterial I have left to the consideration of my reader." (Newton, 2004: 103) Many interpreters think that Newton is alluding here to God's active role in the universe. What is less noticed is that if Newton is alluding to God's activity, he is insisting that God's activity is entirely law-governed ("an agent acting constantly according to certain laws"). So the language of design is

---

[9] Kant also explicitly compares his own effort to offer a mechanical, law-governed ["mechanischen Gesetzen"] account of the development of the universe with Descartes' earlier approach (UNH, preface, 235 (312)).

[10] Perhaps, Kant is drawing on a Leibnizian distinction between chance and indifference; see Leibniz's fifth response to Clarke (70, p. 227). Part of the complication is caused by the fact that the doctrine of necessity can be opposed to both the rejection of providence as well as Lucretian swerve. My argument does not require that Kant read the Leibniz–Clarke correspondence.

[11] In the manner of Newton's *Opticks*: "it may be also allowed that God is able to create particles of matter of several sizes and figures, and in several proportions to space, and perhaps of different densities and forces, and thereby to vary the laws of nature, and made worlds of several sorts in several parts of the universe" (Query 31, Newton, 1979: 403–404; see the discussion in chapter 6 with Biener).

compatible with an immanent God that is in some sense subservient to law. Something akin to this—law-abiding God—doctrine is articulated by Halley in the opening stanza of his very Epicurean ode to Newton at the start of the *Principia* (Albury [1978] is the *Locus Classicus*): "Behold the pattern of the heavens, and the balances of the divine structure/ Behold Jove's calculation and the laws/That the creator of all things, while he was setting the beginnings of the world/ would not violate/Behold the foundations he gave to his works" (Newton, 1999: 379).

My preferred way of interpreting the cumulative impact of these three complications is to suggest that Kant sets up an opposition between Lucretian "chance" and his own law-governed view of nature, to draw attention away from the fact that by his lights his own view is "Epicurean" in non-trivial ways. (Being law-governed is compatible with alternative systems, too.)

In the final section of this chapter I explore key features of the arguments that Kant used to distance his cosmology from the charge of Epicureanism. I argue that these turn Newton's arguments for the same conclusion on their head. But first I analyze an Epicurean interpretation of the *Principia*, in Newton's correspondence with Richard Bentley.

3.2.2 Bentley

The identification between Epicurean innate gravity and Newtonian attraction occurred to an early and significant reader of the *Principia*, Richard Bentley. After Bentley presented his Boyle lectures, Bentley contacted Newton before he produced final revisions to the published text of these lectures. Bentley's side of the correspondence is missing with the exception of one letter.

Even so, we can infer from the responses by Newton (which were appended to the published version of Bentley's Boyle lectures) to the letters from Bentley, that Bentley repeatedly attributed Epicurean innate gravity to Newton.[12] In a passage in the fourth and final letter, Newton strongly resisted the attribution (Newton, 2004: 102–1032).

---

[12] For example, it appears from Newton's first response that in his original letter to Newton Bentley had supposed that "every particle had an innate gravity" (Newton, 2004: 94); in the second letter, Newton writes, "You sometimes speak of gravity as essential and inherent to matter. Pray do not ascribe that notion to me; for the cause of gravity is what I do not pretend to know, and therefore would take more time to consider of it" (Newton, 2004: 100).

The passage has been at the core of recent interpretive disputes over Newton's commitment to action at a distance and his matter theory.[13] Just about the only thing not under dispute is that Newton rejects the attribution of gravity as an essential and inherent property of matter. Here I do not restate my interpretation. Scholars tend to quote the passage selectively. Even a careful scholar such as Schönfeld quotes only part of the first sentence, "It is inconceivable that inanimate brute matter should, without the mediation of something else which is not material, operate upon and affect other matter without mutual contact" and omits the part about Epicurus! (Schönfeld, 2000: 86). So, the question of Epicureanism is not explored in Schönfeld's interpretation of Kant's pre-Critical period. Schönfeld notes that Kant is explicitly engaging with Fontenelle's *Conversations on the Plurality of Worlds* (French: *Entretiens sur la pluralité des mondes*) but not its many not-so-hidden Epicurean (and Spinozist) themes (Martin, 2003).

"Epicureanism" was often used rather loosely during the period as a term of abuse (akin to the way "atheist," "Hobbist," "Pappist," or "Spinozist" could be used). But in the exchange with Bentley, "Epicureanism" has a precise meaning for Newton: It is associated with the system that attributes gravity to an innate and essential (we would say, intrinsic) quality of matter. Moreover, from Newton's response to Bentley's first letter, we can infer that in reading the *Principia*, Bentley had supposed that "every particle had a innate gravity towards all the rest" (Newton, 2004: 94).[14] Bentley was not a naïve reader of the Ancients. He had just established himself as a leading classicist after publishing *Epistola ad Johannem Millium* (1691; see also Westfall [1983: 589]).

So, I explore, first, what in the *Principia* could have made Bentley discern Epicureanism. Second, by drawing on the intellectual and cultural context of Newton's correspondence with Bentley, I argue that Newton had reason to be alarmed by Bentley's attribution of Epicurean doctrine to the *Principia*.

---

[13] In addition to chapters 1–2, see Friedman (2011), Henry (1994), Henry (2011), Janiak (2007), and Janiak (2008), as well as Schönfeld (2000). "For Newton, Matter was passive," while Kant embraces matter as active centers of force (110–111). See Massimi (2011) for more background information on the Newtonian chemical sources that embrace a more active conception of matter.

[14] This differs from how Bentley treats Epicureanism in the published version of his (1692) Sermons preached at Boyle's lecture, remarks upon a discourse of free-thinking; there he treats Epicurean atomism, infinite space, vacuums, and downward gravity (contrasted with mutual gravity) as distinctively Epicurean (157; on p. 48 he also treats the swerve as distinctly Epicurean). I thank Katherine Dunlop for calling my attention to this. I have also consulted Bentley (1838).

3.3 The *Principia*'s Epicureanism

The association of Newton with Epicureanism is prima facie puzzling because the "General Scholium" seems to argue straightforwardly against Epicureanism (the system of "blind metaphysical necessity" [Newton, 1999: 942]).[15] Moreover, Epicureanism was not known for promoting mathematical sciences. To explain this puzzle, I show how the mathematical-physical core of the *Principia* could and was plausibly read as a tract in which Epicurean themes could be discerned (regardless of Newton's intentions).

So, in this section, I explore three related issues: first, I list the dramatic changes to the *Principia* between the first and second edition. Second, I do this in order to motivate a new look at how the first edition of the *Principia* would have been perceived by its early readers (and explain why readers accustomed to the third edition would find such an Epicurean reading surprising). I argue that stripped from its many subsequent changes the *Principia* can plausibly read along Epicurean lines. Third, I argue that a considerable number of these changes might have been motivated by Newton's desire to prevent an Epicurean reading. In particular, I argue that Newton would have every reason to be alarmed by Bentley's attribution of Epicureanism to Newton. Bentley was powerful politically and he prevented Halley from obtaining a Professorship at Oxford a year before the correspondence with Newton. I challenge a recurring theme in Newton scholarship: that Bentley would have been handpicked by Newton to deliver the inaugural Boyle lecture. (The claim is even in Westfall, 1983).

3.3.1 Nontrivial Changes to the *Principia*

The second edition of the *Principia* was published more than a quarter-century after the first under the editorial guidance of Roger Cotes. In addition to a lot of small corrections to the wording and mathematical proofs of the *Principia*, and some comments pertaining to the calculus controversy with Leibniz, we can identify nine significant changes to the *Principia*: (1) there are

---

[15] For the equation of "blind metaphysical necessity" and "Epicureanism" see, besides (as we have seen) Kant's UNH, also Clarke (1705: 227–228), where Lucretius and Epicurus are taken to task for their denial of final causes. Earlier, Clarke had connected Spinoza's view with the position that God was a "mere necessary agent," which is said to be "intelligent." But according to Clarke because Spinoza's God lacks choice it is a "blind and unintelligent necessity" (1705: 102). See also chapter 4 with Domski.

lots of new experimental results, especially in Book II on resistance (Smith, 2001: 264ff), (2) there are new empirical results on planetary orbits and the shortening of pendulum at the equator that strengthen Newton's argument against Huygens' criticism of universal gravity (Smith, 2001), (3) Newton added important material based on his way of measuring the "crookedness" of a curve using an osculating circle (Brackenridge, 1995), (4) Newton largely removes the language of "hypotheses" from the *Principia*. In particular, the nine "hypotheses" that were listed at the start of Book III of the first edition of the *Principia*, were split into three "Rules of Reasoning" (the fourth Rule was added to the third edition) and "six Phenomena." (Newton, 1999: 794)) These four changes will not play a role in what follows.

However, from the point of view of this chapter's argument, there are five other nontrivial changes between the first and second edition of the *Principia*: (5) Halley's ode to Newton was heavily edited by Bentley; (6) Cotes added a new, polemical preface; (7) Newton removed one hypothesis concerning the nature of matter entirely; (8) Newton made significant adjustments in his "matter-theory"; (9) Newton added a new conclusion, the "General Scholium." All these changes can be understood in light of Newton's attempt to distance the *Principia* from the attribution of Epicureanism, although I do not mean to be understood as claiming that all of these are motivated exclusively (or even primarily) with concerns over the attribution of Epicureanism. Debates with Leibniz and his followers also account for some of these changes (Shapiro, 2004). But the issues are not unrelated, because one of the charges that Leibniz pressed against Newton in the correspondence with Clarke (which appeared after the second edition) was that he seemed to embrace Epicureanism. (See Leibniz's letters, 2.1–2 (29–23); 5.130 (227); 5.70 (227), among other allusions.)

In what follows, I explain why the first edition can be read as offering an Epicurean account and along the way discuss how these five alterations respond to that. In particular, there are ten aspects of the first edition of the *Principia* that I will single out as encouraging an Epicurean reading: (i) it embraces infinity of space; (ii) it asserts a vacuum; (iii) it flirts with atomism; (iv) it embraces a strong homogeneity of matter thesis; (v) it appears to embrace action at a distance (and universal gravity); (vi) it is naturalistic, by which I mean that God is almost entirely absent from the book (the exception will be significant); (vii) Halley's ode to Newton is not only modeled on Lucretius but also presents Epicurean (and blasphemous) themes; (viii) the book ends with a mathematical treatment of comets such that comets are law-governed entities (and not portents of the gods) and these play a non-trivial role in cosmic economy; (ix) Newton embraces life on other planets and plurality of worlds (although this will also cause a wrinkle in my argument). In addition (x) together (vii) and (viii) frame the *Principia* such that classically Epicurean themes can be said to bookend the work. Such literary framing is significant because few readers then (or now) would have been able to understand most of the technical claims in the body of the *Principia*. (e.g., Newton suggested to Bentley that he read the first three sections of Book I and then proceed to Book III [Westfall, 1983: 504]). To be clear, these ten aspects do not force an Epicurean reading on the reader of the *Principia*; they are compatible with other readings. But they are also more than highly suggestive of Epicurean themes.

Here my focus is on how the first edition of the *Principia* could have been received. I am not trying to show that Newton was deliberately echoing Epicurean themes. But my argument is, in part, inspired by recent scholarship that has explored Epicurean themes in Newton's unpublished manuscript, De Gravitatione, which has become central to debates over Newton's metaphysics. Jalobeanu (2007) has demonstrated Newton's debts to the English Epicurean Walter Charleton (see also McGuire, 1978; Wilson, 2008). Moreover, my argument builds on an older scholarly literature that called attention to significance of Epicurean language in the *Principia*'s treatment of inertia (Cohen, 1964). So while the main focus here is on the reception of Newton, that reception can help us discern important aspects of the first edition of the *Principia*.

3.4 Ten Epicurean Themes in the *Principia*

In this section, I provide evidence for ten Epicurean themes in the *Principia*. I also discuss the fate of some of these in subsequent editions. First, Newton embraced the infinity of space right at the start of the *Principia*: "the only places that are unmoving are those that all keep given positions in relation to one another from infinity to infinity [*ab infinito in infinitum*] and there always remain immovable" (Newton, 1999: 412).[16] This is a core Newtonian commitment unchanged through all the editions. It is not by itself a distinctively

---

[16] For the classic treatment see Koyré (1957). Koyré emphasizes Platonism in Newton. For the Lucretian roots of infinite space, see Grant (1981, 183ff). Kant also embraced the Newtonian–Lucretian claim that "infinite space teems with cosmic systems" [*der unendliche Weltraum von Weltgebäuden wimmele*] (UNH, Part 1, chapter 1, 257 [247]). In context, Kant refers to Huygens as his source of baseline knowledge.

Epicurean commitment (even Scholastics could embrace varieties of it), but when combined with a commitment to a vacuum, atomism, and homogeneity of matter, it becomes more emblematically Epicurean.

Second, there can be little doubt that Newton embraced the possibility of a vacuum.[17] A doctrine traditionally associated with Epicureanism (as Leibniz points out in his second response to Clarke, paragraph 2 [23]). In the *Principia* some characteristic passages are these: Newton writes, "Hence also it is manifest that the heavens are lacking in resistance" (Newton, 1999: 895). "And thus a vacuum is necessary" (Newton, 1999: 810 note bb. I return to this passage below). This last corollary was heavily rewritten in later editions, but always asserting vacuum. (In the Preface to Book III, Newton argues that "spaces void of all bodies" is one of the general "topics" of the "most fundamental for philosophy" (McGuire, 1995: 104; Newton, 1999: 793). Finally, "And therefore in the heavens, which are void of air and exhalations, the planets and comets, encountering no sensible resistance, will move through those spaces for a very long time" (Newton, 1999: 816). The natural reading of these passages is that space is largely empty. They also constrain quite dramatically the nature of any ether: it must have near negligible mass and be practically frictionless. As Roger Cotes puts it in the "preface": "it must be concluded that the celestial fluid has no force of inertia, since it has no force of resistance" (Newton, 1999: 396).

As a tantalizing aside, Newton's treatment of the vacuum is connected with little-noticed, circumstantial evidence for my thesis about the changes to the *Principia* (as being motivated, in part, by the desire to distance the book from charge of Epicureanism): Robert Boyle gets mentioned explicitly in the *Principia* only from the second edition onward. These mentions occur in two of the notable additions of the Principia. First, Newton writes in the "General Scholium" that "with the air removed, as it is in Boyle's vacuum, resistance ceases, since a tenuous feather and solid gold fall with equal velocity in such a vacuum" (Newton, 1999: 939). This reference to Boyle's air-pump echoes Cotes' "Preface": "Now, that all falling bodies universally are equally accelerated is evident from this, that in the vacuum produced by Boyle's air pump (that is, with the resistance of air

---

[17] Kant also embraces the doctrine in UNH (Preface, 237 [229]). For discussion, see Schönfeld, 2000: 80. McGuire (1995, 107ff) contains a nuanced discussion of Newton's subtle shifts in the doctrine of celestial vacuum (within and outside the solar system) throughout the editions of the *Principia*.

removed), they describe, in falling, equal spaces in equal times, and this is proved more exactly by experiments with pendulums" (Newton, 1999: 387) Boyle is a convenient authority if one is eager to distance one's defense of the vacuum from the potential charge of Epicureanism from a Boyle lecturer (i.e., Bentley; for a nuanced treatment of Boyle and Epicureanism, see MacIntosh, 1991).

Third, Newton's cautious embrace of corpuscularianism is well known from the *Opticks* (see Query 31). But corpuscularianism, which is compatible with infinite division of matter, need not be identical to atomism, which embraces perfectly hard, smallest particles. In the *Opticks*, perfectly "hard bodies" out of which other bodies are composed are presented as possible and likely (Newton 1730: 363ff; 375). So, it is no surprise that Newton's position was associated with atomism (Adam Smith, 1982: 140).

Newton's atomism is more evident in correspondence with Cotes during the editorial activity leading up to the second edition of *the Principia*: "A body is condensed by the contraction of the pores in it, and when it has no more pores (because of the impenetrability of matter) it can be condensed no more" (Quoted from Biener and Smeenk [2012, 133]). But this material was not available during the 18th century.

In the first edition of the *Principia*, Newton's atomism is harder to discern; there is a passage that can be shown to assume it: "And thus a vacuum is necessary. For if all spaces were full, the specific gravity of the fluid with which the region of the air would be filled, because of the extreme density of its matter, would not be less than the specific gravity of quicksilver or of gold or of any other body with the greatest density, and therefore neither gold nor any other body could descend in air" (Newton, 1999: 810 note bb [reworked in later editions, but the core counterfactual claim remains]; Biener and Smeenk, 2012; McGuire, 1995: 105–107; Valentin, 2009, and the reference to Query 28 of the *Opticks*). Newton's argument is by no means obvious to all readers (Chalmers, 2009, chap. 7).

Fourth, in the first edition of the *Principia* Newton embraces a strong homogeneity of matter thesis: "Every body can be transformed into a body of any other kind and successively take on all the intermediate degrees of qualities" (Newton, 1999: 198; see also McGuire [1995: chap. 7]). Commentators see in this hypothesis evidence of Newton's alchemical interests (Dobbs, 2002: 23ff); Newton kept these a well-guarded secret from most of his contemporaries. The hypothesis is one of the most significant Epicurean legacies among a wide variety of New Philosophers

(Wilson, 2008: 52–54 and 80), and a mainstay of its anti-Scholasticism. This is exemplified by the fact that Newton only uses the hypothesis in the proof that shows there are no forms (Newton, 1999: 809 note aa). In UNH Kant also endorses the claim that fundamentally, all material is alike: "I assume that all the matter making up the spheres belonging to our solar system, all the planets and comets, at the origin of all things was broken down into its elementary basic material and filled the entire space of the cosmic structure in which these developed bodies now move around" (Part 2, chapter 1, 275 [263]). All visible differences are evolved from it. In UNH Kant echoes the transformation thesis of Newton's third hypothesis: "And should not other planets be gradually changing into comets by means of a series of intermediate types approximating the composition of comets and linking together the family of planets with the family of comets?" (Part 1, 268; [257–258]).

I call the third hypothesis a 'strong' homogeneity of matter thesis to distinguish it from a thinner homogeneity of matter thesis that remains in all *Principia* editions: 'mass' is a quality that all bodies have in common. (See the postscript to chapter 2 for more on this.) Newton officially dropped Hypothesis 3 in subsequent editions of the *Principia*. All kinds of early modern philosophers could embrace the strong homogeneity of matter, as did Spinoza (*Ethics* II, *Lemmas* I-II; Spinoza, 1994: 125). But this did not entail that they were in other respects Epicurean. Spinoza (whose anthropology and critique of religion has many Epicurean overtones) ridiculed the Epicurean doctrine of seeds in Ethics IP8S2 (Spinoza, 1994: 88). Now, it is well known that when Newton and David Gregory were busy preparing an aborted second edition, Newton studied Lucretius and the Epicurean tradition more generally and intensely. He seems to have thought he could deflect the charge of atheism from Lucretius and Epicurus.[18] Nevertheless, from this period stems a remark by Gregory on Hypothesis 3, "This the Cartesians will easily concede. But not the Peripatetics, who make a specific difference between the celestial and terrestrial matter. Nor the followers of the Epicurean

---

[18] See the landmark paper by McGuire and Rattansi (1966). In his Sermons, Bentley treats Lucretius as a "system of atheism" (p. 375; recall that Kant treats Leucippus and Democritus as systems of atheism); Bentley recognizes that in the ancient world, Epicurus and Democritus were suspected of introducing gods into their system to avoid political trouble, but he also believed that they were forced into accepting the existence of gods on philosophic grounds (pp. 5–6; see also p. 24). In the seventeenth century a number of thinkers—most prominently Gassendi—had promoted a Christianized Epicureanism—and many of Newton's views are consistent with it (Newton's conception of time is indebted to Gassendi's; see Arthur [1995]. The treatment of time may be thought to be an eleventh Epicurean theme in the *Principia* (see chapter 7).

Philosophy, who make atoms and seeds of things immutable" (Newton, 1999: 203). This is evidence that in the aftermath of the exchange with Bentley, some people in Newton's circle wanted to distance the Newtonian project from Epicureanism (on the further significance of Gregory's views in the 1690s, see Schliesser and Smith, forthcoming). Gregory also calls attention to the doctrine of seeds to distinguish Newtonianism from Epicureanism. It is by no means obvious that Gregory can succeed at this distancing. For Newtonian atomism is compatible with Hypothesis 3.

Fifth, the official doctrine of the *Principia* on the cause of attraction is an agnostic one. In all editions of the *Principia*, Newton explains what he means by "attraction":

I use the word 'attraction' here in a general sense for any endeavor whatever of bodies to approach one another, whether that endeavor occurs as a result of [i] the action of the bodies either drawn toward one another or [ii] acting on one another by means of spirits emitted or [iii] whether it arises from the action of aether or [iv] of an air or [v] any medium whatsoever—whether corporeal or incorporeal—in any way impelling toward one another the bodies floating therein. (Newton. 1999: 588; Book 1, Section 11, Scholium)

Newton allows that "attraction" can be caused by many possible mechanisms. So, Newton's claim in the "General Scholium" that "I have not as yet been able to deduce from phenomena the reasons for these properties of gravity" (Newton, 1999: 943)[19] is no change of position and in complete accord with the quote from Book 1, Section 11, Scholium. But it is no surprise that Leibniz and Huygens thought that Newton embraced action at a distance, for [i], [iii], and [v] are all compatible with it. We can infer from Newton's response to Bentley's now lost, first letter that Bentley also ascribed "an innate gravity" to "every particle" toward "all the rest" (Newton, 2004: 94). Bentley's original interpretation embraces the first option listed by Newton.

So despite Newton's official agnosticism, many were inclined to read him as committed to a metaphysical doctrine. I suspect the reason for this is the manner in which the discussion Definition 8 of the *Principia* is worded:

---

[19] However, Newton's further claim in the same sentence, "I frame no hypotheses," is a development in his methodological views. As we have seen, the first edition embraces the (strong) homogeneity of matter hypothesis explicitly.

Accelerative force [may be referred to], the place of the body as a certain efficacy diffused from the center through each of the surrounding places in order to move the bodies that are in those places; and the absolute force [may be referred], to the center as having some cause without which the motive forces are not propagated through the surrounding regions, whether this cause is some central body (such as the lodestone in the center of a magnetic force or the earth in the center of a force that produces gravity) or whether it is some other cause which is not apparent. (Newton, 1999: 407)

Newton seems committed here to the claim that a single particle generates a force field in the places around it (Stein, 1970; see chapter 1 and the post-script to chapter 2). So what Newton and his interlocutors call "innate" gravity seems a reasonable inference from Newton's position. Even so, it goes against the third law of motion, which is cashed out in terms of interactions. That this is true even for gravity becomes fully clear with the third Rule of Reasoning, which was added to the second edition; it shows that the ascription of universal qualities is really—to use David Miller's apt phrase—to systems of bodies (Miller, 2009; my chapter 1). For in that rule, Newton allows that gravity is a "universal quality" of bodies (note the plural throughout the rule), but he explicitly denies that gravity is "essential" (we would say intrinsic) to bodies. (Newton, 1999: 796) It is a contingent fact that gravity is universal.

Sixth, the first edition of *Principia* is a surprisingly naturalistic book. By which I mean that God plays a negligible role in it. This goes against some prominent recent scholarship that emphasizes the theological aspects of Newton (cf. Cunningham, 1988 and 1991; Janiak, 2008; McGuire, 1978). But most such arguments rely on the correspondence with Bentley and the "General Scholium" (as well as Newton's voluminous theological manuscripts). I have found only one mention of God in the first edition of the *Principia*: "Therefore God placed the planet at different distances from the sun so that each one might, according to the degree of density, enjoy a greater or smaller amount of heat from the sun" (Newton, 1999: 814; Book 3, proposition 8, corollary 5). As Bernard Cohen argues, Huygens saw the implications of the argument while commenting on proposition 8 that it showed what kind of gravity "the inhabitants of Jupiter and Saturn would feel" (Newton, 1999: 219).

Despite the fact that the argument was removed after the first editions, Kant also saw the point:

"Newton, who established the density of some planets by calculation, thought that the cause of this relationship set according to the distance was to be found in the appropriateness of God's choice and in the fundamental motives of His final purpose, since the planets closer to the sun must endure more solar heat and those further away are to manage with a lower level of heat." (UNH, Part 2, section 2, 284–285 [271]).[20]

As Kant implies, Newton's claim in the proposition certainly suggests a design argument. Kant rejects Newton's argument on physical grounds (see UNH, Part 2, section 2, 285 [272]). But Newton's claim is not anthropocentric. The antianthropocentric nature of the argument runs through the "General Scholium," which calls attention to the significance of that particular "the diversity of created things, each in its [proper] place and time." (Newton, 1999: 942). Also, the beauty of our solar system is mimicked by countless other solar systems, too far apart to be of interest to us: "This most beautiful System of the Sun, Planets, and Comets, could only proceed from the counsel and dominion of an intelligent and powerful being. And if the fixed Stars are the centers of other like systems, these, being form'd by the like wise counsel, must be all subject to the dominion of One" (Newton, 1999: 940). Moreover, natural diversity is suited to times and places regardless of human interest. Kant's UNH echoes this antianthropocentrism design argument (Schönfeld, 2000: 101ff) but—as I show below—with a Spinozistic twist (see also chapter 8).

Proposition 8 was reworded without mention of God in the second edition in the *Principia*. Interestingly enough, in the original draft version of Newton's system of the world, the posthumously published *Treatise*, the same claim is treated in conditional form: "If God has plac'd different bodies at different distances from the Sun, so as the denser always possess the nearer places, and each body enjoys a heat suitable to its condition, and proper for its nourishment" (Newton, 1728: 34)[21]

Now what makes these not-so-subtle changes significant is that through all editions of the *Principia* Newton called attention to the existence of this

---

[20] While we cannot rule out that Kant was familiar with the first edition of the *Principia*, he may have become familiar with Clarke's argument (Clarke, 1998). That argument may have prompted readers to go back to the first edition. According to Watkins (2015), Kant owned the second edition of the *Principia*. So, perhaps he was unfamiliar with Newton's fourth Rule of Reasoning.

[21] Because Kant echoes the law-like claim, Schönfeld (p. 272 n. 60) discusses the *Treatise* passage, but fails to recognize the conditional nature of the mention of God, and he fails to note the version of the argument in the first edition of the *Principia*.

suppressed draft. In the published introduction to Book III of the *Principia*, Newton writes he suppressed his *Treatise* in order to "avoid lengthy disputes" (Newton, 1999: 793). So, when the *Treatise* did finally appear shortly after Newton's death, inquisitive readers could use it as a guide to what they took to be Newton's original views.

Seventh, the first thing that people would encounter when reading the *Principia* is Halley's ode to Newton. It is clearly modeled on Lucretius' praise of Epicurus (Albury, 1978); Wilson, 2008: 33). As Albury has shown, the ode was heavily edited and revised in the second edition of the *Principia* by Bentley (before being restored in the third). As Albury demonstrates, much of the content of the ode is also Epicurean. Consider the following lines:

> Behold set out for you the pattern of the Heavens, and the balances of divine Mass, and indeed the Calculation of Jove; Laws which the all-producing Creator, when he was fashioning the first-beginnings of things, wished not to violate and established as the foundations of his eternal work. . . . No longer does error oppress doubtful mankind with its darkness: the keenness of a sublime Intellect has allowed us to penetrate the dwellings of the Gods [*Superum*] and to scale the heights of Heaven. (Halley's ode, quoted in Albury, 1978: 27).[22]

In Halley's rendition, God obeys the (eternal?) laws of nature. Throughout the Ode, there is no sign that God the creator [Demiurge] needs to intervene in his own creation. More important, he uses the plural "Gods." I single out these lines because in 1691 they were used to create the charge against Halley "for asserting eternity of the world." Halley's candidacy for the Savilian Chair of Astronomy at Oxford was rejected (Albury, 1978: 37). As Albury points out, Halley's Epicureanism was not disguised and it was used against him. Leading the charge were Bentley and his mentor Stillingfleet, who is known today among philosophers for his correspondence with Locke, but who was one of the most powerful clergyman in the aftermath of the Glorious Revolution (1688). As a young man, Bentley had lived with Stillingfleet for several years (Guerlac and Jacob, 1969). The front page of the published

---

[22] In the second edition these lines were rewritten as follows: "Laws which the all-powerful [*omnipotens*] Creator himself, when he was forming the first-beginnings of all things, pronounced to himself, and indeed the foundations which he layed down for his work. . . Newton now permits us to penetrate the dwellings above and to reach the high Temples of Heaven" (quoted from Albury, 1978: 40). Note, that (the singular) God is now all powerful and the source of nature's laws.

version of Bentley's first Boyle Lectures identifies Bentley as "Chaplain to Right Reverend Father in God, Edward, Lord Bishop of Worcester"—that is, Stillingfleet.

Halley's commitment to the eternity of the world (i.e., "the unchanging order of things") is revealed in the following, blasphemous passages in the Ode:

Arise, Mortals, throw off earthly cares; and discern herein the powers of a heaven-born Mind, far and away remote from the life of the beasts. He who commanded men, by written Tablets, to curb Murder, Theft and Adultery, and crimes of perjured Deceit; or who Counseled nomad peoples to establish walled Cities; or who blessed the nations with the gift of Ceres; or who pressed from the grape relief from cares; or who showed how to combine pictured sounds upon a Nile reed, and to exhibit Voices to the eyes—each of these improved the lot of mankind less, in only looking to a few benefits for the unhappiness of life. But now we are truly admitted as table-guests of the Gods; we are allowed to examine the Laws of the high heavens; and now are exposed the hidden strongholds of the secret Earth, and the unchanging order of things." (Halley's Ode quoted in Albury, 1978: 27)

Now Newton's divinity might be brushed off as a rhetorical flourish. But Halley comes close to suggesting that Revelation seems to have human origins; he clearly suggests that the Gospel is not mankind's greatest gift.

Eight, in his "Ode," Halley writes "Now is revealed what the bending path of horrifying comets is; no longer do we marvel at the Appearances of the bearded Star" (Albury, 1978: 27). As Albury has documented, this line was removed by Bentley in the second edition. In it, Halley draws on the classic Epicurean argument that religious superstition is caused by and causes ongoing fear. In the "Ode," Halley thus calls attention to one of the most significant achievements by Newton: the last propositions of the *Principia* provide a (rather complex) procedure to calculate and predict the orbits of comets. Thus, in the first edition, the *Principia* closes on a series of propositions that in Halley's way of framing them allow one to tame the causes of superstition.[23] In Newton's hands, comets are not supernatural portents, but a "kind of planet

---

[23] Cohen and Whitman's 1999 translation of the *Principia*, fails to note that the first edition of the *Principia* ends on p. 930 (with Newton's Q.E.I. in proposition 42, problem 22). I thank Roger Ariew and Dan Garber for helping me establish this.

revolving about the sun in very eccentric orbits" (Newton, 1999: 928; accepted by Kant in UNH, Part 2, Section 3, 292ff [277ff]).

Moreover, in the closing pages of the first edition of the *Principia*, Newton inserted the following connected speculations:

> [F]or the conservation of the seas and fluids on the planets, comets seems to be required, so that from the condensation of their exhalations and vapors, there can be a continual supply and renewal of whatever liquid is consumed by vegetation and putrefaction and converted into a dry earth . . . .I suspect that it is chiefly from the comets that spirit comes, which is indeed the smallest but the most subtle and useful part of our air, and so much required to sustain the life of all things with us. (Newton, 1999: 926)

Here Newton is committed to multiple worlds (as we have seen this doctrine is emphasized in the "General Scholium"). In Newton's view, comets play an indispensable part in, what one may term, the economy of nature. They transfer essential material throughout the solar system. Echoing the providential argument of corollary 5 to Proposition 8 Book 3 (quoted above), comets play a purposive role in maintaining nature's equilibrium on various "planets." Again, this is an antianthropocentric argument. Most importantly, comets play a role in circulating the building blocks of life (see postscript to chapter 2).[24] This has a very Epicurean ring to it; Lucretius also insists that the infinite celestial heavens are the source of earthly life (see the end of Book II of *De Rerum Natura*). So it is no surprise that discerning readers would have interpreted Newton as an Epicurean![25]

Now, Newton is silent on the particulars of the Lucretian theory of seeds, so one can understand why Gregory would have been tempted to explore it as a way to distinguish Newton from Epicureans.[26] Either way, Gregory could

---

[24] This is not Kant's position in UNH (Kant appears to recognize Newton's argument, but he denies its empirical adequacy—he also thinks comets are inhospitable to advanced life forms [something that Newton is quiet about]. Part 2, Section 8, 362–363 [338]).

[25] Steve Snobelen has objected that Newton's views are entirely compatible with a Christian Epicurean treatment of comets. My approach turns on reading Newton's treatment of comets in light of Halley's unchristian Epicurean Ode.

[26] Having granted that, the following is the last—very Epicurean—sentence added to the second edition of the *Principia* proper (before the "General Scholium"): "And the vapors that arise from the sun and the fixed stars and the tails of comets can fall by their gravity into the atmospheres of the planets and there be condensed and converted into water and humid spirits [*spiritus humidos*], and then—by a slow heat—be transformed gradually into salts, sulphurs, tinctures, slime, mud, clay, sand, stones, corals, and other earthly substances" (Newton, 1999: 938). What is to say that these other earthly substances are not seed-like? Incidentally, in private correspondence (May 10, 2011), Ted McGuire points out that here Newton presupposes the discarded third hypothesis, after all.

point to other aspects of the *Principia* that definitely seem un-Epicurean; Newton seems to have no interest in moral philosophy (in the body of the *Principia*).

None of the points listed in this section by itself would settle the case for an Epicurean reading of the first edition of the *Principia*. But it should be clear that, especially in light of Halley's opening Ode and the closing pages on comets, their cumulative effect would have made the *Principia* seem like an (perhaps innovatively mathematical, modern) Epicurean tract to a reader more interested in metaphysics and theology than in solving the sophisticated mathematical-empirical problems that Newton bequeathed to future generations.

3.5 The Correspondence with Bentley, Reconsidered

The aim of the previous section was to try to explain why knowledgeable readers would have discerned Epicurean commitments in the first edition of the *Principia*. But they also are meant to lay the ground for a reconsideration of Newton's correspondence with Bentley and its significance for the reception of and changes to future editions of the *Principia*. This correspondence has played two significant roles in recent scholarship of Newton: (i) to motivate interpretations of Newton's views on action at a distance, (ii) to motivate interpretations of Newton's views on theology.[27]

There has been surprisingly little attention to the circumstances of the exchange between Newton and Bentley. As I indicated above, Bentley was not just an ambitious young intellectual who contacted Isaac Newton for some feedback on his arguments between the spoken and printed version of his Boyle lectures. He was the protégé of powerful clergy of England after the Glorious Revolution. While Stillingfleet and his Latitudinarian comrades promoted interest in natural philosophy and were relatively tolerant about doctrinal differences, their successful opposition to Halley's appointment at Oxford also shows the limits of their tolerance.* In particular, the circle around Stillingfleet promoted a providential interpretation of the Glorious Revolution and a natural religion in accord with it (Guerlac and Jacob, 1969).

So when Newton received letters from Bentley with the repeated suggestion that a central part of the *Principia* sounded like Epicureanism, he

---

[27] There is a third role: to explore Newton's cosmology. See, especially Kerszberg (2012).
* William Whiston's fate at Cambridge University is another example.

would have had every reason for concern. In particular, while Epicureanism was not unfashionable before the Glorious Revolution, even Christian Epicureanism—first promoted by Gassendi and popularized in England by Charleton—became impolitic after 1688 on my interpretation. Bentley could be dangerous to him. The first sentence of his initial response to Bentley's now lost first letter reads, "When I wrote my *Treatise* about our system, I had an eye upon such principles as might work with considering men, for the belief of a Deity, and nothing can rejoice me more than to find it useful for that purpose" (Newton to Bentley, December 10, 1692; Newton, 2004: 94). Commentators have interpreted this sentence as sincere autobiography. We need not attribute to Newton duplicity, however, in order to recognize that the claim—however sincere and carefully crafted—is almost entirely political in nature. It is designed to curry favor with a powerful and dangerous interlocutor.

More important, the first sentence of Newton's response to Bentley does very little justice to the content of the *Principia* (Guerlac and Jacob, 1969: 311–312). As I have argued, the tenor of the first edition of the *Principia* is largely naturalistic. It is hard to see how it can be read as promoting "principles as might work with considering men, for the belief of a Deity." (This is not to deny that indirectly the *Principia* may be conducive to such an end). By contrast in the book, there is an alternative statement that gives an entirely adequate account of the aims of the *Principia*:

> But in what follows, a fuller explanation will be given of how to determine true motions from their causes, effects, and apparent differences, and, conversely, of how to determine from motions, whether true or apparent, true causes and effects. For to this was the purpose for which I composed the following treatise. (Scholium to the Definitions; Newton, 1999: 415).[28]

Philosophically, the crucial part of Newton's claim is that against the dominant hypothetical approach prevalent in his time, his method would be able to uncover causes in nature. But theologically, the crucial part of Newton's

---

[28] Against purely instrumental readings of Newton's enterprise, Newton's self-understanding of his enterprise is couched in causal terms. (See also the wording of the first two rules of reasoning). See also Schönfeld (2000: 90), who claims without offering any textual evidence from Newton (or even secondary literature) that Newton had "scoffed" at the discovery of laws, and he claims that Newton "had maintained" that "causal investigations" should "not be the business of 'experimental philosophy.'" The best recent defense of the instrumentalist reading is McMullin, 2001. See chapter 8 on Toland's reading of Newton.

statement in the Scholium to the definitions is that he promises to settle the Copernican controversy once and for all; he will teach how to distinguish apparent from true motion. As we have seen, the theology of even the first edition of the *Principia* is compatible with the commitment to the existence of a deity. Previously I have called attention to the ways in which Newton argued for a providentialist deity. But Newton's deity is antianthropocentric.

Newton's arguments in the *Principia* were traditionally viewed as belonging to an Epicurean tradition (which did not deny the existence of deities), and inimical to Christian religion. Once alerted to this reading of his book, Newton prevented it from gaining currency: first, in his adamant rejections of it to Bentley; later (through the 1690s) in private attempts to develop a reading of Epicurean tradition more in accord with the new orthodoxy of his time;[29] eventually, by reframing the second edition of the *Principia* in dramatic ways by letting Bentley rewrite Halley's ode, by the addition of the "General Scholium," which Voltaire called "the little treatise on metaphysics" (1927: 172), and by letting Cotes argue for Newton's orthodoxy in the new "preface." Cotes directly confronts the Epicurean hypothesis (see Newton, 1999: 397). The intended target here is compatible with a broad range of views, including Spinozist.

My interpretation is speculative. Many will prefer to avoid a political interpretation of these letters. But it is worth pointing out that the current historiography accepts uncritically the claim that Newton had somehow handpicked Bentley to be the first Boyle lecturer (even Westfall, 1983: 489 and 589 implies this). But there is no reason to claim that Newton had anything to do with the selection of Bentley (the claim seems to originate in Guerlac and Jacob [1969]). Newton was present at Boyle's funeral (Guerlac and Jacob, 1969: 317–318). But there is no other evidence linking Newton to the selection of Bentley as a Boyle lecturer.[30]

[29] See McGuire and Rattansi (1966). In McGuire 1995a, he shows how Newton's engagement with Lucretius in the 1690s led Newton to embrace Lucretian views on the nature of existence. So on my reading, the changes to the *Principia* are motivated to make the book more Christian (which is compatible with some Epicurean doctrines), but less obviously Epicurean.

[30] Guerlac and Jacob call attention to a note by Gregory from December 1691 as evidence that even before Bentley was selected as Boyle lecturer, "Newton was obviously suggesting that his discoveries in celestial physics would serve the argument from design" (1969, 317). They think Newton was proposing a cosmic argument from design that would be simpler than the then recently published John Ray's *Wisdom of God manifested in the Works of Creation*. It is by no means obvious that, in Gregory's note, Newton is endorsing design arguments. Rather he is making the methodological point that natural philosophy ought to start with the most general and universal empirical phenomena.

The upshot of this is that treating the correspondence with Bentley as evidence for Newton's views in the first edition of the *Principia* is fraught with difficulty not sufficiently appreciated by those who routinely appeal to the correspondence to settle interpretive debates. In particular, given that Newton's views on gravity are singled out as "Epicurean," Newton's main points in the passages in the exchange with Bentley and subsequently in writings available to a wide audience that deal with gravity should be first and foremost read as Newton distancing himself from the potential charge of blasphemy (Henry, 1999, 2007).

3.6 Understanding Kant's Response to Newton

In conclusion, in this section, I argue first that UNH as a whole responds to a number of specific arguments by Newton in the "General Scholium." I provide considerable textual evidence that Kant was aware of this. Second, I show that Kant's argument flirts with Spinozistic themes.

3.6.1 Kant and Newton's General Scholium

Here I offer a new interpretation of aspects of the relationship between the General Scholium and Kant's UNH. It is well known that the "General Scholium" offers arguments for the existence and the nature of our knowledge of God. Newton first argues that while the orbits of celestial bodies are governed by laws, neither such laws nor the "mechanical causes" of his philosophic opponents can be the cause of the orbits themselves. Newton barely gives an argument for his claim that the laws of motion cannot be the cause of the orbits. He does claim that it is "inconceivable" that the laws of nature could account for such "regular" orbits (Newton, 1999: 940; see also Kant UNH Part 2, section 8, 363–364 [338–339]). But in what follows, Newton packs quite a bit into this claim. In particular, it turns out that for Newton the regularity consists not merely in their being law-governed, but also that the trajectories and mutual attractions of the planets and comets hinder each other least. This culminates in Newton's conclusion that "This most beautiful System of the Sun, Planets, and Comets, could only proceed from the counsel and dominion of an intelligent and powerful being" (Newton, 1999: 940). Without further argument, Newton rules out the possibility that these particular three features—(i) law-governed orbits that (ii) hinder each other minimally and that (iii) are jointly beautiful—could be caused by other causes than God. Newton then offers the "immense distances" among the planetary systems, which thus avoid the possibility of gravity-induced mutual collapse, as another, empirical phenomena that supports his argument from inconceivability (Newton, 1999: 940). Moreover, as Chris Smeenk emphasized to me, given that Newton could put no constraint on the mass of comets, he must have found it striking that these do not disrupt the motions of the solar system through which they pass.

Newton's position rules out two contrasting, alternative approaches, both discussed later in the "General Scholium," the first being that God is constantly arranging things in nature. He rejects this quickly: "In him all things are contained and move, but he does not act on them nor they on him. (Newton, 1999: 941).[31] No further argument is offered against a hyperactive God. Second is that everything is the product of "blind metaphysical necessity" (Newton, 1999: 942). This second view is associated with the system of Spinoza. Newton offers an independent argument against that approach, namely that given that necessity is uniform it cannot account for observed variety. (Newton, 1999: 942) Now this is only a limited objection against Spinozism; for it is committed to there being sufficient reason for infinite variety in the modes (E1p16 [Spinoza, 1994: 97] and E1p28 [Spinoza, 1994: 103]). At best Newton has shifted the burden of proof (see the discussion in chapters 4 and 8) Because Newton has the better physics, he can claim to have constrained any possible explanation that will account for the observed variety. But it is not insurmountable: All a necessitarian needs to show is how the laws and the "regular" orbits are *possible* given some prior situation.

Moreover, in the absence of a discussion of initial conditions of the universe Newton's claim begs the question. One can understand Kant's UNH as taking up the challenge of accounting for the universe in light of Newton's laws of motion. In particular, in Part 2, Section 8 of UNH is Kant's refutation of Newton's argument. That section starts with the following claim:

> We cannot look at the planetary structure without recognizing the supremely excellent order in its arrangement and the sure marks of God's

---

[31] As Ted McGuire pointed out, this passage may be thought problematic for my claim that Newton's God is immanent in nature. But the sentence is compatible with the denial that God stands outside nature (e.g., God "will not be never or nowhere" [Newton, 1999: 941]).

hand in the perfection of its interrelationships. After reason has considered and wondered at so much beauty and excellence, it rightly grows indignant at the daring foolishness which permits itself to ascribe all this to chance [*Zufalle*] and a happy contingency [*Ungefähr*]. There must have been a Highest Wisdom to make the design, and an Infinite Power must have produced it. Otherwise it would be impossible to encounter in the planetary structure so many purposes cooperating in a single intention. (UNH, Part 2 Section 8, 355, (331–332))

In what follows Kant offers two opposing options: first, Newton's approach in the General Scholium (God as an "alien hand" that produces "restriction and coordination which permit us to see the perfection and beauty in it"; Kant also calls it the thesis that introduces "the immediate hand of God") and, second, his preferred proposal: "the plan for the structure of the universe is already set in the fundamental composition of eternal natures by the Highest Understanding and implanted in the eternal laws of motion, so that they develop themselves freely from them in a manner appropriate to the most perfect order." He remarks ruefully: "An almost universal prejudice [*Vorurtheil*] has made most philosophers oppose the capability of nature to produce something ordered through its universal laws, just as if it meant that we were challenging God's rule over the world, when we seek the primordial developments in the forces of nature, as if these forces were a principle independent of the Godhead and were an eternally blind fate" (UNH, Part 2, Section 8, 356 [332]).

Consider one of Kant's arguments against Newton's position:

What then will this curious method of demonstrating the certain existence of a Highest Being out of the fundamental incapacity of nature prove by way of an effectively counter to Epicurus? If the natures of things bring forth by the eternal laws of their being nothing but disorder and absurdity, then they will show in that very manner the nature of their independence from God. What sort of an idea will we be able to create for ourselves of a divinity whom the universal natural laws obey only through some sort of compulsion and in and of themselves act against the wisest designs of the Divinity? Will the enemy of providence not win just as many victories from these false basic principles, when he can point to harmonies [*Übereinstimmungen*] which the universally effective natural laws produce without any special limitations? (UNH, Part 2, Section 8, 357 [332ff])

According to Kant, Newton's argument in the "General Scholium" fails as an argument against Epicureanism! It makes the laws of nature too independent from God.

This does not address the burden-shifting argument that Newton has offered. But UNH as a whole meets the burden of evidence. As Kant writes (emphasis in the original) *"the world gives evidence of a mechanical development from the general natural laws as the origin of its arrangement and, secondly, that the manner of the mechanical development which we have presented is the true one"* (UNH, Part 2, Section 8, 359; A.149). Even if one denies that Kant has succeeded in providing a true account, his speculative, Newtonian cosmogony shows how the world *could* have developed such that the orbits of the planets can be explained by Newtonian laws. This is not to suggest that Kant's account is fully convincing. In particular, Marius Stan has suggested that "Kant wants to get angular momentum from nothing" (Personal correspondence, June 28, 2012). Even if we grant Stan's point, Kant has transformed a decisive burden-shifting argument into an outstanding technical, localized research problem. In that respect, certainly Kant has met Newton's burden-shifting argument.[32]

3.6.2 Kant and Spinozism

But a question remains: does Kant's success in meeting Newton's burden-shifting argument against (neo-Epicurean) Spinozism reopen the door to the system of blind metaphysical necessity? Now given all of Kant's talk about providence, it might seem perverse to even pose this question.[33] Yet, as we have seen, Kant's (Newtonian) providence is resolutely antianthropocentric.

---

[32] I leave open how unwelcome this result would have been to Newton. Leaving aside the question about Kant's use of a vortex as a stage in cosmic development, here I take for granted that Kant's physical principles are properly Newtonian.

[33] Given the prevalent providential language in UNH, it is no surprise that most scholars have linked Kant's UNH to Leibniz's metaphysics (e.g., see Schaffer, 1978: 191; the significance of Spinoza on the young Kant seems largely unchartered territory, but see Boehm [2014 and 2016] and Schliesser [2015b]). Watkins (2013) uses later Kant to argue against a Spinozistic reading of UNH, but this cannot shift the burden of evidence. In particular, a Spinozist reading of young Kant can accept that God's understanding is essential to the grounding of all possibility (understood as the reality of eternal return of all variations in the universe as implied by E1p16).

In a single striking passage (UNH, Part 3, Appendix, 379–381; [353–354]), Kant embraces several aspects of this antianthropocenticism that are distinctly Spinozistic. In particular, (i) Kant endorses the "satire" of the "creatures who live in the forests of a beggar's head" and (ii) in which men are compared to lice by a Fontennelle-like character (as noted before Fontenelle is not an innocent reference). In his own voice, Kant explicitly endorses the analogy between "humans" and "insects" (Schönfeld, 2000: 101ff) in order (iii) to combat the anthropocentric idea that nature is designed for human purpose (final causes aimed at humans). Moreover, Kant insists that (iv) "The limitlessness of creation contains within itself, with equal necessity, all natures which its superbly fecund richness produces" while arguing that (iv) "the beauty of the whole . . . consists in the interrelatedness" and (vi) "everything is determined by universal laws which nature effects through the combination of forces originally planted in it." Even (vii) the surprisingly moralistic conclusion of the passage "has the possessor of those inhabited forests on the beggar's head ever created greater disasters among the races of this colony than the son of Philip brought about among the race of his fellow citizens, when his wicked genius gave him the idea that the world was created only for his sake?" echoes Spinoza's denunciations of Alexander the Great (i.e., the son of Philip) in chapter 17 of the *Theological Political Treatise*.

The first six of these are all recognizable Spinozistic themes, and echo Spinoza's famous letter to Oldenburgh where he compares man to a worm in the blood (Spinoza, 1994: 82–84). One may, perhaps, argue that the fifth point about beauty is un-Spinozistic. In the Appendix to *Ethics* 1, Spinoza had insisted that cosmic beauty is a subjective, imaginative projection (Spinoza, 1994: 114). But here Kant's way of understanding beauty as consisting in the interrelatedness of the universe is Spinozistic. For in the letter to Oldenburgh, Spinoza had propounded a kind of cosmic harmony of the sort that Kant explicitly calls beautiful. Kant's UNH is suffused with appeals to this kind of harmony. He never names Spinoza, but he recognizes such appeal is not innocent— "The defender of religion fears that the harmony [*Übereinstimmungen*] which can be explained by a natural tendency of matter would demonstrate the independence of nature from divine providence" (Preface, 229; [223])—and shortly thereafter he calls such a system, in which nature is independent from divine providence, "Epicurean" in the passage [II] quoted in section 1A above. We have already encountered Kant's response to this charge: it contrasts the "blind collision" of Lucretius with his own embrace of "matter bound to certain necessary laws."

Then follows Kant's main argument for the existence of God that is supposed to distinguish him from the Epicureans: "There is a God for just this reason, that nature, even in a chaotic state, can develop only in an orderly and rule-governed manner," (UNH, Preface, 235 [228]). One might think this is not Spinozistic because in his letter to Oldenburgh, Spinoza writes: "I attribute to Nature neither beauty nor ugliness, neither order nor confusion. For things can only be called beautiful or ugly, orderly or confused, in relation to our imagination" (see also *Ethics*, Appendix 1; Spinoza, 1994: 82; Spinoza, 1994: 114). But all Kant is saying here is that nature is law-governed and therefore there is a God. Spinoza would not deny this.

In context, Kant's argument seems to be something like this:

P1) the uniformity of matter is evidence of shared, "primordial origin"
P2) nature can develop from chaos in an orderly and rule-governed manner (as has been shown by the whole argument of UNH)
P3) matter is not intrinsically self-organizing[34]
P4) only a "self-sufficient Highest Reason" [*allgenugsamer höchster Verstand*] can be the original source of matter's ordering and rule-governed tendency (UNH, Preface, 234 (227))
P5) from chaos nature cannot develop by chance (ruling out Lucretius)
∴: Therefore, there must be a God (if understood as self-sufficient highest reason).

Now, Kant's argument to design 'advances' over the version presented in Newton's "General Scholium" in two respects: first, with the help of a Newtonian cosmogony and cosmology, God's role is pushed back to the origin of matter; second, God has been cleansed of anthropomorphic qualities.[35] (Kant is not the first to make these moves; hence, the scare quotes around "advances.") God has been turned into a "self-sufficient highest reason." Kant's position rules out a voluntaristic God. It looks as if even God could not have done otherwise because from chaos nature could have developed only in one way. Thus, Kant's God can neither perform miracles (because these would violate the one and only possible development) nor have a choice in adding different qualities to the nature of matter (because this, too,

---

[34] This is compatible with the empirical fact that matter encountered in nature is self-organizing.
[35] Abe Stone and Aaron Koller have both urged me to see a Platonic provenance originating in the *Timaeus* in this.

would violate the one and only possible development). This is to say, while the argument is not Spinozistic, the argument's conclusion is. In particular, Kant ends up endorsing key Spinozistic positions (such as necessitarianism, uniformity of matter, no miracles, and antianthropomorphism).

In explaining his argument for the existence of God, Kant writes, "All things connected together in a reciprocal harmony must be united among themselves in a single being on which they collectively depend. Thus there is present a Being of all beings, an Infinite Intelligence and Self-Sufficient [*Also ist ein Wesen aller Wesen, ein unendlicher Verstand und selbständige Weisheit, vorhanden*]" (UNH, Part 2, Section 8, 358 [334]). This does echo Spinoza's *causa sui* and the infinite intellect more generally, as well as the argument of the letter to Oldenburgh, in particular. For after describing how all the parts of nature are in harmony with each other, Spinoza describes the existence of "since it is of the nature of substance to be infinite" and "each part" of corporeal nature "pertains to the nature of corporeal substance, and, can neither be nor be conceived without it" (Spinoza, 1994: 84).

Newton's anti-Epicurean burden-shifting arguments of the General Scholium leave this particular argument as a viable alternative available to Kant. Kant concludes UNH as follows, "The collective essence of creatures, which has a necessary harmony with the pleasure of the Highest Original Being, must also have this harmony for its own pleasure and will light upon it only in perpetual contentment" (UNH, Conclusion, 395 [367]). It is hard to shake the feeling that Kant is vividly describing Spinoza's vision of the eternal mind's contented, intellectual love of God in the act of knowing (E5p33–37; Spinoza, 1994: 259–261).[36]

[36] I thank audiences at Bucharest, especially Dana Jalobeanu, Brussels, Utrecht; and especially Paul Ziche and Ernst-Otto Onnasch, and Groningen for comments on presentations where I discussed earlier incarnations of this chapter. I also thank Ted McGuire, Katherine Dunlop, Eric Watkins, and Steve Snoeblen, for terrific comments on a penultimate draft. Finally, I am very grateful to Michela Massimi for encouraging this chapter; her detailed comments and, especially, her calling attention to the puzzling role of Epicurus in Kant's *Universal Natural History* (UNH). She has also kindly shared her (2011) on the Newtonian-chemical sources to Kant's matter theory in UNH. The usual caveats obtain.

4

Newton and Spinoza

On Motion and Matter (and God, of Course)

with Mary Domski

4.1 Introduction

Rosalie L. Colie's (1963) work has been indispensable for understanding the depth of the hostile responses to Spinoza's *Ethics* (1677) by Cambridge Platonists, especially from Ralph Cudworth and Henry More[1]. She emphasizes, in particular, that More's *Confutatio* (1678) is as much an attempt to refute Spinoza as it is an attempt to restate More's own position as distinct from Spinoza's. My goal in what follows is to extend Colie's important work on British anti-Spinozism by connecting More's criticisms to later arguments against Spinoza from Samuel Clarke and Colin Maclaurin.[2] Tracing this peculiar line in the course of British anti-Spinozism shows how the authority derived from the empirical success of Newton's enterprise was used to settle debates within philosophy (see Schliesser [2011] on "Newton's Challenge").

In the arguments on which we focus, More, Clarke, and Maclaurin aim to establish the existence of an immaterial and intelligent God precisely by showing that Spinoza does not have the resources to adequately explain the origin of motion. What we emphasize is that in the progression from More to Clarke to Maclaurin, key Newtonian concepts from the *Principia* (1687) are introduced and exploited in order to challenge the account of matter and motion that is presented in Spinoza's *Ethics*. Namely, Clarke's arguments can be seen as innovations over More's arguments insofar as Clarke adopts a Newtonian

---

[1] This chapter first appeared as Eric Schliesser, "Newton and Spinoza: On motion and matter (and God, of course)." *The Southern Journal of Philosophy* 50.3 (2012): 436–458.

[2] Recent French scholarship has tended to read Clarke's engagement with Spinoza through his controversies with Leibniz, Toland, and Wachter. See the brilliant work by Tristan Dagron (2009). This approach tacitly projects Diderot's historiographical views back onto a previous generation and, more crucially, ends up assuming that we have nothing to learn about Spinoza from some of his "Newtonian" critics. We thank Mogens Laerke for calling attention to Dagron.

Newton's Metaphysics. Eric Schliesser, Oxford University Press. © Oxford University Press 2021.
DOI: 10.1093/oso/9780197567692.003.0006

conception of motion to buttress More's general critique of Spinozism. Maclaurin later appeals to other elements of Newton's mechanics—namely, atomism and the existence of a vacuum—to strengthen the charges leveled against the "blind necessity" that characterizes Spinoza's natural philosophy. Building on this treatment, we use the arguments from More and (especially) Clarke to help discern the anti-Spinozism that can be detected in Newton's "General Scholium" (1713). Ultimately, the Newtonian criticisms that we detail offer us a more nuanced view of the problems that plague Spinoza's philosophy, and they also challenge the idea that Spinoza seamlessly fits into a progressive narrative about the scientific revolution (cf. Schliesser, 2018).

Before proceeding to the arguments from More, Clarke, and Maclaurin, we begin with a brief overview of Spinoza's position on motion. This will set the stage for understanding why British critics found the Spinozistic natural system so worrisome and the weaknesses they discerned in it.

4.2 Motion in Spinoza's *Ethics* (1677)

It is an axiomatic fact for Spinoza that "All bodies either move or are at rest" (E2p13A1).[3] Motion, in fact, is one of the distinguishing conditions of simple bodies (see especially the demonstration included with E2p13L3), and, as presented by Spinoza, a compound individual is an entity (or nature) that maintains the same ratio of motion to rest among its parts (E2p13L5).[4]

Spinoza also appeals to the existence of "absolute" motion (E2p13L2) and (presumably), thus, distinguishes between merely apparent (or relative) and absolute motion. While motion clearly plays a crucial role in Spinoza's fundamental metaphysics, neither in the context of E2p13 nor anywhere else in the Ethics does *Spinoza* define 'motion' or 'rest' (or even 'speed', for that matter). In a letter to Tschirnhaus dated January 1675, Spinoza admits that his observations on "motion . . . are not yet written out in due order, so I will reserve them for another occasion" (Spinoza, 2002: 913).

However, in a letter dated May 5, 1676, Spinoza offers some elaboration on the origins of motion and writes to Tschirnhaus that "For matter at rest,

---

[3] Our approach to Spinoza's views on mathematical physics is an elaboration of Savan (1986) and a corrective to Israel (2002, 242ff). This present chapter is a companion to Schliesser (2017). Our treatment of the English Newtonian response to Spinoza runs parallel to the treatment of Dutch Newtonians in Jorink (2009).

[4] For a very good treatment of how bodies are individuated in Spinoza's system, see Section 5 of Manning (2012).

as far as in it lies, will continue to be at rest, and will not be set in motion except by a more powerful external cause" (Spinoza, 2002: 956). Initially, this sounds like a Cartesian argument for the existence of God, who (like an infinite billiard-ball player) sets in motion unmoving matter, which is governed by an inertia-like law.[5] But given that in this context Spinoza explicitly rejects Descartes' conception of extension and Cartesian natural philosophy more generally ("Descartes' principles of natural things are of no service, not to say quite wrong" [Spinoza, 2002: 956]), we should be cautious in pressing the analogy between Spinoza and Descartes here. In particular, while Spinoza does embrace an inertia-like principle in his corollary to E2L3 ("a body in motion moves until it is determined by another body to rest; and that a body at rest also remains at rest until it is determined to motion by another"),[6] Spinoza indicates that the only cause that can determine matter to move is some other (larger or more forceful) body. This follows, in fact, from a core commitment that can be traced back to the start of the Ethics, especially E1D2, according to which modes terminate, co-constitute, and delimit each other only within an attribute.[7] So in Spinoza's system it is nonsensical to think of motion as somehow originating outside of the attribute to which it properly (as a common notion) 'belongs.'

Moreover, in Spinoza's system, unlike in Descartes, God is not external to physical nature; for Spinoza, God is immanent (E1p18), so is in no way to be thought of as an external cause.[8] Now Spinoza does distinguish between God as a free cause (*Natura Naturans*), by which Spinoza refers to the fact that God is the only thing that exists and acts from the necessity of his nature (E1p17C2 and E1p16), and those things which follow from the necessity of God's nature (*Natura Naturata*). For Spinoza bodies "are in God, and neither be, nor be conceived without God" (E1p29S). Now, while it is not easy to understand what Spinoza means by 'in' or 'inherence' (see Melamed [2006] and Della Rocca [2008], 61ff.), he is very careful to make clear that it is a mistake to understand God as being somehow outside nature. As he infamously puts it, "*Deus Sive Natura*" (E4Intro).

---

[5] For God as the general cause of motion in Descartes, see *Principles of Philosophy* (1984), Part 2, sec. 36.

[6] For the significant differences between Descartes' account of inertia and Spinoza's see Schliesser (2011b). For an opposing view that treats the conatus doctrine as inertia-like, see Viljanen (2008).

[7] So Spinoza is not merely committed to attribute parallelism but also to a kind of attribute "closure."

[8] This is not to deny that God plays no causal role for Spinoza; God is the cause of the being of things, including the material ones (E1p28). We understand Newton's doctrine of substantial omnipresence in the "General Scholium" along similar lines (chapter 2).

Thus in Spinoza's system if matter 'starts' at rest, no motion will be generated; therefore, given the existence of motion, there must be some motion in the universe from the infinite past. And this is what Spinoza seems to claim at E1p28, where he says: "this cause, or this mode [of an attribute] ... had also to be determined by another, which is also finite and has a determinate existence; and again, this last (by the same reasoning) by another, and so always (by the same reasoning) to infinity."[9]

Now this is not to deny that God plays some role as the source of natural motion: "From the necessity of the divine nature [who has absolutely infinite attributes by D6] there must follow infinitely many things in infinitely many modes" (E1p16), including, presumably, motion. So in Spinoza's system there is what we may call sufficient reason for the existence of motion. However, from our human vantage point, it is not clear that this is a very impressive explanation, which is precisely the point that More and Clarke forward in their attack on the Ethics.

4.3 More's *Confutatio* (1678)

Shortly after Spinoza's *Ethics* appeared posthumously in 1677, Henry More published a blistering attack on it in his *Confutatio*.[10] In this section we explore two arguments presented in this work, one that targets Spinoza's claim that there is an infinite succession of motions and the other that aims at Spinoza's rejection of final causes. Our main reasons for considering these arguments is that they offer a clear sense of the problems that British opponents found with Spinoza's conception of motion and also give us an instructive starting point for understanding the Newtonian elements that are introduced in similar arguments that are later offered by Clarke.

According to More, Spinoza is a materialist ("matter is God"; Jacob, 1991: 70–72, 77ff.).[11] Against this position, More attempts to show that a (spiritual) God

[9] In order to avoid misunderstanding: it need not follow that, for Spinoza, bits of matter, which mutually codetermine, must be passive. Spinoza often repeats that we do not have adequate knowledge of body (e.g., E2p24–25) and so leaves the source of its activity open.

[10] Colie provides useful political and intellectual context for the reception of Spinoza in England and also calls attention to the significance of Spinoza's views on motion to the debates over his reception by Toland and Wotton (Colie, 1959: 44). Hutton (1990) is an indispensable guide to More's thought. All citations to the *Confutatio* refer to the translation by Alexander Jacob (1991).

[11] More's interpretation of Spinoza here is controversial insofar as he appears to suppose that the other infinite attributes of Spinoza's God are irrelevant. However, it is by no means entirely implausible. In fact, Curley (1988) is the contemporary standard-bearer for a very influential set of arguments that treat Spinoza as a sophisticated materialist.

is required to explain certain pertinent facts about our world, including the existence of motion. Here we focus on two such attempts, the first of which comes from More's self-described "third argument" against substance monism. The argument is presented as follows:

it is manifest that that which at any time was not present was not even for a moment past, in the succession of the world. Whence it is plainly proved, since all the moments of its succession were at some time present, and many do not at any time follow at the same time as one, but single moments always follow one after the other, that it was at some time, since all things, at any rate apart from one, were in the process of becoming present. And thus perforce we will be led back to the head or principle [*ad caput sive principum*] of all successive durations, of whatever extent, and suppose it to be extended, and think and declare, what is equally contradictory, that there can be an infinite successive duration, and a figured infinite magnitude. When it plainly follows that this corporeal world, with all its motions and revolutions of changes, has not existed nor can exist from eternity, and [A] matter cannot be by itself or [A⁺] at least moved by itself, and so [B] it is necessary that some other substance exists before matter, which communicates motion to matter in some way. (Jacob, 1991: 96; bracketed symbols added to facilitate discussion)

Now, on the whole, More does not argue from premises shared with Spinoza. Consider, for instance, More's assumption that time successively unfolds, as it were, moment by moment and that this is an "objective" feature of the world. For Spinoza, this is not the case. Rather, time is imaginary or merely abstract (e.g., E2p45S and E5p29), where to imagine something does not always mean it is false. Even so, imaginings, or confused knowledge, can never yield adequate knowledge (see the long Scholium at E2p49C), because from the point of eternity, things that are not fully adequate (such as duration) do not have full existence.[12] To put this anachronistically, these imaginings

---

[12] This follows from the way Spinoza applies E1A4 in E2p7. Spinoza is committed to different things having more or less reality (E1p9). It is interesting to note that at one point in Newton's career, in De Gravitatione (Newton, 2004), the ontological difference between space and body is one of different degrees of being; bodies have more reality than space and "whatever has more reality in one space than in another space belongs to body rather than to space" (Newton, 2004: 27). Newton also affirms the position that bodies have a "degree of reality" that "is of an intermediate nature between God and accident" (32). (See chapter 5.) For more on the notion of space in De Gravitatione, see Domski (2013).

should not be thought to belong to the fundamental ontology of the world. Whatever specific premises Spinoza might accept or reject (and certainly there are others in the passage above), the argument as a whole promotes further inquiry into the problems surrounding Spinoza's account of motion. Specifically, reflection on [A], [A⁺], and [B] raises important questions about the nature and origin of motion in Spinoza's system. In particular, we will have to give more consideration to how Spinoza blocks the inference from [A] and [A⁺] to [B], namely, how he can accept that both the existence and motion of matter depend on some cause without also committing to the existence of an external, nonmaterial cause.

What is at stake here is the origin of matter and, in particular, the origin of matter's motion. In Spinoza, their origin follows from the divine nature; God serves as the sufficient reason. But Spinoza is frustratingly silent on the details, and this leaves "God" acting like an empty placeholder rather than a specific cause for the origin of an infinite succession of motion. We can recognize this point even if we are not very impressed by either More's blanket denial that the material universe can be eternal or More's conclusion that there must be "some other substance [that] exists before matter" that is the original source of motion.

That More considers that this external cause for the existence and motion of matter must be immaterial is made clear in his earlier anti-Cartesian *Enchiridion Metaphysicum* (1671), where More presents the same argument against the existence of an infinite succession of motions. He writes:

> this corporeal world with its motions and revolutions of changes neither existed nor could exist from eternity, and [I] matter does not arise from itself or [I⁺] at least cannot be moved by itself; and so [II] it is necessary that there exists a certain immaterial principle or incorporeal substance which impresses motion on matter. (Jacob, 1991: 87n36)

Now if one accepts [I⁺] that matter cannot be self-moving (because, say, it is taken to be passive), then even if More's [I] is false or thought to be question begging against Spinoza, then [II] may be accepted. As already noted, it is not clear that Spinoza can block the inference, given his reticence in discussing the nature of matter and the origin of motion.

In the *Confutatio*, More offers a separate and extended argument against the Spinozistic claim that there is an infinite succession of motions (in the context of his criticism against what he calls the "fourth" of Spinoza's seven arguments against final causes):

> Indeed he supposes an infinite succession of motions. . . . For Spinoza here [E1p32C] speaks of God in the same way as of some infinite matter, whose parts are pushed and pulled one by the other from eternity. But if not so, then motion begins in matter from within at some time, unless God is, such as He doubtless is, distinct from matter, who . . . from a mind that is never incapable of foresight and counsel (since that mind is eternal and infinite, which, as it were in flash of the eye, can see what is best in each thing) according to His eternal Ideas which include the cause and the end of all things has produced the entire creation as soon as He was capable of creating it. (Jacob, 1991: 87)[13]

Here More provides a framework for the way Newtonians will later argue against Spinoza: either motion is eternal, which, as we have already seen, More takes to be absurd; or if not, then either matter generates motion or a wise God generates motion. Crucially, More presupposes that matter is necessarily passive, which renders absurd the notion that matter generates motion. Thus, only one option remains: a wise God generates and is the ultimate source of motion.[14]

It is worth remarking here that the passivity of matter was a standard commitment of the mechanical philosophy (and even a Platonist like More relies on it in his arguments in favor of the existence of active spirits), but it raises complications in the context of Newtonian claims about action at a distance. To put this tersely: the moment Newton opens the door to the possibility that matter is active, he undercuts this peculiar anti-Spinozistic

---

[13] When More writes that "Spinoza here [E1p32C1] speaks of God in the same way as of some infinite matter," on our reading, More subtly reinterprets Spinoza: Spinoza's God acts in a law-like manner (e.g., in E1p32C1 Spinoza denies that God produces "any effect by freedom of the will"), but that does not prevent God from being distinct from matter in some nontrivial sense. For Spinoza, matter, or extension, is an infinite attribute of substance, but Spinoza's God also encompasses infinitely more attributes.

[14] There is a line of argument that insists that for Spinoza matter is not passive but is naturally active. For example, Nadler (2006, 196n7) has pointed to the Conatus doctrine as evidence that Spinoza rejects the passivity of matter and accepts that it has innate active powers. Diderot seems to have read Spinoza this way (see Wolfe, 2010a and 2010b).

strategy for the existence of (a providential) God.[15] And what we see below in Clarke's arguments, in fact, is that he is able to enhance More's strategy precisely by appealing to the Newtonian notion of motion without making any commitments about the nature of matter.

The second Morean strategy we focus on is employed in one of his arguments in favor of final causes (something famously denied by Spinoza, as in *Ethics* Part I, Appendix).[16] This particular argument from the *Confutatio* also gives a nice flavor of More's generally (hostile) tenor toward Spinoza:

> O the intolerable petulance and haughty virulence of the completely blind and stupid philosophaster, who since he represents the descent of a stone to the earth as a mechanical cause, most indolently and boastfully holds forth as if he has certainly realized that the structure of the human body was made not by counsel or providence but by blind mechanical necessity. We, on the other hand, hold for certain that mechanical power cannot extend so far. (Jacob, 1991: 79)

The core of More's argument here is that (i) mechanical power is limited; (ii) the human body exhibits design; and therefore, (iii) there must be a designer. Now, Spinoza would deny (i), so the argument does not seem very compelling. Moreover, given that Spinoza thinks we know very little about the nature and capacities of the human body (E2p24S; E3p2S), claims such as (ii) reveal mostly wishful thinking (see especially E1, Appendix; see also Schliesser, 2017). Regardless of the premises Spinoza might reject here, we mention More's argument for two reasons. First, he forwards the idea that if one can properly delimit the scope of mechanical causes—and the meaning of these causes shifts before and after Newton—then one has room to marshal evidence for the existence of alternative, nonmechanical causes (including God), which (as we will see) will be a recurring strategy among the Newtonians. Second, the association of Spinoza's position with "blind mechanical necessity" will become a trope that we see recurring in later anti-Spinozistic arguments and will alert us to possible anti-Spinozism even when Spinoza is not named. In fact, both of these themes emerge in Clarke's arguments, to which we now turn.

---

[15] For evidence of the possibility of active matter in Newton, see chapters 1–2 and 8.

[16] According to Colie (1963), More's argument probably has roots in Boyle's arguments for design.

4.4 Clarke's Demonstration (1705)

Samuel Clarke is, of course, very well-known from his celebrated correspondence with Leibniz and Collins. These are extremely relevant to a more thorough study of Clarke's critical engagement with Spinozism, but here we focus more narrowly on some arguments in Clarke's 1704 "Boyle's lectures," collected as *A Demonstration of the Being and Attributes of God* (1705), with the subtitle "More particularly in answer to Hobbs, Spinoza, and their followers." We follow Colie's assessment that "Alone of the [Boyle] lecturers, Clarke treated Spinoza as an important philosopher, not just as an atheist whose foolish arguments could be met with stock physico-theological orthodoxy" (Colie, 1963: 207).

However, although Colie calls attention to the importance of motion to Clarke's argument against Spinoza, our approaches to their debate are orthogonal to each other. What we emphasize here is that Clarke engages with the *Ethics* as offering a natural philosophical system and, in particular, that he introduces a distinctively Newtonian conception of motion into his arguments against Spinoza's natural philosophy and thereby offers important innovations over More's earlier arguments against Spinoza's account of motion.

In a lengthy and very important passage, Clarke begins by telling us that there are three possible origins of motion: (i) an eternal intelligent being is the cause of motion, (ii) the motion is necessary and self-caused, or (iii) "an endless successive communication" is the cause of motion. Clarke then offers explicit criticisms of Spinoza in order to rule out (ii) and (iii) and thereby leave us with (i), namely, the position that an eternal intelligent being is the cause of motion. He writes:

[I] If [motion] was of itself necessary and self-existent, then it follows that it must be a contradiction in terms to suppose any matter to be at rest. And yet, at the same time, [II] because the determination of this self-existent motion must be every way at once, the effect of it could be nothing else but a perpetual rest. Besides . . . [III] it must also imply a contradiction to suppose that there might possibly have been originally more or less motion in the universe than there actually was, which is so very absurd a consequence that Spinoza himself, though he expressly asserts all things to be necessary, yet seems ashamed here to speak out his opinion, or [IV] rather plainly contradicts himself in the question about the origin of motion[+] [[+]The accompanying footnote refers to E1p33 with E2p13L3]. But [V] if it be said, lastly, that motion, without any necessity in its own nature and without any external necessary cause, has existed from eternity merely be an endless successive communication as Spinoza inconsistently enough seems to assert (+E2p13L3), this I have before shown in my proof of the second general proposition of this discourse to be a plain contradiction. (Clarke, 1998, Part VIII: 45; bracketed numerals added to facilitate discussion)

Clarke's discussion explicitly attacks the argument of the Ethics and shows significant textual command over details of Spinoza's position. Moreover, a cursory reading of this discussion shows agreement between Clarke and Spinoza on some key points. In particular, Clarke and Spinoza agree that "it is manifest" that matter at rest "could never of itself begin to move" (recall Spinoza's letter to Tschirnhaus, dated May 5, 1676).[17] Also, Clarke and Spinoza agree (without argument) that there is "now such a thing as motion in the world" (recall Spinoza's commitment to "absolute" motion at E2p13L2). However, it is the points on which they disagree (numbered [I]– [V] in the passage earlier) that reveal Clarke's Newtonian innovations over More's initial attack on the Ethics from the Confutatio. We treat each of these five innovations in turn.

Clarke's first innovation is the claim "[I] If [motion] was of itself necessary and self-existent, then it follows that it must be a contradiction in terms to suppose any matter to be at rest." Such a claim, of course, seems question begging against Spinoza, who, as we saw, claims it is axiomatic that "All bodies either move or are at rest" (E2p13A1) and, thus recognizes no blatant contradiction between self-existing, or necessary motion, and matter at rest. That said, Clarke is right to suggest that if we accept something like a PSR, then it does at least seem arbitrary to accept that, at any given time, matter can be at rest if motion is self-existent and if, as Spinoza claims, there must have always been some motion in the universe.

Something like this intuition informs Clarke's best arguments against Spinoza, and this intuition reveals how Newtonian mechanics subtly enters the debate. The basic charge against Spinoza is that he cannot sufficiently

---

[17] Of course, Clarke's position is a bit more ambiguous. The argument here does not address what intelligent matter could do, only how we should conceive of "unintelligent matter." In what follows we ignore how Clarke might conceive of intelligent (or minds that are superadded to) matter because this option is irrelevant to the debate with Spinoza (for whom intelligence/intellects/minds would belong to the attribute of thought, not extension).

distinguish motion from rest, or the existence of motion from the existence of rest, as the natural philosophy of Newton's *Principia* does.

That Newton sets the standard for Clarke's criticism of Spinoza is also revealed in his second innovation, according to which "[II] because the determination of this self-existent motion must be every way at once, the effect of it could be nothing else but a perpetual rest." With the claim that the determination of self-existent motion "must be every way at once," Clarke suggests here that any body will have tendencies (or determinations) to motion in every direction. Thus, for any determination, d1, in a body there will be an opposing determination, −d1, that cancels it out. The effect is that the body does not move, that is, it will be in perpetual rest. Clarke assumes, of course, that the determinations, or tendencies, to motion are acting with the same force—that is, are of equal degree, as it were. And if this holds, then what Clarke has ultimately shown is that Spinoza cannot designate a single, well-defined determination of motion for each body, which is precisely what Newton does with his system of laws of motions and definitions.

Clarke's third innovation is the most revealing of the Newtonian commitments that underwrite his criticisms of Spinoza. Recall the argument: "[III] it must also imply a contradiction to suppose that there might possibly have been originally more or less motion in the universe than there actually was, which is so very absurd a consequence that Spinoza himself, though he expressly asserts all things to be necessary, yet seems ashamed here to speak out his opinion." Now, it is important to first point out that all Spinoza actually requires is that the ratio of motion to rest in the universe remains the same, where the universe is taken as an infinite individual that keeps its nature or form (see E2p13L4 and E2p13L7). Keeping this ratio fixed is entirely compatible with variation in the actual motion of the universe. All the same, given that we are dealing with a ratio here, Clarke is right that at any given time there is a sense in which "there might possibly have been originally more or less motion in the universe than there actually was." This, in turn, violates Spinoza's claim that "things could have been produced by God in no other way" (E1p33) and gives merit to the charge leveled above.

From our present point of view, what is even more important is the role that quantity of motion plays in Clarke's criticisms. Considering motion as a quantity is at the heart of Newton's mechanics, and it is a conception that Spinoza would have severe problems addressing. Namely, while adequate knowledge of motion and rest as such is clearly possible for Spinoza—it is one of the common notions of extension—quantities of motion do not enter into such knowledge, since common notions are about qualitative, not quantitative, properties of extension. The manner or magnitude of such properties is extrinsic and, thus, not a common notion. This becomes clear by reflecting on how Spinoza characterizes common notions: common notions are qualities, as it were, that all bodies share regardless of their state (see, especially, E2p38–39). Second, these properties do not just have a high degree of generality—they are common to all bodies (E2L2, cited in E2p28C)—but the manner in which they are present within each and all bodies is also equal (E2p39Dem). The best way to make sense of common notions is, therefore, to suggest that they are intrinsic properties of modes within an attribute (in Spinozistic terms they share an "affection") and that they reflect the peculiar modal qualities of such a mode within an attribute. For example, all bodies are equally capable of motion and of rest, and of moving slower and quicker (E2L2) (see Schliesser, 2011c).

At the level of quantities of motion, Spinoza is always going to claim that we are dealing with confused knowledge by way of the imagination.[18] At best, it seems we can say that Spinoza will deflect this sort of argument by relegating the natural science of motion to the realm of inadequate knowledge. Thus, in advancing from More, Clarke has not only shifted the burden from explaining the nature and origin of motion in general to explaining the nature and origin of Newtonian motion (understood as a physical quantity), he has also underscored Spinoza's inability to meet the standards of Newtonian mechanics and to explain the success and authority of mathematical physics as a body of adequate knowledge about natural motions.

The final two arguments from Clarke address worries already suggested by More—namely, that Spinoza is unable to give a general explanation for what causes motion and, thus, that Spinoza does not have the resources to identify any natural motions at all. However, as in the *Natura Naturata* case of the third innovation, Clarke is implicitly adopting a Newtonian conception of motion as a physical quantity and a peculiar conception of God in order to strengthen the case against Spinoza. Consider the fourth innovation, where Clarke cites two passages (E1p33 and E2p13L3) and claims that in these, Spinoza "[IV] rather plainly contradicts himself in the question about the origin of motion." The two texts state (1) "things could have been produced by God in no other way" (E1p33), and (2) "A body which moves or is at rest must be determined to motion or rest by another body, which has

---

[18] See especially Spinoza's "The Letter on the Infinite" (discussed in Schliesser, 2017).

been determined to motion or rest by another, and that again by another, and so on to infinity" (E2p13L3 relying on E1p28). Certainly, from Spinoza's point of view, there is no obvious contradiction that emerges here; it is possible to have a deterministic system in which bodies cause other bodies to move. But from Clarke's point of view, there is a contradiction that stems from Spinoza's account of God because it looks as if E1p33 understands God in terms of *Natura Naturans*, that is, as a free cause (E1p29S). By contrast, E2p13L3 seems to be describing God's activity in terms that come closer to following from some other cause (E1p29S). Even if it is too strong to call Spinoza's position a contradiction—*Natura Naturans* and *Natura Naturata* are complementary, not contraries—here Clarke successfully puts his finger on Spinoza's ambiguities and lack of determinateness on the source of motion.

Finally, consider the contradiction cited in Clarke's fifth innovation: "[V] if it be said, lastly, that motion, without any necessity in its own nature and without any external necessary cause, has existed from eternity merely be an endless successive communication as Spinoza inconsistently enough seems to assert [Clarke adds a note to ⁺E2p13L3], this I have before shown in my proof of the second general proposition of this discourse to be a plain contradiction." As in the previous case, the "plain contradiction" is not easy to detect: from Spinoza's point of view, there is no contradiction in claiming the existence of an infinite and eternal succession of motion—this simply follows from "the absolute nature" of God's attribute of extension (see E1p28Dem, which Spinoza relies on in the demonstration of E2p13L3). Spinoza simply asserts that there is some connection between (a) the absolute nature of the attribute of extension, (b) the infinite and eternal succession of motion that "follows" from it, and (c) the particular motion(s) we find in the world. But from Clarke's point of view, Spinoza can say nothing about the kind of motion we experience. In particular, from Clarke's point of view, a chain of motions is conserved by Newtonian mechanical causes. These are so-called passive principles, but the motions require some active principles to be generated.

To see the point, consider the full argument that Clarke had presented earlier in the book,

an infinite succession . . . of merely dependent beings without any original independent cause is a series of beings that has neither necessity, nor cause, nor any reason or ground at all of its existence either within itself or without. That is it is an express contradiction and impossibility. It is supposing something to be caused (because it is granted in every one of its stages of succession not to be necessarily and of itself), and yet that, in the whole, it is caused absolutely by nothing, which every man knows is a contradiction to imagine done in time; and because duration in this case makes no difference, it is equally a contradiction to suppose it done from eternity. And consequently there must, on the contrary, of necessity have existed from eternity some one immutable and independent being. (Clarke, 1998, Part II: 10–11)

Here Clarke is making a Newtonian move insofar as he is explaining the origin of mechanical causes—or explaining the cause of mechanical causation—in terms of an active, immaterial agent. While this is not the only option available within a Newtonian framework—an "independent" God is not the only active principle in Newton's system[19]—it is an important strain in Newtonian anti-Spinozism. As already seen in More's arguments, it is by properly delimiting the scope of mechanical causes that Clarke makes room to marshal evidence for the existence of an immaterial cause of motion, that is, for the existence of an "immutable and independent" God.

While several of Clarke's innovations over More's may seem question begging against Spinoza (insofar as accepting Spinoza's notions of God/Nature and body seem to render many of the arguments moot), on balance Clarke succeeds in putting the spotlight on the problematic features of motion within Spinoza's system precisely by adopting a generally Newtonian framework for his analysis. In particular, he has successfully raised the suspicion that when it comes to thinking about motion in terms of a physical quantity Spinoza has very few resources at his disposal, other than to dismiss mathematical physical science as at best offering inadequate knowledge of natural motions.

We now turn to Maclaurin's use of two further Newtonian elements—the vacuum and atomism—to critique Spinoza's philosophy of nature. Then, in final section, by returning to Clarke, we examine how the framing of Newton's natural philosophy in the *Principia* is taken to subvert the Spinozistic commitment to nature governed by "blind necessity."

[19] See chapters 1 and 2.

4.5 Maclaurin's Account (1748)

Colin Maclaurin was the most sophisticated Newtonian of the generation following Clarke (and Newton). As documented elsewhere (Schliesser, 2011a & 2012a), in Maclaurin's widely read (posthumous) *An Account or Sir Isaac Newton's Philosophy*, one of the main targets is Spinoza's claim that there is "one substance in the universe, endowed with infinite attributes, (particularly, infinite extension and cogitation) that produces all other things, in itself, necessarily, its own modifications; which alone is, in all things, cause and effect, agent and patient, in all respects physical and moral" (Mclaurin, 1748: 78). Here we focus on just one aspect of his criticisms of Spinoza:

> So Spinoza represented [the universe] as infinite and necessary, endowed always with the same quantity of motion, or (to use his inaccurate expression⁺) always having the same proportion of motion and rest in it, and proceeding by an absolute natural necessity; without any self-mover or principle of liberty. . . . Sir Isaac Newton's Philosophy, . . . altogether overthrows the foundation of Spinoza's doctrine, by showing that there may be, but actually is a vacuum; and, instead of an infinite, necessary, and indivisible, plenitude, matter appears to occupy but a very small portion of space, and to have its parts actually divided and separated from each other. (Maclaurin, 1748: 77)[20]

Notice the two elements of Newton's natural philosophy that are marshaled against Spinoza: the vacuum and an atomist conception of matter. Now, there can be little doubt that Newton embraced the possibility of a vacuum; he says explicitly in Book 3, Proposition 6, Corollary 3 of the Principia that "a vacuum is necessary" (Newton, 1999: 810; see also Book 2, sec. 9, Scholium; Book 3, Lemma 4, cor. 3; Book 3, Proposition 10, Theorem 10). The natural reading of these passages is that space is largely empty. They also constrain quite dramatically the nature of any ether that may pervade space, namely, consistent with the remarks in the Queries of the *Opticks* (1704), any such ether must have near negligible mass and be practically frictionless.

---

[20] In the accompanying footnote, Maclaurin quotes Spinoza in Latin: "⁺*omnia [enim] corpora ab aliis circumcinguntur, et ab invicem determinantur ad existendum et operandum, certa ac determinate ratione, servata semper in omnibus simul, hoc est, in toto universe, eadem ratione motus ad quietem, Epist.—Corpus motum vel quiescens ad motum vel quietem determinari debuit ab alio corpore, quod etiam ad motum vel quietem determinatum fuit ab alio, et illud iterum ab alio, et sic in infinitum*" (E2p13L3). (The same Lemma targeted above in Clarke, 1998, Part VIII: 45.)

Newton's cautious embrace of atomism is well known from the *Opticks*, where he writes in Query 31 that "the small Particles of Bodies" that have "certain Powers, Virtues, or Forces, by which they act at a distance, not only upon the Rays of Light for reflecting, refracting and inflecting them, but also upon one another for producing a great Part of the Phenomena of Nature" (Newton, 1952: 375–376). But of course corpuscularianism, which is compatible with the infinite division of matter, need not be identical to atomism, which embraces perfectly hard smallest particles. In the *Opticks*, perfectly "hard bodies" out of which other bodies are composed are presented as possible and likely (Newton, 1952: 364, 370, 375–378). So it is no surprise that Newton's position was associated with atomism.[21]

In the quoted passage, Maclaurin is as coy about atomism as Newton became from the second edition of the *Principia* onward (Biener and Smeenk, 2012). Maclaurin (correctly) attributes to Newton the position that there is a vacuum in nature and that matter can be separated from each other. He opposes these positions to the "foundation of Spinoza's doctrine." Now few readers of Spinoza will have thought that Spinoza's claim that "there is no vacuum in Nature" (E1p15S) is fundamental to the program of the *Ethics*, even if one allows that it may be the only place in the *Ethics* where Spinoza is vulnerable to empirical refutation.[22]

So why would Maclaurin call this a "foundation"? Echoing Clarke, Maclaurin appears to be homing in on Spinoza's commitment to the universe having "the same quantity of motion or (to use his inaccurate expression) always having the same proportion of motion to rest, and proceeding by an absolute natural necessity; without any self-mover or principle of liberty." But unlike Clarke, Maclaurin neither analyzes Spinoza's argument carefully nor argues against Spinoza. Here Maclaurin simply asserts the authority of Newton's natural philosophy as offering the more legitimate source of explanation on these matters.* But this does not mean Maclaurin is being merely dogmatic here. For while Maclaurin cites Letter 4 to Oldenburgh from 1661,

---

[21] Newton's atomism is more evident in his correspondence with Cotes during the editorial activity leading up to the second edition (1713) of the *Principia*: "A body is condensed by the contraction of the pores in it, and when it has no more pores (because of the impenetrability of matter) it can be condensed no more" (Biener and Smeenk, 2012: 133). But this material was not available during the eighteenth century. See Biener and Smeenk, 2012 for a useful analysis of the atomism presupposed at *Principia*, Book 3, Proposition 6, Corollary 3.

[22] For an excellent treatment of Spinoza on the vacuum, see Schmaltz (1999).

* This is an example of what Schliesser calls "Newton's Challenge to philosophy" (Schliesser 2011a; 2012a).

in the *Ethics* Spinoza also states the conservation doctrine in E2p13L7S. If we trace the sources of the doctrine through earlier commitments in the *Ethics* (via the proofs of E2p13L7, E2p13L4, and E2p13L1), Spinoza refers back to E1p15S and the arguments for the denial of a vacuum in nature! This suggests that Maclaurin has read the Ethics carefully. More crucially, it also suggests that by explaining away a vacuum, a post-Newtonian Spinozist must respond to the authority that follows from the unprecedented success of empirical mathematical philosophy.

4.6 The General Newtonian Argument Against Spinoza and the General Scholium (1713)

In Clarke and Maclaurin, we see the use of specific elements of Newton's mechanics to buttress the original arguments against Spinoza made by More. Beyond that, we notice in Clarke especially that Newton's natural philosophy in general is taken to challenge Spinoza's account of nature and motion. What Clarke argues is that the Newtonian natural system and the findings that stem from it are incompatible with the "blind necessity" that characterizes both the Epicurean and Spinozistic world picture, precisely because this system implies the existence of an immaterial and wise Creator.[23] Consider these remarks from Demonstration (1705) where Clarke claims that recent Newtonian science offers "additional strength" to an a posteriori argument that "the supreme cause and author of all things must of necessity be infinitely wise":[24]

What would [Cicero] have said [against the Epicureans], if he had known the modern discoveries in astronomy? The immense greatness of the world (I mean of that part of it, which falls under our observation), which is now known to be as much greater than what in his time they imagined it to be, as the world itself, according to their system, was greater than Archimedes' Sphere? The exquisite regularity of all the planets' motions without epicycles, stations, retrogradations, or any other deviation or confusion whatsoever? The inexpressible niceties of the adjustments of the primary

---

[23] Colie (1963, 207) claims that Spinoza's association with Epicureanism in the Boyle lectures starts with John Hancock, who preached his sermons in 1706.
[24] The classic treatment of these arguments is by Hurlbutt (1965). For the significance of Spinoza in the Boyle lectures, see Colie, 1963: 204.

velocity and original directions of the annual motions of the planets, with their distance from the central body and their force of gravitation towards it? . . . The wonderful motions of the comets, which are now known to be as exact, regular, and periodical as the motions of other planets? Lastly, the preservation of the several systems and of the several planets and comets in the same system from falling upon each other, which infinite past time (had there been no intelligent governor of the whole) could not but have been the effects of the smallest possible resistance made by the finest ether and even by the rays of light themselves to the motions (supposing it possible there ever could have been any motion) of those bodies? (Clarke, 1998, Part XI: 81–83; for more on this argument see chapters 8–9)

The negative force of these remarks is clear: the very findings of recent science and astronomy cannot be explained by a purely mechanical and deterministic system. Such a Spinozistic/Epicurean system lacks explanatory force because, as Clarke suggests, it cannot explain what Newtonian science has made manifest: the regularity, greatness, exactness, and "niceties" of the natural world. This reinforces the argument by Clarke that we had analyzed above; even if Spinoza can offer a sufficient explanation of motion, he lacks the resources to account for the particular exacting details we find in nature. For a Spinozist to meet this burden, she must articulate a full-fledged mechanics from first principles.

One might object that Clarke is discussing Epicureans and not Spinoza here. But this ignores that Clarke tends to equate the two. Indeed, later in the text, Clarke equates "blind metaphysical necessity" and "Epicureanism" (1705, 227–228), where Lucretius and Epicurus are taken to task for their denial of final causes. And prior to that Clarke had connected Spinoza's view with the position that God was a "mere necessary agent." But according to Clarke, because Spinoza's God lacks choice, his is a "blind and unintelligent necessity" (1705: 102).

With a "blind necessity" unable to account for the world as described by Newtonian science, we are left to embrace the existence of a wise and governing creator:

Certainly atheism, which [at the time of Cicero] was altogether unable to withstand the arguments drawn from this topic, must now, upon the additional strength of these later observations which are every one an unanswerable proof of the incomprehensible wisdom of the creator, be utterly ashamed to show its head. We now see with how great reason the author of the book of Ecclesiasticus after he had described the beauty of the Sun and stars, and all the then visible works of God in heaven and Earth, concluded, ch. 43, v 32, (as we after all the discoveries of later ages may no doubt still truly say) 'there are yet hid greater things than these, and we have seen but a few of his works.' (Clarke, 1998: Part XI: 81–83; see chapter 9 for more on this argument.)

Now, the main point of Clarke's argument is to deploy the amazing precision and empirical success of modern mathematical physics, which reveals the greatest amount of regularity and apparent design in nature as an a posteriori argument for a designing God. God is not only a necessary origin for the very possibility of motion (motion would be impossible "had there been no intelligent governor of the whole"), but also the particular kinds of motions exhibited in the world. (Spinoza was, of course, the greatest critic of such arguments for design, as we see in *Ethics*, Part 1, Appendix.) In the quote, Clarke does not mention Spinoza explicitly. But, for those who doubt that Spinoza could be treated as a latter day follower of Epicurus and Lucretius in the period, there is an oblique reference to Spinoza here nevertheless.

Clarke insists that when it comes to knowledge of nature, rather than using clear and distinct intellectual perception, we need to see "clearly and distinctly" with telescopes! Spinoza was of course a great craftsman of telescopic lenses. He was also the proponent of the method of clear and distinctness, which carries its own self-evidentness (E1p8S2).[25] According to Clarke, the only thing we may not be able to learn is the "nicety" of God's "Adjustment of the Primary Velocity and Original Direction of the Annual Motion of the Planets" (recall the quote from Clarke, 1998: Part XI: 81) at the origin of the universe.[26] But the main point of the passage is revealed by Clarke's optimistic use of Ecclesiasticus 43:32. For Clarke this is the predictive assertion that there is still much to learn about nature. In context, Clarke

[25] Maclaurin explicitly targets the method of clear and distinctness to buttress support for an "experimental philosophy": "In all of these, Spinoza has added largely from his imagination, to what he had learned from Descartes. But from a comparison of their method and principles, we may beware of the danger of setting out in philosophy in so high and presumptuous a manner; while both pretend to deduce *compleat* systems from the clear and true ideas, which they imagined they had, of eternal essences and necessary causes. If we attend to the consequences of such principles, we shall the more willingly submit to experimental philosophy, as the only sort that is suited to our faculties..." (1748: 77). For discussion, see Schliesser (2012a).

[26] In a general sense this position, too, has its provenance in Boyle, who had emphasized that while we know God exists, we know almost nothing of him (Colie, 1963: 201). There are affinities with Newton's position on our near complete lack of knowledge of God's substance in the "General Scholium" (see the postscript to chapters 2 and 9.)

treats Newton's discoveries as evidence of our progress in knowledge and as a promise that there is much more to learn.

Now Clarke's arguments are not merely an argument for the existence of a wise designer. The significance of Clarke's arguments rests, in part, in their ability to marshal the authority of Newton's unprecedented novel, precise, and accurate discoveries to settle an argument within philosophy. (This move is "Newton's Challenge to Philosophy," Schliesser [2011a]). Importantly, Clarke's approach anticipates the contours and much of the specific details of a similar set of arguments that Newton presents in the "General Scholium,"[27] which was added to the second edition (1713) of the *Principia*:

But though these bodies may indeed persevere in their orbits by the mere laws of gravity, yet they could by no means have at first deriv'd the regular position of the orbits themselves from those laws. The six primary Planets are revolv'd about the Sun, in circles concentric with the Sun, and with motions directed towards the same parts and almost in the same plan. Ten Moons are revolv'd about the Earth, Jupiter and Saturn, in circles concentric with them, with the same direction of motion, and nearly in the planes of the orbits of those Planets. But it is not to be conceived that mere mechanical causes could give birth to so many regular motions: since the Comets range over all parts of the heavens, in very eccentric orbits. For by that kind of motion they pass easily through the orbits of the Planets, and with great rapidity; and in their aphelions, where they move the slowest, and are detain'd the longest, they recede to the greatest distances from each other, and thence suffer the least disturbance from their mutual attractions. This most beautiful System of the Sun, Planets, and Comets, could only proceed from the counsel and dominion of an intelligent and powerful being. And if the fixed Stars are the centers of other like systems, these, being form'd by the like wise counsel, must be all subject to the dominion of One; especially since the light of the fixed Stars is of the same nature with the light of the Sun, and from every system light passes into all the other systems. And lest the systems of the fixed Stars should, by their gravity, fall on each other mutually, he hath placed those Systems at immense distances from one another. (Newton, 1999: 940)

---

[27] There is, of course, rich scholarship on the "General Scholium": in theological matters see, for example, Snobelen (2001) and Ducheyne (2006). For useful treatment on Newton's sources see McGuire and Rattansi (1966) and De Smet and Verelst (2001).

Now this is not the place to discuss all the features of this argument (see Smeenk and Schliesser [2017]). For our present purposes, what is important is that Newton rules out the possibility that three particular features of planetary orbits—that they (i) are law-governed, (ii) hinder each other minimally, and (iii) are jointly beautiful— could be caused by any other cause than God. Newton then (iv) offers the "immense distances" among the planetary systems, which thus avoids the possibility of gravitationally induced mutual collapse, as another empirical phenomenon that supports his argument from inconceivability. Moreover, given that Newton could put almost no constraint on the mass of comets, (v) he must have also found it striking that these do not disrupt the motions of the solar system through which they pass. Comets provide a further hint of providential design, in that (vi) at aphelia they are sufficiently far apart so as not to disturb each other's motion.

Newton's position rules out two contrasting, alternative approaches, both discussed later in the "General Scholium": (i) that God is constantly arranging things in nature. As he writes, "In him are all things contained and moved; yet neither affects the other" (Newton, 1999: 941). No further argument is offered against a hyperactive God. (ii) Everything is the product of "blind metaphysical necessity" (Newton, 199: 942). As we have documented above, in the writings of More and Clarke, this system of blind necessity is associated with the system of Spinoza. So while Newton does not mention Spinoza by name, readers familiar with Clarke's and More's rhetoric and arguments would have recognized the target. Thus, in the new second and third editions' capstone of the *Principia*, Newton lends his authority to the project of refuting Spinoza. (For more on these arguments, see chapters 8–9.)

In addition to a refined version of the argument to design, Newton offers an independent argument against the system of blind necessity, namely, that given that necessity is uniform, it cannot account for observed variety (Newton, 1999: 943; see chapter 8 for more discussion). Now, this is only a limited objection against Spinoza, for as we have seen, Spinozism is committed to there being sufficient reason for infinite variety in the modes (E1p16 and E1p28). At best Newton has shifted the burden of proof. Because Newton has the better physics, he can claim to have constrained any possible explanation that will account for the observed variety (see chapter 8). Of course, this charge is not insurmountable: all a Spinozist needs to show is how the observed laws and the "regular" orbits are possible given some prior situation.[28] Even so, by focusing on the origin and nature of motion, the

---

[28] Kant's UNH *should* be understood as accepting this challenge; see chapter 3 above.

Newtonians do call attention to the Achilles heel of Spinoza's system. Given their objectives, this is no small achievement.

4.7 Conclusion: Reading Spinoza

More, Clarke, Newton, and Maclaurin were not dispassionate respondents to Spinoza. But their motivated, sophisticated criticisms help us more clearly recognize tensions in Spinoza's *Ethics* and other writings. In particular, they were in a position to exploit the increasingly high intellectual status of Newton's natural philosophy to successfully press the case against Spinoza's treatment of motion and, consequently, Spinoza's unprovidential God. A progressive narrative that situates Spinoza at the center of the scientific revolution (as offered, for instance, in Israel [2002]) misses the fact that the authority of triumphant, mathematical-empirical science is deployed against Spinozism. Such a progressive narrative is also unsatisfying philosophically because it leaves us wondering how Spinoza could meet the challenge of the Newtonians to address and account for the crucial elements of Newton's *Principia*. More important, the progressive narrative has made us overlook that many eighteenth-century thinkers—Mandeville, Hume, Diderot, Buffon, and others—drew upon Spinozistic resources (as developed in the "Letter on the Infinite") to contest the authority of mathematical natural philosophy and support their insistence that it was useless for the knowledge worth.*

Finally, the story told here has an unexpected afterlife in the historiography of analytic philosophy. In the middle of the nineteenth century, in chapters 13 and 14 of his classic *An Investigation of the Laws of Thought*, George Boole rationally reconstructed the Spinoza–Clarke exchange in order to showcase the utility and significance of his new symbolic language. Boole had no doubt that:

> The analysis of its [Spinoza's *Ethics'*] main argument is extremely difficult, owing not to the complexity of the separate propositions which it involves, but to the use of vague definitions, and of axioms which, through a like defect of clearness, it is perplexing to determine whether we ought to accept

---

* The existence of this strain of anti-mathematicism is discussed in Nelson (2017); Schliesser (2018, 2021); Demeter and Schliesser (2019).

or to reject. While the reasoning of Dr. Samuel Clarke is in part verbal, that of Spinoza is so in a much greater degree; and perhaps this is the reason why, to some minds, it has appeared to possess a formal cogency, to which in reality it possesses no just claim. (Boole, 1854: 145)

This is not the place to evaluate Boole's judgment or its reflection in Maxwell's treatment of the relative merits of Newton and Spinoza (Maxwell, 2010: 18). We just note that one influential reader of Boole, Russell, praises Spinoza's moral vision but explicitly discounts the argumentative value of Spinoza's works in the context of his attacks on the British Idealists and Bergson (Russell, 1986: 64). But the details of these stories must be told elsewhere.[29]

[29] We have benefited from Noa Shein's working paper "Between Physics and Metaphysics— Samuel Clarke's Response to Spinoza," and our subsequent, ongoing discussions with her. The ideas of this chapter were tried out on audiences at Oxford, where we received very helpful comments from Martine Pécharman, Sarah Hutton, Daniel Garber, Alex Douglas, and Jasper Reid, and at Cal State Long Beach, where we received splendid criticisms from Marcy Lascano, Larry Nolan, Alex Klein, and the incomparable Wayne Wright. We are grateful to two anonymous referees and Daniel Schneider and Allison Peterman for insightful comments on an earlier draft.

5

Newtonian Emanation, Spinozism, Measurement, and the Baconian Origins of the Laws of Nature

5.1 Introduction

The main aim of this chapter is to get clear on the relationships among God, space, existence and the science of motion in Newton's famous manuscript De Gravitatione[1]. In the first two sections, I recast the ongoing scholarly debate over the status of Newton's emanation doctrine in De Gravitatione. In the third section I focus on Newton's claims about existence.

A major theme of this chapter is that De Gravitatione offers us two ways of thinking about God's relationship to nature. Following Whitehead (1933) and Oakley (1961), I call these an "immanent" and an "imposed" conception.[2] The first, immanent, conception is the God of the philosophers while, the second, more anthropomorphic and voluntarist God is familiar from the Judeo-Christian tradition. I argue that the composition of De Gravitatione largely segregates these two conceptions, but that the immanent one accounts for more of the metaphysically significant features of Newton's account. I argue that the immanent conception brings Newton rather close to Spinozism in De Gravitatione.[3] In Newton's published works the voluntarist strain is more prominent (see Harrison [2004] and Henry [2009]), but the immanent strain in Newton's thought was perceived by Leibniz who (while unfamiliar with De Gravitatione, of course) raised the specter of Spinozism in his second letter to Clarke in paragraph 7 (Leibniz and Clarke, 2000: 9).

---

[1] This chapter originally appeared as Eric Schliesser, "Newtonian emanation, Spinozism, measurement and the Baconian origins of the laws of nature." *Foundations of science* 18.3 (2013): 449–466.
[2] Oakley is self-consciously developing Whitehead (1933). Both are writing before the publication of De Gravitatione in the twentieth century. I thank Zvi Biener for calling my attention to these.
[3] This is not to claim identity between the two. On my reading of Spinoza, he is uninterested in, even hostile to, developing a conceptual framework for the mechanical sciences (Schliesser, 2017).

Newton's Metaphysics. Eric Schliesser, Oxford University Press. © Oxford University Press 2021.
DOI: 10.1093/oso/9780197567692.003.0007

Before I summarize my chapter, I offer three caveats, two of which regard my stance toward De Gravitatione and one on methodology. First, the De Gravitatione manuscript was (a) probably put together from earlier reflections in order (b) to start a large ambitious work. The case for (b) is fairly straightforward. The Halls report that the notebook that contains it was left blank after the abrupt ending of De Gravitatione (Newton, 2004: 39). This suggests that Newton intended to add more to it. Below I call attention to other various ways in which Newton's plans for the work betray considerable ambition. The case for (a) is more speculative, but there are two further arguments on its behalf: (i) unlike many other manuscripts in Newton's hand De Gravitatione does not contain numerous (obsessive) false starts and new beginnings. This suggests, as George Smith pointed out to me, that Newton was copying from preselected pieces.[4] The corrections in the manuscript of De Gravitatione are all corrections to an already fairly polished draft, and (ii) if (b) is true then it would help explain the difficulty in dating the manuscript that has bedeviled discussion of it. The piece reflects Newton's thought at different stages of his intellectual development. (One can be committed to this without thinking Newton's development is always linear.) I assume, following a suggestion of Karen Verelst, that De Gravitatione was brought together in the first half the 1680s when Newton decided to write an ambitious anti-Cartesian tract in the science of motion. For my argument the dating is irrelevant. But second, I resist a close identification between De Gravitatione and *Principia*, especially the scholium on space and time and the "General Scholium." In recent years, there has been a tendency in Newton scholarship to resolve interpretive difficulties in one of these texts in light of each other (this is as true of Stein [2002], as it is of McGuire [1978]). But this does no justice to the fact that the metaphysical commitments of the *Principia* changed significantly with the second edition. It also tends to ignore the evidence we have of the *Treatise* (dating from the mid-1680s and presumably composed after De Gravitatione was put aside) and posthumously published (see chapters 1–3 above).

My third caveat is methodological. Below I draw on writings of Bacon to illuminate some of Newton's concepts. My contribution should not be understood in terms of the history of ideas; I am using Bacon to offer an alternative interpretation of the meaning of "emanation." I am not claiming that Bacon's

---

[4] Feingold (2004: 25–26). But see Henry (2011) for a reaffirmation of an early dating. See also Smith (2020).

writings influenced Newton (although the claim would not be far-fetched). In general, my argument resists the temptation to assimilate Newton with any particular school of thought—I view him as an eclectic, conceptual alchemist, who is constantly trying out subtle (and sometimes dramatic) innovations on a wide range of views available to him.

The first two sections of this chapter investigate what Newton could have meant in a now famous passage from De Gravitatione that "space is as it were an emanative effect of God" (21). First I examine four key passages within De Gravitatione that bear on this. I argue that the internal logic of Newton's argument permits several interpretations, and I call attention to a Spinozistic strain in Newton's thought.

Second I sketch four options: (i) a neo-Platonic approach that builds on work by Christia Mercer on Leibniz's views of emanation; (ii) an approach associated with the Cambridge Platonist, Henry More, which was recently investigated by Ed Slowik, and a variant on this (ii⁺) articulated by McGuire and more recently by Dana Jalobeanu emphasizing that Newton mixes Platonist and Epicurean themes; (iii) a necessitarian approach associated with Howard Stein's influential interpretation, recently reaffirmed by Andrew Janiak, and (iv) an approach connected with Bacon's efforts to reformulate a useful notion of form and laws of nature.

Hitherto only the second and third options have received scholarly attention in scholarship on De Gravitatione. I offer new arguments to treat Newtonian emanation as a species of Baconian formal causation as articulated, especially, in the first few aphorisms of part two of New Organon. If we treat Newtonian emanation as a species of formal causation then we can reconcile Stein's necessitarian reading with most of the Platonist elements that others have discerned in De Gravitatione, especially Newton's commitment to doctrines of different degrees of reality as well as the manner in which the first existing being 'transfers' its qualities to space (as a kind of causa sui). This can clarify the conceptual relationship between space and its formal cause in Newton as well as Newton's commitment to the spatial extended-ness of all existing beings.*

My interest is not exclusively driven by exegetical concerns. In particular, we can appreciate that one of Newton's most important decisions in recasting the material for Principia was to drop the language of formal causation that was still present in De Gravitatione and replace it with the language of

---

* In this chapter, I mostly ignore the relationship between God and duration, but this, too, is recast in the Principia (see chapter 7 and its postscript).

law—ironically, this was a concession to Cartesian terminology. I argue that in De Gravitatione (and also the queries of the *Opticks*) there is, thus, overlooked evidence for Thomas Kuhn's old speculative claim that formal causes are replaced by laws of nature during the scientific revolution (see also chapter 6).

While the first two sections of this chapter engage with existing scholarly controversies, in the final section I suggest that the recent focus on emanation has obscured the importance of Newton's very interesting claims about existence and measurement in the same passage(s). Newton writes:

> Space is an emanative effect of the first existing being, for if any being whatsoever is posited, space is posited. And the same may be asserted of duration: for certainly both are affections or attributes of a being according to which the quantity of any thing's existence is individuated to the degree that the size of its presence and persistence is specified. So that the quantity of the existence of God is eternal in relation to duration, and infinite in relation to the space in which he is present; and the quantity of the existence of a created thing is as great in relation to duration as the duration since the beginning of its existence, and in relation to the size of its presence, it is as great as the space in which it is present." (Newton, 2004: 25–26)

McGuire and Slowik are the only commentators that I am familiar with to have remarked on Newton's focus on "the quantity of existence." Recently, Slowik writes it is "a fairly mysterious and undefined notion in the De Gravitatione, so it is difficult to draw a specific conclusion based on this use of terminology."[5] McGuire (1978), by contrast, makes sense of it in light of Newton's Epicurean turn of the 1690s. I am greatly indebted to McGuire's brilliant study, but in the final section of this chapter I deviate from his reading and argue that in De Gravitatione God and other entities have the same kind of quantities of existence. My overarching position is that in De Gravitatione, Newton is concerned with how measurement clarifies the way of being of entities. Newton is not claiming that measurement reveals all aspects of an entity. But if we measure something then it exists as a magnitude in space and as a magnitude in time. This is why in De Gravitatione Newton's conception of existence really helps to "lay truer foundations of the mechanical sciences" (Newton, 2004: 21), which is one of Newton's ambitions for De Gravitatione that I wish to recover here.

---

[5] Slowik (2009) originally appeared as Slowik (2008); if I mention page-numbers it will be to Slowik (2009). See also Slowik (2012).

5.2 Four Passages from De Gravitatione on Emanation and *Causa Sui*

In this section, I introduce and analyze four passages from De Gravitatione. The first passage, [A], reads:

> [N]ow it may be expected that I should define extension [space–ES] as substance, accident, or else nothing at all. But by no means, for it has its own manner of existing which is proper to it and fits neither substances nor accidents. It is not substance: on the one hand, because it is not absolute in itself, but it is as it were an emanative effect of God and an affection of every kind of being; on the other hand, because it is not among the proper affections that denote substance, namely actions, such as thoughts in the mind and motions in body. (Newton, 2004: 21)

The second, [B], reads,

> space is eternal in duration and immutable in nature because it is emanative effect of an eternal and immutable being." (Newton, 2004: 26)

In these two passages Newton claims that space is eternal in duration and immutable. This much appears clear, although my use of 'is' might not do full justice to the claim that it has its "own manner" of existence. Space is an emanative effect of an eternal and immutable being. It is tempting to identify the eternal and immutable being with God, but Newton's use of "as it were" in passage [A] should give us pause. What is the eternal and immutable emanative cause of space? In this section I offer some preliminary answers. We shall also come to understand what it might mean to say that space "is an affection of everything."

The plot thickens when we consider two further passages. The third passage, [C], reads:

> Space is an affection of a being just as a being. No being exists or can exist which is not related to space in some way. God is everywhere, created minds are somewhere, and body is in the space that it occupies; and whatever is neither everywhere nor anywhere does not exist. And hence it follows that space is an emanative effect of the first existing being, for if any being whatsoever is posited, space is posited. (Newton, 2004: 25)

The fourth passage, [D], reads,

[I]est anyone should ... imagine God to be like body, extended and made of divisible parts, it should be known that spaces themselves are not actually divisible and furthermore, that any being has a manner proper to itself of being present in spaces. (Newton, 2004: 26)

That space is an affection of everything is clarified by these two passages: it means that all existing entities are located in some spatial structure (Stein, 2002). A more Kantian way of saying this is that space is (part of) the condition of possibility for any entity. Or one might say that: spatiality is a quality that belongs to everything (see also McGuire, 1978: 456ff.).

Moreover, because of passage [C] we can specify something more explicit about the eternal emanative cause of space; it is the first existing being. Most readers would be inclined to call the first existing, eternal, and immutable being "God." But this is too easy because it leaves unexplained Newton's use of "as it were." Of course, we are also still left clarifying what it means that space is an emanative effect of the first existing eternal and immutable being.

One way to approach Newton's use of "as it were" is to interpret it as his attempt to indicate that he is discussing a philosophical God, one that has no anthropomorphic qualities. For emanation as a form of divine causation is traditionally distinguished from conceptions that refer to God's will.[6] Newton is clearly signaling that his God does not stand outside nature; even God exists spatially. In De Gravitatione, Newton plainly rejects a soul of the world with God standing outside nature (Newton, 2004: 30; Slowik, 2009: 443; Newton, 2004: 124–125). This fits nicely with what he wrote later in life in his "Account of the Commercium Epistolicum," in which he rejects Leibniz's view of God as "an intelligence above the bounds of the world; whence it seems to follow that he cannot do anything within the bounds of the world, unless by an incredible miracle" (Newton, 2004: 125). In context Newton has just affirmed that, while God is not the soul of the world, he is omnipresent, so this accords with the view expressed in De Gravitatione (recall "no being exists or can exist which is not related to space in some way" [Newton, 2004: 25]).

---

[6] See Internet Encyclopedia of Philosophy, http://www.iep.utm.edu/e/emanatio.htm, accessed on October 24, 2008.

Nevertheless, before we start exploring how Newton's contemporaries use "emanation" (Sect. 3), we have not exhausted all the reasonable options in trying to establish the identity of the emanative cause of space based on Newton's text alone. It is probably safe to assume that for Newton ordinary material entities and (maybe less likely) ordinary minds (which, however, are more "noble" than bodies [Newton, 2004: 20; Slowik, 2009: 438 n. 10]) are not eternal and immutable, and can, thus, be ruled out as the first existing beings.

But despite the recent philosophic interest in De Gravitatione, it has not yet been noted that Newton's wording in these four passages is compatible with the position that space is the first existing being. The emanative cause of space could be space itself! Or to be more precise God and space would be related by way of self-causation.[7] Newton's space would then be a Godlike *causa sui*. This is not as crazy as it sounds: Newton is certain that space is eternal, immutable, immobile, indivisible, and infinite ("space is extended infinitely in all directions"—[Newton, 2004: 23]); it is the condition of possibility of all beings. So, given these claims why not call space a philosophic conception of God? This could then explain Newton's "as it were," especially if we see the phrase as modifying emanation (and not God). The relationship between the first eternal cause and its eternal effects (space, time, etc.) is then emanative-like, and that would be a way to capture *causa sui*. Newtonian space and Newtonian time are then, in a certain sense (and more precisely), attribute-like aspects of a self-causing God.[8]

Before one rejects this Spinozistic reading out of hand, one should recognize that Newton seems aware of this option when he writes, "I see what Descartes feared, namely if he should consider space infinite, it would perhaps become God" (Newton, 2004: 25; see also the treatment of Descartes and the atheists at 31–32).[9] In context, Newton is discussing Descartes'

---

[7] These funny sounding phrases are meant to echo Descartes' responses to Caterus and Arnauld. For excellent discussion, see Lee Jr. (2006).

[8] In order to avoid confusion, with most recent commentators I recognize that De Gravitatione drops the traditional substance-property/attribute structure (McGuire, [1978: 474]). For Newton entities can exist without inhering in a substance. However, I have argued in chapter 2 that Newton probably was a substance monist, with God being the only entity with full substantial reality. Gorham (2011) wishes to argue against the recent consensus in order to explain that space is a principle, generic attribute of God; Gorham draws on a rich Cartesian framework to argue this. Without claiming to have done justice to all of Gorham's subtle arguments I claim to have captured the same insight by stressing that the emanative relationship between God and space is a form of *causa sui* (something Gorham also notes) but that Newton can say so without accepting the traditional substance-attribute structure in De Gravitatione.

[9] Concern over Descartes' flirt with atheism seems to have been something of a trope; it also shows up in MacLaurin (1748: 77).

Principles, but we know from his early notebooks that he was also familiar with the *Meditations*. In fact, Newton interpreted the ontological argument of the fifth Meditation in terms of self-causation: "A Necessary being is ye cause of it selfe or its existence after ye same manner yt a mountaine is ye cause of a valley . . . (wch [sic] is not from power or excellency, but ye peculiarity of theire natures" (Quaestiones, folio 83r, quoted in McGuire, 1978: 485). So even if Spinoza was never on Newton's mind in such passages, Cartesian ideas on *causa sui* were available to Newton.[10] It might appear unlikely that in De Gravitatione, Newton would endorse this reading, because he insists that it is "repugnant to reason" that God created "his own ubiquity" (Newton, 2004: 26). Newton's striking appeal to reason here offers independent evidence for my claim that we are here dealing with the God of the philosophers. (It also should also make us cautious about reading Newton's later strict methodological empiricism back into De Gravitatione. See chapter 2.) One might be tempted to claim that this "repugnance" rules out any *causa sui*. Unfortunately, this conclusion cannot be established because emanation is a doctrine that avoids creation in time. And there is no trace that Newton is using emanation to indicate activity (Gorham, 2011; Stein, 2002). All we can say is that is that the first emanative cause and its effect are both eternal, and this is compatible with *causa sui*.[11]

In the next section, I canvas four usages of "emanation" that were available in the seventeenth century. My argument will proceed independently from my remarks about self-causation in De Gravitatione. But they can be fruitfully presupposed when I treat more fully of Newton's immanent or philosophic conception of God below.

5.3 Four Kinds of Emanation

There can be no doubt that in De Gravitatione according to Newton space is something like an emanative effect of something (eternal, immutable, etc.). I am familiar with four possible meanings of the word 'emanation' in the

---

[10] As noted in chapter 4, Henry More published his *Confutatio* directed against Spinoza in 1678.

[11] Gorham (2011) has argued for what he calls an assimilationist interpretation in which space is an attribute of God in Newton. Gorham sharply distinguishes Newton from Spinoza, because according to Gorham Newton has a voluntarist conception of God. As I argue more fully below in De Gravitatione Newton's voluntarism is very attenuated.

seventeenth century. Let me survey these and comment on the plausibility that Newton might have had any one of them in mind.

First, there is what I call a traditional neo-Platonic version of emanation, which relates a pure or perfect cause with an impure or imperfect imitation of it. The perfect cause emanates a property to the imperfect effect so that the effect "participates in" or has an inferior version of the property or accident (Mercer, 2001: 189–190). Something like this concept is clearly presupposed in passage [B]. For in it immutability and eternality are transferred from the cause to the effect. This is characteristic of neo-Platonic emanation. It is a bit unfortunate that Newton does not explain the causal transference of immobility, indivisibility, and infinitude from God to space in a similar matter. Newton's silence frustrates our search (but see chapter 8). If Newton had fully intended this use of "emanation," it is a bit surprising that in passage [D] he does not mention God as the source of space's immobility, indivisibility, and infinitude; this would have been a natural place for him to do so in clarifying the relationship between the emanative cause and effect.

Nevertheless, we should not entirely forego this option because it fits with another aspect of Newton's view: while space is "uncreated" (Newton, 2004: 33), it is somehow not quite absolute (recall passage [A]). In fact, Newton is committed to things having different degrees of reality (see also Spinoza's *Ethics* 1p11; because of the shared commitment to different degrees of reality, this is one further reason why I connect Newtonian emanation with Spinozistic *causa sui*—space/extension is dependent on God, while being, in some sense, self-same with it).[12] This is partly indicated by Newton's claim that things have their "own manner of existing which is proper" to them (Newton, 2004: 21). Later in the piece Newton also claims: "whatever has more reality in one space than in another space belongs to body rather to space" (Newton, 2004: 27). Perhaps only "God" has full ("absolute") reality and space has less of it; as Newton writes, space has "some substantial reality" (Newton, 2004: 33), but not complete reality. This is a doctrine familiar from Descartes, who employs an "emanative cast" in the third Meditation; different degrees of reality are crucial in the fourth and fifth Meditations (Grene, 1999: 102).

Second, there is the definition of emanation by the Cambridge Platonist, Henry More, who was well known to Newton. Even by contemporaries

---

[12] I thank Maarten Van Dyck and Ted McGuire for discussion, although I fear my discussion does not satisfy either.

Newton was taken to be following More in various ways. For example, in his "Account of the Commercium Epistolicum," Newton notes that the editors of *Acta Eruditorum* point to More's potential influence on Newton (Newton, 2004: 124–125). For More, an emanative cause is an immediate cause which is co-present with its effect. As More writes, "a Cause as merely by being, no other activity or causality interposed, produces an Effect" (quoted in Jalobeanu, 2007).[13] No doubt some of Newton's comments point to this definition. Indeed this notion of causation must have been very influential, because as late as the 1730s it is still Hume's target in his eight Rule of Reasoning (without mention of More): "an object, which exists for any time in its full perfection without any effect, is not the sole cause of that effect, but requires to be assisted by some other principle, which may forward its influences and operation. For as like effects necessarily follow from like causes, and in a contiguous time and place, their separation for a moment shews, that these causes are not complete ones" (Hume, 1739: 1.3.15.10).[14]

Nevertheless, while there are ways in which Newton's approach harmonizes with Cambridge Platonism (Slowik, 2009: 2012; cf. Stein, 2002: 269), More's particular definition of emanation is uninformative as an aid to understanding the details of Newton's claims (in passages [A] through [D]). It is at best consistent with these passages, but even Slowik's splendid papers do not help account for all the rich details in them. Ted McGuire and Dana Jalobeanu, in particular, have argued that these passages mix Platonist imagery with Epicurean themes that Newton would have derived from Charleton (Jalobeanu, 2007; McGuire, 1978: 471ff). Their approach in which Newton is an eclectic in its original sense (i.e., choosing what is best) has much to recommend to it. It is, thus, a bit surprising that so much scholarly attention is focused on mining the More-Newton connection.[15]

A third conception, diametrically opposed to the Platonizing readings, has been defended by Howard Stein. No one has done more to rehabilitate the reputation of Newton's analysis of space among the community of space-time theorists and historians of philosophy than Stein (with the classic Stein,

---

[13] Slowik (2009) quotes the relevant passage from More's *The Immortality of the Soul* as follows: ""an Emanative Effect is coexistent with the very Substance of that which is said to be the Cause thereof," and explains that this "Cause" is "the adequate and immediate Cause", and that the "Effect" exists "so long as that Substance does exist" (438).

[14] For more on Hume's Rules of Reasoning and their relationships to Newton, see Sect. 4.5 in Demeter and Schliesser (2020).

[15] But see also Slowik (2012) for an attempt to integrate the Morean influence with strands coming from Gassendi and Charleton.

1967). Part of his argument turns on a revisionary analysis of emanation in Newton (Stein, 2002). By relying on passages [A] and [C], in particular, and on linguistic contextual evidence, Stein argues against a noncausal reading of emanation. Instead Stein reads Newton as employing "emanation" in the sense of a "necessary consequence" (Stein, 2002: 269).

I have relied on aspects of Stein's argument, and moreover I have agreed with Stein's focus on the empirical basis of Newton's doctrines, including those pertaining to God, and Stein's emphasis on Newton's probabilism, even fallibilism. Stein's arguments have recently been affirmed by Janiak (2015).

Nevertheless, Stein and Janiak are silent on some crucial details of these passages. In particular, they seem to miss the fact that passage [B] purports to explain how qualities of space are caused by the emanative cause (see also Slowik, 2009: 445–446). We could put the argument against Stein as follows: He ignores the fact that even if we posit any being, we need not posit space as "eternal in duration and immutable in nature." So, Stein's argument seems to attribute to Newton conceptual question begging. To be clear: Newton does offer a separate argument from our ability to imagine a rotating triangle to the conclusion that space is infinite (Newton, 2004: 23). So, it does seem that if we posit any (geometric) object we (can) posit infinite space.[16] But I cannot see how Stein's resolutely anticausal approach can account for these crucial details in passage [B]. This is not to say that Stein's focus on emanation as necessary consequence is entirely misguided. In Section I, I agreed with his claim that Newton is committed to the view that all beings presuppose space as a condition of possibility. It is Stein's focus on the logical structure of Newton's position that exhibits this fact most clearly. Thus we should not ignore the possibility that emanative causation may in many respects be more akin to a conceptually necessary consequence than efficient causation, but this cannot be the whole story. We can illustrate and make more precise what I mean if we focus on an unlikely source: Bacon, which frames the fourth conception.

Bacon has been hitherto ignored as a possible source in this discussion. There was a time, from the middle of the eighteenth century onward, and especially since the writings of Thomas Reid, that Bacon and Newton were viewed as the twin sources advocating a shared and proper method of philosophizing (a view especially popular among early nineteenth-century thinkers like Whewell and Mill). Particularly highlighted were Bacon's and

---

[16] This is not to deny that the point of the example "is to give us a handle for grasping [space's] actual, not potential, infinity" (McGuire, private communication, May 16, 2011).

Newton's empiricism in the service of the discovery of the true causes of nature, especially if accompanied with an *experimentum crucis*. In recent years commentators have been more eager to differentiate between Bacon's natural history and Newton's mathematical experimental approaches (Feingold, 2001). While I endorse the recent trend, we should not be blind to Bacon's significance to Newton. In the *New Organon*, Bacon sometimes uses "emanation" in an innocent sense, as when he writes that the rays of light "emanate" from the sun (Bacon, 1870: 129).

But at other times "emanation" has a more technical meaning in Bacon. Consider the following two paragraphs of *New Organon*, 2.I–II:

On a given body, to generate and superinduce a new nature or new natures is the work and aim of human power. Of a given nature to discover the form, or true specific difference, or nature-engendering nature, or source of emanation (for these are the terms which come nearest to a description of the thing), is the work and aim of human knowledge. Subordinate to these primary works are two others that are secondary and of inferior mark: to the former, the transformation of concrete bodies, so far as this is possible; to the latter, the discovery, in every case of generation and motion, of the *latent process* carried on from the manifest efficient and the manifest material to the form which is engendered; and in like manner the discovery of the *latent configuration* of bodies at rest and not in motion.

In what an ill condition human knowledge is at the present time is apparent even from the commonly received maxims. It is a correct position that "true knowledge is knowledge by causes." And causes again are not improperly distributed into four kinds: the material, the formal, the efficient, and the final. But of these the final cause rather corrupts than advances the sciences, except such as have to do with human action. The discovery of the formal is despaired of. The efficient and the material (as they are investigated and received, that is, as remote causes, without reference to the latent process leading to the form) are but slight and superficial, and contribute little, if anything, to true and active science. Nor have I forgotten that in a former passage I noted and corrected as an error of the human mind the opinion that forms give existence. For though in nature nothing really exists besides individual bodies, performing pure individual acts according to a fixed law, yet in philosophy this very law, and the investigation, discovery, and explanation of it, is the foundation as well of knowledge as of operation. And it is this law with its clauses that I mean when I speak of *forms*, a name which I the rather adopt because it has grown into use and become familiar (Bacon, 1870: 119; emphasis in original).

Bacon is no Scholastic, as can be clearly seen by his claims that (a) "final cause rather corrupts than advances the sciences, except such as have to do with human action," and (b) "that forms give existence" is "an error." Yet Bacon embraces the three other Aristotelian causes (material, formal, and efficient) as entirely appropriate to his proposed science of nature. In particular, here I focus on Bacon's willingness to embrace a reformulated version of formal causation; he writes that "the aim of human knowledge" is to discover the "nature-engendering nature" of bodies in order to create new "natures." Bacon's forms are materialistic. In labeling his causes in this fashion, Bacon is even willing to risk confusion with the Scholastics. To be clear: a Baconian formal cause has nothing to do with hylomorphism. (See Mancosu [1999] for discussion of nonhylomorphic formal causation in mathematics.)

Two aspects are relevant for my treatment of De Gravitatione. First Bacon glosses a "nature-engendering nature" as "a source of emanation." Second the cause or source of emanation is associated with the form. As I understand Bacon here, he is relying on the traditional meanings of words in the first paragraph, and then begins to offer innovative uses in the second. If this is so, then those commentators that have tried to treat emanative causation as a species of efficient causation (e.g., Slowik, 2009: 437–438; Slowik, 2012) are mistaken. We need not saddle the Platonizing reading with this mistake; one may question the modern identification of, for example, More's claim about God as "the adequate and immediate Cause" of space with an efficient cause. Why not treat More as explicating a formal cause? I can allow that Bacon is innovating linguistically in both paragraphs. While I do not claim that Bacon is the source of Newton's doctrines about space, it is worth noting that for Bacon, too, space is infinite (Bacon, 1870: 162). Moreover, according to Bacon "forms [. . .] are (in the eye of reason at least, and in their essential law) eternal and immutable" (Bacon, 1870: 126). So if Newton's "first existing being" were in Bacon's sense the formal cause of space it would have to be eternal and immutable. Thus Newton's first existing being could fit the characteristics of a Baconian form as a source of emanation.

Now while Bacon is clearly an admirer of Galileo's discoveries and willing to speculate about interplanetary travel on the basis of them (Bacon, 1870: 193), his views on space and motion are too undeveloped for them to count as a source of the details of Newtonian doctrine on space. Here,

I insist only that Bacon offers helpful clues for understanding Newton's use of "emanation."

The crucial point is that we should consider treating Newton's "emanative" causation along the lines of nature-engendering nature, that is, as a species of formal causation in Bacon's sense.[17] Interestingly, in De Gravitatione formal causation shows up twice explicitly: first "that product of the divine will is the form or formal reason of the body denoting every dimension of space in which the body is to be produced" (Newton, 2004: 29); second we should "distinguish between the formal reason of bodies and the act of the divine will" (Newton, 2004: 31—for the sake of argument, allow that for Newton a "reason" is a "cause"). Now the context is Newton's thought experiment about how God could have created bodies. So, despite the fact that it appears that the thrust of Newton's treatment of body is to emphasize a voluntarist conception of God, who could have created bodies differently (see also the late query 31 in Opticks), Newton is warning his readers not to confuse God's will for the formal reason of bodies. That is to say, the "nature-engendering nature" of bodies is not God's will, but (presumably) the essential (we would say, intrinsic) qualities of body: (e.g., (i) mobility; (ii) impenetrability, which Newton articulates in terms of law-like behavior in collision; and (iii) that they can "excite various perceptions of the senses and the imagination in created minds, and conversely be moved by them" [Newton, 2004: 28–29)). This list helps us appreciate that Newton's formal reason(s) should not be conflated with an efficient cause. It is regularly underestimated that even in the most voluntarist part in De Gravitatione, much of the metaphysical work is being done by a more immanent conception of God.

One might object that given that Newton did not shy away from the language of formal causation/reason, why would he not use that language in describing the relationship between the first existing (eternal, immutable, etc.) thing and space? The response to this is straightforward: if Newton is following Bacon's usage then emanation picks out a particular species of formal causation.[18]

---

[17] McGuire (1978: 480), uses the language of "ontic dependence" in order to avoid the language of causation. Given our post-Humean condition this may be said to attempt to capture the insight that 'as it were emanation' understood as Baconian formal causation is not quite our notion of causation.

[18] Marc Lange has called my attention to a very interesting passage in Hooker: "Whereas therefore things natural which are not in the number of voluntary agents . . . do so necessarily observe their certain laws, that as long as they keep those forms which give them their being, they cannot possibly be apt or inclinable to do otherwise than they do; seeing the kinds of their operations are both constantly and exactly framed according to the several ends for which they serve, they themselves in the meanwhile, though doing that which is fit, yet knowing neither what they do, nor why: it followeth that all which they do in this sort proceedeth originally from some such agent, as knoweth, appointeth,

What else can be said in favor of this option? First, if we treat the emanative cause as a species of formal causation then we can make sense of Stein's qualms about treating it as a species of efficient causation; we can capture Stein's insight that there is a necessary inference between the first existing being and the qualities of space. That is to say, it is not merely that space is a necessary consequence of any entity, but various particular qualities (immutability, eternal duration, and perhaps also infinitude and indivisibility) of space are also necessary in this way. One might capture this with the claim these are essential or intrinsic qualities of space for Newton. Spaces would not be space without them. Second, we can accommodate the evidence in favor of the neo-Platonizing reading of passages [A] to [D]; we can allow that "emanation" signals Newton's commitment to a version of the doctrine of different degrees of reality, as well as the claim (in passage [B]) that the first existing being 'transfers' (somehow) its qualities to space. We can do this if we reject the claim that emanation is a species of efficient causation.

So Bacon shows us the way to reconciling the first three options.[19] I extend my reading by offering a speculation inspired by Thomas Kuhn, who argued that during the scientific revolution formal causes morphed into laws of nature (Kuhn, 1977: 21–30). But I have presented evidence that one of Bacon's innovations appears to reinterpret the formal cause, or nature-engendering nature, in terms of a law of nature.

My proposal is that we should treat De Gravitatione as offering evidence of Newton's willingness to try out a radical new idea: the emanative source of space could be better conceived as a formal cause of a decidedly modern

holdeth up, and even actually frameth the same" (Hooker, 1888, Of the *Lawes of Ecclesiastical Politie*, Book I.iii.4.). As Lange remarks: "Shades of Hempel and Oppenheim from 350 years later!" Lange (2009) discusses the Hooker passage briefly in chapter 1. Hooker is claiming that natural things are law-like in virtue of God's unknowable general providence. (Cf. Descartes in *Meditations*!) The form is an imposed manifestation (if I may use that term) of God's providence. What's especially interesting is that Hooker's nature is very much knowable because it is part of maker's knowledge; yet Hooker transforms that traditional (medieval) doctrine because he does not seem to be interested in discovering 'local' final causes. So, not unlike Spinoza (who probably was familiar with the Bacon passage quoted in the body of the text), Hooker has an account of what one might call 'blind' or, less able-ist, 'masked' forms ('blind' because divorced from final causes) that are responsible for the law-following order we find in nature.

[19] Slowik's very fine papers do one disservice to the dialectic surrounding the meaning of Newtonian "emanation." Stein's treatment of Newton's use of "emanation" is not itself an argument for a 'third way' between the debate over "substantivalist" and "relationist" interpretations of "absolute" space, although it can certainly be slotted into such a one (as it is in Slowik's reconstruction of Stein's argument). Stein's treatment of emanation is really designed to put claims about the importance of Newton's theology for his physics on the defensive (see Stein, 2002: 268, quoted in Slowik, 2009: 436).

kind: a law of nature. To be clear: I have very little positive evidence for the view that I am defending. I am certainly not claiming that while writing De Gravitatione Newton conceived of the emanative cause as a law of nature. But reflection on the thrust of De Gravitatione could have pushed him to recognizing this. Let me offer some arguments, some new and some reiterating earlier points.

First, laws of nature can play the logical role that Stein assigns to the emanative cause. They provide the necessity that is required for his arguments to be persuasive. Second, when we conceive of emanative causation as a species of formal causation, then laws of nature fit the function that "as it were emanation" has in the discussion of space in De Gravitatione; Baconian laws of nature are eternal and immutable (etc.).[20]

Third, the laws of motion in Newton's *Principia* have been treated as constitutive principles (by neo-Kantians) or deeply entrenched empirical working assumptions (by empiricists), but we can also read them as formal causes in Bacon's sense. They are the natures that help define the nature of motion. (See McGuire [2007] for the importance for Newton of identifying natures; I have extended this in chapter 1.)

Fourth the Baconian aspect of Newton's conception of a law of nature shows up in a decidedly surprising place. In one of the most voluntarist passages in Newton's oeuvre, he writes: "it may be also allowed that God is able to create particles of matter of several sizes and figures, and in several proportions to space, and perhaps of different densities and forces, and thereby to vary the laws of nature, and made worlds of several sorts in several parts of the universe" (*Opticks*. Query 31, 403–404; see also chapter 6 with Biener). It is the relationships among the densities and forces of matter to space that account for the varying laws and worlds. That is to say, in the *Opticks* for Newton the nature-engendering nature of our world is simply the way in which bodies are organized. This picks out a Baconian form. In Newton's conception the laws of nature vary if the way bodies are organized varies. With different laws of nature or Baconian emanative causes we have different worlds. The whole point of Query 31 is to deny that the laws of nature have a separate causal standing apart from the way matter is organized (see chapter 1 above and chapter 6 with Biener). This is the Baconian form.

[20] As Katherine Brading has pointed out to me, this argument works best for force laws (such as law of universal gravitation) because "they make matter clump together into the kinds of bodies that there are" (personal communication, June 17, 2011).

Of course, Newton is no Baconian because he is also willing to consider "densities and forces" as basic components.

There is an objection to this Baconian reading: I have treated the emanative cause of space as a kind of philosophic, unanthropomorphic God. This God is immanent in the world. But it looks as if De Gravitatione has a very anthropomorphic conception of God; Newton repeatedly appeals to the "divine will"; "the power of God"; God's "action of thinking and willing" (27); "divine constitution" (30); "the work of God" and "creation" (33). In the *Principia* this kind of anthropomorphism is more strictly confined to the General Scholium added to the second edition; in *Opticks* it shows up in Queries 28 and 31 (of the final edition) especially. Yet it appears to be a central conception of De Gravitatione. If correct, it is hard to accept a reading of Newton that treats emanation as a Baconian form or nature-engendering nature.

Yet matters are not so simple. To see this we must investigate a little-noticed peculiarity of the structure of De Gravitatione. It contains two explicit appeals to the utility of Newton's arguments in De Gravitatione: one is methodological in character, the other theological. It turns out that Newton's presentation of God shifts between the two. First, at the start of De Gravitatione Newton introduces his "twofold method" (Newton, 2004: 12), which combines mathematical demonstration that abstracts from physical consideration with experimental confirmation. The twofold method is said to be mirrored in Newton's text: the mathematical demonstration is supposed to be confined to lemmas, propositions, and corollaries, while the "freer method of discussion" associated with the experimental confirmations are "disposed in scholia" (Newton, 2004: 12). Newton explicitly justifies his introduction of experimental confirmations to make clear the "usefulness" of his mathematical demonstrations (Newton, 2004: 12).

Second, later in De Gravitatione Newton sums up his treatment of the idea of body by calling attention to its "usefulness" because "it clearly involves the principal truths of metaphysics and thoroughly confirms and explains them" (Newton, 2004: 31). In the very next sentence Newton makes explicit the advantages: "we cannot posit bodies of this kind without at the same time positing that God exists, and has created bodies in empty space out of nothing, and that they are beings distinct from created minds, but able to be united with minds" (Newton, 2004: 31). Newton's target is Descartes and the atheists (Newton, 2004: 32).

There need be no conflict between the methodological and theological utility. Nor do I mean to suggest that the whole of De Gravitatione is organized around a distinction between a methodological and theological section. In fact, most of what we have of De Gravitatione is explicitly a digression (Newton, 2004: 36) meant to dispose of Cartesians' "fictions" (Newton, 2004: 14; in Janiak's 2004 edition the digression is most of the document). But the first half of the digression—on the flaws of Cartesian conception of motion (Newton, 2004: 14–21) and Newton's positive conception of space (Newton, 2004: 21–27)—is not theologically useful in the way the more speculative treatment of body (Newton, 2004: 27–35) is. Newton's employment of "emanative" causation (as indicated in the four passages, [A]–[D] above), which offers us an unanthropomorphic God, is confined to the treatment of space—that is, in the part that is designed to "lay truer foundations of the mechanical sciences" (Newton, 2004: 21). Presumably these foundations are the way in which his methodology is useful (see also the use of "foundations" in Newton, 2004: 12 and 21). But when Newton turns to his theologically useful treatment of body, emanative causation is absent (except, perhaps, hypothetically at Newton, 2004: 31). The treatment of the creation of body, which is useful to metaphysics, relies on an anthropomorphic God. So my response to the objection is: do not conflate the first half of De Gravitatione's digression (Newton, 2004: 21–27) with the second half (Newton, 2004: 27–35; see also Slowik, 2009, 438ff). It is only the first half of the digression that fits more clearly the general methodological aims of De Gravitatione; there emanative causation figures in the ways that resemble Bacon's notion of a formal cause. Strikingly, Newton is more confident of the status of the "exceptionally clear idea of extension" (22) than he is of the "more uncertain" part about "the limits of the divine power" (27).

Finally, Geoff Gorham uses these remarks as a source of an objection. Gorham accepts my claim that in De Gravitatione Newton separates God's will from the formal reason of bodies. But he is more skeptical about my claim that one can use Baconian emanation as a way to interpret the relationship between God and space. As Gorham (2011a: 304) writes: "Space and bodies are very different sorts of effects according to De Gravitatione: the former is necessary, eternal, uncreated, imperceptible, infinite, indivisible, immobile and causally inert while the latter are, as Newton says, 'opposite in every respect'(Newton, 2004: 33). Indeed, laws are only mentioned in De Gravitatione after the account of space has been completed and the more 'uncertain' account of body is introduced."

I agree with Gorham that one should not conflate Newton's treatments of space and body in De Gravitatione. In fact, my argument relies on the assumption that these need to be kept separate not only because they have a different epistemic status (as Gorham notes) but also because they presuppose largely different conceptions of God (although I have also argued that even the treatment of the essential features of body is less voluntarist than ordinarily supposed in the scholarly literature).

So to clarify my position and sum up: I use the evidence from Bacon to argue, first, that when Newton speaks of "emanation" he could be thinking of *causa-sui*-like, formal causation. If we treat emanation as a species of formal causation then we can reconcile Stein's necessitarian reading with most of the Platonist elements that others have discerned in De Gravitatione, especially Newton's commitment to doctrines of different degrees of reality as well as the manner in which the first existing being 'transfers' its qualities to space. This is the main point of my argument. Second, I use the evidence from Bacon to claim that Newton shares with Bacon a privileging of material composition as (part of) the ontic source of lawlike regularity in the world, and this is another, more speculative, way to understand the role of emanation in De Gravitatione.

We can also appreciate that one of Newton's most important decisions in recasting the material for *Principia* was, first, to drop the language of formal causation that was still present in De Gravitatione and replace it with the language of law—a concession to Cartesian terminology—and, second, to separate the theologically useful material from the main argument.

While it is clear that I advocate the fourth option, my main aim has been to suggest the shortcomings of all the current readings. In particular, with possible exception of McGuire (1978), no reading I am familiar with has attempted to do full justice to the intricacy of all of the details of Newton's treatment of emanation. My appeal to Bacon is meant to show the fruitfulness of treating Newton's use of "emanation" as a species of formal causation.[21] But I do not claim that I have accounted for all the peculiarities of Newton's position. I turn to one of these in next section of this chapter.

---

[21] Goldish (1999: 148 and 162ff), points out that later in life Newton rejected emanation theories which he associated with Leibniz (e.g., *Monadology; Discourse on Metaphysics*, 14). Moreover, on Goldish's account emanation conflicts with Newton's embrace of a voluntarist conception of God. On my account the voluntarism only becomes predominant with the "General Scholium" introduced in the second edition of the *Principia*. I thank John Henry for reminding me of the significance of Goldish's piece.

5.4 Newton on Existence and Measurement

Recall passage [C]:

"Space is an affection of a being just as a being. No being exists or can exists which is not related to space in some way. God is everywhere, created minds are somewhere, and body is in the space that it occupies; and whatever is neither everywhere nor anywhere does not exist. And hence it follows that space is an emanative effect of the first existing being, for if any being what-soever is posited, space is posited."

It continues with [E]:

"And the same may be asserted of duration: for certainly both are affections or attributes of a being according to which the quantity of any thing's ex-istence is individuated to the degree that the size of its presence and per-sistence is specified. So that the quantity of the existence of God is eternal in relation to duration, and infinite in relation to the space in which he is present; and the quantity of the existence of a created thing is as great in re-lation to duration as the duration since the beginning of its existence, and in relation to the size of its presence, it is as great as the space in which it is present." (Newton, 2004: 25–26)

It is to Ed Slowik's credit that he has tried to fit [E] in his Platonizing inter-pretation of De Gravitatione. Nevertheless, even he has to admit that the "'quantity of existence' is a fairly mysterious and undefined notion in the De Gravitatione, so it is difficult to draw a specific conclusion based on this use of terminology" (Slowik, 2013: 440). But this does not mean we cannot say anything at all. McGuire, by contrast, relies on Newton's 1692–1693 man-uscript (which he has dubbed "Tempus et Locus") to argue that Newton is committed to the claim that "'what is never and nowhere, it is not in the nature of things [rerum natura].'" (McGuire, 1978: 465, quoting "Tempus et Locus," CUL Ms. Add. 3965, Sect. 13, folio 545r.40.) In one sense McGuire's interpretation of Newton is correct. As we have already seen above, all things must exist in space and time. To put this in Steenbergen's apt phrase: "an en-tity exists only insofar as its quantity of existence is specifiable."[22]

---

[22] See Steenbergen unpublished ms. Here I cannot do justice to Steenbergen's very interesting re-construction of the relationship among this point, Newton's claims about affection, and emanation.

But as McGuire's argument develops he associates this doctrine with a further claim that "Newton must deny that God exists in space in the manner of an extended being" (505). McGuire's argument rests on the claim that "any being has a manner proper to itself of being in spaces" (McGuire, 1978: 506; recall passage [D] above). The upshot for McGuire is that "quantity of existence" means something different for God and bodies. For the former is a "permanent thing" and the latter a "successive" thing (McGuire, 1978: 496ff; McGuire is relying on the opening lines of "Tempus et Locus" and the "General Scholium" for his argument). Crucially this is the case for McGuire's reading of Newton because the successive things rely "upon Divine will" (McGuire, 1978: 497). But in context (that is, of the claim "that any being has a manner proper to itself of being in spaces") Newton wishes to deny that God is "like a body, extended and made of divisible parts." (This is stressed by Gorham [2011a]) Newton is not claiming that there are two ways of having a quantity of existence in space or two ways of having a quantity of existence in time. As Steenbergen notes, Newton is "distinguishing between the fact of a thing as existing in space and the manner in which it does so." So, what is Newton's position in De Gravitatione?

First, Newton indicates that he is treating space and time symmetrically; this carries over in *Principia* (Gorham, 2011a; but for complications and exceptions, see chapter 7 and its postscript). This means that time is also a condition of possibility of all existing entities. Time is presumably also indivisible, infinite, etc. Whatever we say about the emanative relation between "God" and space also holds between God and time.

Second, the "geometric richness" of space (and by analogy of time) is sufficient for all things to have determinate quantities of extension and duration.[23] The significance of this is that Newton's twofold method is not exhausted by mathematical propositions and experimental confirmations; measurement is the method to establish *a way of being of entities*.

Before I explicate this (and what Newton means by "quantity of existence") I need to make an important point to prevent misunderstanding (and distinguish my view from McGuire's): being in space and in time is a necessary condition for existence of an entity, but these (for lack of a better word) dimensions of being do not exhaust the way of being for Newton in De

---

[23] See, especially, Newton (2004: 22–23). It would have been nice if Newton had also mentioned these geometric qualities in passage [B] above; now their source is left unexplained. I thank Katherine Brading for discussion.

Gravitatione. Recall that Newton also allows that entities can have different degrees of reality. As he writes, bodies have a "degree of reality" that "is of an intermediate nature between God and accident" (Newton, 2004: 32). Now Newton asserts the doctrine of different degrees of reality to help explain (as a kind of error-theory) the Scholastic "prejudice" in applying "the same word, substance . . . univocally. . . to God and his creatures."[24] Bodies have some substantial reality but are not themselves substances. This makes sense in light of Newton's claim that a substance "is absolute in itself," while extension "is not among the proper affections that denote substance, namely actions, such as thoughts in the mind and motions in bodies" (Newton, 2004: 21). To have full reality is to be the source of activity (see also chapter 2).

Yet, third, when in the mechanical sciences we individuate quantities, which exist in some space-time structure, they can then have "substantial reality" without "inhering in a subject" (Newton, 2004: 32–33; see also Stein [2002], which also defends the use of the slightly anachronistic space-time language).[25] Newton insists against Descartes that bodies must also have "the capacities" to "stimulate perceptions in the mind by means of various bodies" (Newton, 2004: 35). For a body to be a body in the mechanical sciences it must be susceptible to measurement.* (The criticism of Descartes' treatment of motion boils down to the claim that Descartes' concepts cannot yield a "determinate motion"—see Newton [2004: 20!]) What is true of body is true of all entities as a subject of mechanical science: they must have the ability to be perceived by minds "by means" of other bodies. We can now appreciate the relevance of [E].

In [E], Newton is concerned with the way of being of entities as appropriate to mechanical science. This is why Newton focuses on quantities and sizes in [E]: quantities matter because they are what is measured. Newton is not claiming that measurement reveals all aspects of an entity. But if we measure an entity then it exists as a magnitude in space and as a magnitude in time.[26] This is why in De Gravitatione Newton's conception of existence really helps to "lay truer foundations of the mechanical sciences," (Newton, 2004: 21).

---

[24] This is one of the few places in De Gravitatione, where Newton is in agreement with Descartes (Principles of Philosophy, 1.51). For more on such agreement, see Gorham (2011).

[25] In earlier draft I had claimed that for Newton we individuate entities. But as Steenbergen notes, "Newton does not say we can individuate an entity. Rather he says we can individuate a quantity."

* My argument from the fact that a body must able to "stimulate perceptions in the mind by means of various bodies" to being a subject of measurement clearly skips a step. The crucial step is that bodies are also used to measure. See the postscript to chapter 7.

[26] In private correspondence (May 16, 2011), McGuire puts it as follows, "that being in a space-time nexus confers actuality on things; but it also indicates the magnitude of a thing's existence. Thus, having magnitude is a crucial part of what it is to have actuality, even for God."

If God exists, he has (spatial and temporal) magnitude. If we read De Gravitatione carefully we see it has a very radical message: If God is going to be susceptible to analysis within natural philosophy (see the "General Scholium" or Query 31 in *Opticks*), then it, too, must be susceptible to measurement. No wonder Newton suppressed its publication.[27]

[27] Acknowledgments The author thanks Zvi Biener, Sarah Brouillette, Geoff Gorham, John Henry, Geoff McDonough Mogens Laerke, Maarten Van Dyck, and Karin Verelst for very helpful comments on earlier drafts. Moreover, I have benefited from Gordon Steenbergen (ms) "The Role of Measurement in Newton's De Gravitatione," which has caught numerous ambiguities in my formulation of my position, as well as Katherine Brading and Ted McGuire who provided extremely insightful comments on the final draft of this chapter.

In fact, in the unpublished McGuire (2011), McGuire revisited many of his earlier views that are critically discussed here. Our remaining differences on these matters are, I believe, now more a matter of emphasis than deep disagreement except for his ongoing misgivings over my attempt to link Newtonian emanation with Spinozistic *causa sui*. The usual caveats apply.

6

The Certainty, Modality, and Grounding
of Newton's Laws

with Zvi Biener

6.1 Introduction

Isaac Newton began his *Principia* with three now-famous laws of motion[1].
He also called these "axioms," a fact that is often forgotten. It is not obvious why he used both terms. In this chapter, we offer an interpretation of
Newton's dual label, *Axiomata sive leges motus* (Axioms or Laws of Motion).
We claim that by using the terms Newton signals that his foundational principles can be understood as certain premises, shared with the broader natural philosophical community of his time. We are not the first to note this.
George Smith (2014) has proposed to treat the laws as deeply entrenched
working assumptions, while Kantian accounts like Michael Friedman's
(2001, 2009) have treated the laws as constitutive principles.

To shed light on both sorts of claims, we investigate two tensions inherent in
Newton's account of laws. The first arises from the juxtaposition of Newton's confidence in the certainty of the laws and his commitment to their variability and
contingency. The second arises because Newton ascribes fundamental status to
both the laws of nature and to the bodies and forces they govern. We argue that
the first tension is resolvable, while the second is not. However, the second tension
shows that Newton conceives the laws of nature as formal causes of the bodies
and forces which they govern and in which they are grounded. We argue that this
neo-Aristotelian conception goes missing in Kantian accounts of Newton's laws,
as well as accounts that stress the laws' grounding in powers and capacities.

To show this, we draw on the prehistory of Newton's treatment of 'laws' in
the drafts leading to the *Principia* (section 2). We show that Newton introduced

---

[1] This chapter first appeared as Zvi Biener and Eric Schliesser. "The certainty, modality, and
grounding of Newton's laws." *The Monist* 100.3 (2017): 311–325.

Newton's Metaphysics. Eric Schliesser, Oxford University Press. © Oxford University Press 2021.
DOI: 10.1093/oso/9780197567692.003.0008

Axiomata sive Leges Motus as he grew confident in the certainty of his laws and the fact that they underlay all previous mechanical practice. Newton's confidence in the certainty of his laws gives rise to the first tension described. In section 3, we discuss this tension, as well as Newton's various statements concerning the modality and grounding of the laws. We conclude that Newton understands laws on the model of formal causation. We close the chapter by comparing this conception of laws to several contemporary accounts.

6.1.1 *Axiomata Sive Leges Motus* as Shared, True Premises

To determine the meaning of 'axioms' and 'laws' for Newton, it is instructive to uncover the considerations that prompted their use in the *Principia*. The *Principia* was the result of a series of increasingly sophisticated works known as the "De motu drafts," initially aimed at determining the shape of planetary orbits, supposing an inverse square attraction toward the sun.[2] Strikingly, while the *Principia* began with three *Axiomata sive Leges Motus*, the initial De motu drafts only began with a series of hypotheses ("Hyp" or "Hypoth"). Newton's decision to switch terminology is our concern here. We can only rehearse part of the story. The evolution of Newton's thought in 1684–1685 is complex and involves some of the most significant conceptual leaps taken in the latter half of the seventeenth century. Newton himself did not see their full implications for decades (Biener, 2017).

The use of 'hypotheses' in physico-mathematical tracts was common in Newton's time. For example, the deductive portion of Christiaan Huygens' *Horologium Oscillatorium* (1673)—a work Newton admired— began with three 'hypotheses.' Huygens employed the term for principled reasons. He believed that the best we could achieve in natural philosophy was an explanatory intelligibility founded on conjecture and the mechanical principles of shape, size, and motion. Empirical evidence could rule out certain conjectures and give others a high degree of probability, but natural philosophical truth was beyond human ken (Huygens, 1690: 125).

Newton did not agree that natural philosophy was confined to hypothetical reasoning, but neither was he ready to claim in his first tract—"De motu corporum in gyrum"—that his initial premises were certain or even

---

[2] All references to De motu are to John Herivel (1965).

particularly important.[3] At this stage of composition, Newton was merely answering a mathematically challenging and astronomically noteworthy question brought to him by Edmond Halley, but was yet to realize that its solution entailed a reconceptualization of the basic principles of natural philosophy (Cohen, 1971: 47–54).

'Hypotheses' also began the following version of the tract—"De motu corporum in fluidis." They included forerunners to the first two laws, as well as theses concerning resistance, the relativity of motion, and the idea that "By the mutual actions between bodies the common center of gravity does not change its state of motion or rest" (299). In this tract, Newton also engages more explicitly with questions of space, time, and cosmology, but cautiously. He notes that since we cannot determine from the motions of the planets whether the center of gravity of the solar system is moving, we may take it to be at rest and so "the Copernican system is proved a priori" (301). This is the first clear indication that Newton sees his tract as more than a mathematical exercise, but as answering the central question of seventeenth-century natural philosophy. However, Newton also recognizes that since the center of the solar system is not identical with the sun, the planets "neither move exactly in ellipses nor twice in the same orbit." In what has been dubbed the "Copernican Scholium," the complexity of planetary orbits leads Newton to despair that: [T]o consider simultaneously all these causes of motion and *to define these motions by exact laws* allowing of convenient calculation exceeds, unless I am mistaken, *the force of the entire human intellect* (301, emphasis added). At this point, Newton believes that we could not go beyond a rough description of "the system of the world" (see section 2 of Smith (2007) for the wider significance of the so-called Copernican Scholium). Echoing Descartes, Newton seems resigned to the idea that empirical complexity will make it impossible to gain detailed knowledge of real motions and their causes (Garber, 2000).

We come now to our focus. Sometime after writing the earlier hypotheses, Newton crossed out the word "Hypoth" and replaced it with "Lex." This is the first instance of 'law' (as a heading) in the De motu composition sequence (Newton, 1989: 13). Since the tract also concludes with the claim that laws lay beyond the reach of the human intellect, it is almost certain that the

---

[3] For Newton's antihypothetical stance of the 1670s, see Alan Shapiro (1989). Newton's ardent refusal in the 1670s to accept hypothetical reasoning makes clear that his use of 'hypotheses' in 1684 is not unconsidered. Rather, it indicates that he believed the work had not yet passed the epistemic bar that turned 'conjectures' into 'certainties.' See also Howard Stein (1990).

cross-out occurred after the tract was complete. Consequently, the reasons for Newton's introduction of 'law' are likely connected to the developments made explicit in the subsequent tract, "De motu in regulariter cedentibus."

In this tract—which begins with six *Leges Motus*—Newton suggests that his previous attitude was too pessimistic. He distinguishes absolute and relative quantities and begins to work out a method that can determine, at least in principle, 'true' from 'apparent' motions.[4] The tract is striking in its recognition of the deep conceptual connections between Newton's laws of motion, particularly the third law, and previously accepted principles of mechanics, including the relativity of motion and the claim that the motions of bodies among themselves do not change the state of motion of their center of gravity. The distinction between absolute and relative quantities also shows that Newton began believing he could determine how motion really is, not merely how it appears to us. This is surely one of the reasons that led him to cross out 'hypotheses' and replace them with 'laws' in the very same tract where he previously wrote that to define motions by exact laws exceeds the force of the human intellect. In fact, the removal of the pessimistic "Copernican Scholium" suggests he came to believe he found a means of determining the system of the world despite the ineliminable complexity of planetary orbits and the ineliminable epistemic limits set by the relativity of motion (Smith, 2008).

Moreover, it seems that Newton came to believe that the principles that allowed him to describe the true world system were intertwined not just with astronomical tenets (like Kepler's 'area law'), but with the foundations of traditional terrestrial mechanics.[5] Newton makes the connection between his laws and the mechanical tradition explicit. In the same tract, after the first two laws, he writes that "By means of these two laws now widely acknowledged Galileo discovered that projectiles under a uniform gravity acting along parallel lines described parabolas . . ." (312). Newton had not

---

[4] For a discussion of the method, see Smith (2002). The distinction between absolute and relative quantities also suggests that Newton begins thinking about mechanics as tracking abstract quantities, a feature that will be constitutive of his work in the final *Principia* (see Smeenk and Schliesser, 2013).

[5] Of course, what counts as 'mechanics' is a fraught question. It is clear that the scope of 'mechanics' was different for different seventeenth-century authors. The field was only loosely defined by a set of (neither necessary nor sufficient) concerns. These included the use of mathematics, the distinction between natural and artificial construction, a broadly non-Aristotelian matter theory, the rejection of teleological explanation, etc. We thank an anonymous referee for stressing this point. For further discussion, see Gabbey (1992), Domenico Bertoloni Meli (2006b), and Kochiras (2013). We use 'mechanics' here to broadly denote the field of investigation concerned with projectile and pendular motion, as well as the five simple machines.

previously drawn the connection between his exercise in mathematical astronomy and terrestrial motion. Here, however, he asserts that his answer to Halley's question was based on principles that were endorsed by Galileo himself and "widely acknowledged" by the previous half century of research into projectile motion (Rupert Hall, 1952). Moreover, in a later development of the same passage, Newton joins together the study of projectile motion and collision theory: "From the same laws . . ., Sir Christopher Wren, Dr. John Wallis, and Mr. Christiaan Huygens, easily the foremost geometers of the previous generations, independently found the rules of the collisions and reflections of hard bodies . . ." (Newton, 1999: 424).

This scholium to the laws is worth commenting on. Its main task is to defend the laws' validity: it begins by asserting that "the principles I have set forth are accepted by mathematicians and confirmed by experiments of many kinds" (Newton, 1999: 424). Newton's mentions of Wren, Wallis, and Huygens in this context are no coincidence. In 1666–1668, the Royal Society held a highly publicized competition to determine the "Laws of Motion" (see Jalobeanu, 2011). Wren, Wallis, and Huygens took part, and responded to the challenge with mathematically equivalent formulations. Each account, however, was based on different foundational principles and concepts, a fact of which Newton was well aware.[6]

On the face of it, it thus seems that Newton misrepresents the history of his own field, falsely suggesting that his laws were already agreed upon. Why? It seems to us that his appeal to the (invented) consensus of virtuosi of his time is meant to recruit their authority and the authority of their independent investigations, to show that his principles are not arbitrary hypotheses he had manufactured ad hoc. Of course, Newton was not being entirely disingenuous. As he shows in the scholium, due to his recovering basic results regarding simple machines, pendulums, and collision, his principles were implicitly presupposed in previous mechanical works, even if not explicitly recognized.[7] Newton's dual appeal to the authority of, and genuine confirmation through, previous mechanical research is clear in the scholium's first

---

[6] For the significance of this claim, see Schliesser (2011a). Newton himself wrote a tract titled "Lawes of Motion" in the 1660s. It is a work very much in the spirit of the Royal Society competition. It makes clear that Newton's decision not to begin the De motu drafts with 'laws' and only adopt it later is significant. After all, he was not averse to the term and even used it himself in appropriate contexts.

[7] The project is further expanded in Section X of the published *Principia*, where Newton shows that he can exactly recover results that in the Galilean-Huygensian tradition were derived assuming a uniform, parallel gravity, but assuming his own laws and an inverse-square, centripetally directed gravity. See Smeenk and Smith (ms., unpublished).

sentence: "the principles I have set forth are accepted by mathematicians and confirmed by experiments of many kinds" (Newton, 1999: 424). Importantly, Newton first made this appeal—initially only to Galileo—in the first De motu tract that fully replaced 'hypotheses' with 'laws'; this seems like no coincidence, although this must remain a matter of speculation.

Newton next authored two works titled *De motu corporum, Liber Primus*. The first use of 'Axiomata' of which we are certain is in the second *Liber Primus*.[8] In the early modern period, an axiom was often understood as an established principle, one that is widely, if not universally, accepted (e.g., "the whole is greater than a part" and [counterintuitively to modern ears] "God does exist" [Hobbes, 1656: 228; More, 1662: 22]).[9] Such axioms were often left implicit, precisely because their truth was beyond dispute, although they were more likely to be made explicit in deductive contexts. Newton used axioms in this way in the *Opticks*. The *Opticks'* axioms include claims like "The angle of Reflexion is equal to the Angle of Incidence" and "An Object seen by Reflexion or Refraction, appears in that place from whence the Rays after their last Reflexion or Refraction diverge in falling on the Spectator's Eye" (5, 18). Immediately after them, Newton notes: "I have now given in Axioms and their Explications the sum of what hath hitherto been treated of in *Opticks*. For what hath been generally agreed on I content my self to assume under the notion of Principles . . . " (Newton, 1952: 19–20). We believe Newton uses 'axioms' in the *Principia* in the same way. In fact, the second *Liber Primus*— the first extant work where 'axioms' are used—is also the first extant work that contains the already discussed appeal to Wren, Wallis, and Huygens, as well as the extended discussion that justifies the laws (particularly the third) with a wide range of examples from the domain of simple machines and collision theory.[10] In other words, 'axioms' were likely introduced by Newton as he was making an even more extended case for the connection between his principles and the entirety of mechanical practice before him.

To sum up, we claim that Newton turns to 'laws of motion' and then 'axioms or laws of motion' from 'hypotheses' for three main reason: (1) he gains confidence that they provide a means of distinguishing true and apparent

---

[8] While it is possible that Newton added the label 'Axiomata' to '*Leges Motus*' in the first Liber Primus, the term is missing from Halley's commentary on the work (Cohen, 1971: 336–344)

[9] In the *Lexicon Technicum* (1708), John Harris wrote: "AXIOM, is such a common, plain, self evident, and received Notion, that it cannot be made more plain and evident by Demonstration . . . ."

[10] Newton singles out the third law, noting that he wished to show "the wide range and the certainty of the third law of motion" (Newton, 1999: 430).

motions, and thus capture fundamental features of reality; (2) they are not only applicable to celestial motion, but to the analysis of traditional mechanical problems and collisions; and (3) that they have been presupposed by his predecessors' treatment of mechanical problems and collisions, even if not formulated as clearly. This is not to say that these were Newton's only motivations for the terminological change. For example, (4) in the preface to the *Principia*, Newton is clearly concerned to defend the status of his mechanics as properly mathematical, and there is good reason to believe that his use of 'axioms' is meant to recruit the term's mathematical connotations.[11]

Still, Newton's introduction of 'laws' and 'axioms' predates his explicit concern with his work's mathematical status. In fact, we've argued that the timing of the terms' introduction coincides with Newton's articulation of a different set of concerns, ones concerning the truth of and broad agreement regarding his principles.[12] A reader may well wonder how the use of 'laws' connects to Newton's theology. After all it is often thought that laws presuppose a lawgiver. In order to get clear on this issue, we now turn to Newton's treatment of the metaphysics and modality of laws.

6.2 The Modality and Grounding of Newton's Laws, and Contemporary Accounts

While Newton's confidence in the truth of his laws is clear, his position on their metaphysics is harder to determine. There are only a few key passages in his corpus that address the topic directly. The most significant was introduced in the 1706 Latin edition of the *Opticks* and ultimately ended up in Query 31 of the later English edition. We focus on it because Newton published it under his own name. It reads: "Since Space is Divisible in infinitum, and Matter is not necessarily in all places, it may also be allowed that God is able to create Particles of Matter of several Sizes and Figures, and in several Proportions to Space, and perhaps of different Densities and Forces, and thereby to vary the Laws of Nature, and make Worlds of several sorts in several Parts of the Universe" (Newton, 1952: 403–404).

---

[11] We thank an anonymous referee for urging us to stress this point. See also Gabbey (1992) and Bertoloni Meli (2010).

[12] These decisions reflect only one attitude toward 'laws,' from the various possible in the late seventeenth century. For discussion of the concerns embodied in the debates about 'laws,' see Steinle (2002) and Ott (2009).

In the *Principia* itself, Newton did not offer a definitive, or even speculative, cosmogony. In the first edition (1687), he made the (Epicurean) suggestion that comets bring material for the production of life from one solar
system to another, and intimated that nature is cyclical, with its cycles insti-
tuted by a providential, but not especially Christian God. In the second edition (1713), he appealed to a more heterodox Christian theology (in the
"General Scholium"), and argued that a Designer-God must be posited in
order to explain the stability of celestial orbits (see chapter 3). The *Opticks*
passage above provides more detail on this cosmogony.

A terminological point is in order. Newton sometimes uses 'world' in a
metaphysically neutral sense, to mean a solar system or, more generally, a
closed (or virtually closed) system of interacting celestial bodies.[13] In the
above passage, however, 'world' is a metaphysically richer notion. A world
in this sense is constituted by the kinds of particles and forces it contains,
particles and forces that are ontologically prior to the world they constitute. Moreover, the above passage suggests that within 'worlds' of this kind
particles and forces are prior, in some sense, to the laws of nature.

We will return to issue of priority shortly. First, however, note that for Newton
laws are clearly contingent. They are not grounded in God's immutability (as
Descartes' were), but depend on God's will. Their contingency stands in stark
contrast to the necessity of Newtonian space and time. For Newton, space and
time could not have been otherwise and are a consequence of God's necessary
existence. As he writes in the "General Scholium," "by existing always and every
where, [God] constitutes Duration and Space . . . 'Tis allowed by all that the supreme God exists necessarily; and by the same necessity he exists always and
everywhere" (Newton, 1999: 942).[14] Physical laws do not possess such necessity.[15]

---

[13] In the scholium to the definitions of the *Principia*, Newton shows that in order for the First Law
to hold for the center of mass of a closed system of interacting bodies, the Third Law must hold for
the interactions among the bodies. In addition, in Corollary 5 to the Laws he makes it clear that the
relative motions of a closed system of bodies are not affected if the entire system moves uniformly
without rotation (for details see Smeenk and Schliesser [2013]).

[14] There has been a tendency to ignore Newton's commitment to God's necessary existence
by both those that tend to read Newton as a thoroughgoing empiricist (e.g., Stein, 2002; Di Salle,
2002) and those that tend to emphasize inductive design arguments when discussing Newton's theology (Hurlbutt, 1965). For a useful, Kantian corrective to this tendency see Janiak (2008).

[15] In the "General Scholium" Newton deploys a principle that if some effect Y is universal temporally and spatially, then we can infer or posit as the ultimate cause something that itself is necessary in
a way that accounts for Y (this is one of the arguments for God's necessary existence and the principle
is also used in a criticism of Spinozism that need not concern us here). For more on this see chapter 8
below. Furthermore, we should make clear that although Newton's laws are contingent, they are necessary in relation to God's will. Given God's edicts, they could not be otherwise. See Massimi (2014).
We thank Michela Massimi for stressing this to us.

As Roger Cotes wrote in the editor's preface to the second edition: "From this source [God], then, have all the laws that are called laws of nature come, in which many traces of the highest wisdom and counsel certainly appear, but no traces of necessity" (Newton, 1999: 397; the denial of necessity distances Newton from the charge of Spinozism and its attack on design arguments).

In the above passage, the contingency of Newton's laws is also coupled to the claim that they are not universal. They may hold of some worlds (in the metaphysical sense), but not all. Their variability has at least two consequences. First, a universe with metaphysically distinct worlds entails causal disconnectedness among (some of) its parts.[16] This disconnectedness appealed to Newton, because it could prevent different solar systems from collapsing in on each other. It thus supports the argument from design of the "General Scholium" (Newton, 1999: 940).

Second, the variability of laws raises a possible (but ultimately only apparent) conflict with Newton's third Rule of Reasoning. The rule reads, "Those qualities of bodies that cannot be intended and remitted and that belong to all bodies on which experiments can be made should be taken as qualities of all bodies universally" (Newton, 1999: 795). Newton's position relies (as his long commentary on the third Rule shows) on a distinction between universal and essential qualities (see chapter 1). The latter are necessary for a thing to be the kind of thing it is, while the former are metaphysically optional. The third rule states that universal qualities, despite their inessentiality, are present in every bit of matter, whether in our region or elsewhere. The tension between this universality and the variability endorsed in the *Opticks* passage can be put thus: the third Rule recommends that certain features of the physical world ought to be taken as universal, while the *Opticks* passage suggests that such features may vary from region to region.

However, the tension can be resolved by taking the third Rule as a claim about how qualities should be taken in the context of natural philosophical research, as distinct from the context of speculative cosmogony.[17] The third Rule offers a bold generalization from what is empirically available to domains beyond our experimental grasp, both from the visible to

[16] At the end of the vortex theory promoted in his posthumous *Cosmotheoros* (1698), Huygens explicitly posits causal disconnectedness between solar systems, which entails a denial of a universe-wide vortex. We thank Chris Smeenk for calling our attention to this.

[17] That is, 'universally' means something akin to (adopt Leibnizian terminology) 'for all particles (in the actual world)' and not 'for all particles (in all possible worlds).' For Newton, of course, (some of) those 'possible worlds' may obtain in our own universe. We thank Michela Massimi for discussion.

the microscopic (so-called transduction; see McGuire [1970] and Belkind [2012]) and from this world to other worlds. George Smith (2014) has explained the epistemic payoff of such boldness in light of Newton's subtle methodology. His account relies on Newton's fourth Rule of Reasoning, which states that:

> In experimental philosophy, propositions gathered from phenomena by induction should be considered either exactly or very nearly true notwithstanding any contrary hypotheses, until yet other phenomena make such propositions either more exact or liable to exceptions. This rule should be followed so that arguments based on induction may not be nullified by hypotheses. (Newton, 1999: 796)

According to George Smith, the fourth Rule teaches us that the payoff to inductive boldness is the ability to find phenomena that make induced propositions better qualified, either by making those propositions more and more accurate, or by revealing that they only hold accurately in limited domains or under restricted conditions.

The Rules' emphasis on the epistemic attitude to be taken in the context of natural philosophical research suggests that Newton's position on the variability of laws is compatible both with (a) the universe having more than one kind of law of nature, each holding for a different world (or system of worlds) and (b) the universe having one set of laws of nature that cover all the systems of worlds, as long as only (b) is held as an effective position in the context of natural philosophical inquiry. Newton recognizes that the empirical support for his laws is strong, but limited. Consequently, when he reflects more speculatively on ontology, he can allow something like (a). But when writing for methodological purposes (and in the context of natural philosophical inquiry) he adopts (b) as the engine of future research. Put differently, within the context of research, Newton advocates taking as true (in accordance with the fourth Rule) the idea that the universe has one set of laws that cover all the systems of worlds; and that within that context the possibility of (a) cannot be used to nullify (say) the universal character of gravity. But Newton allows that, in a different register, one can speculate about alternative possibilities that may turn out to be fruitful in the future. The possibility of multiple laws thus may exist if empirical inquiry were to encompass areas of the universe God has populated with particles and forces different from our own. And so, although the laws are deeply entrenched in inquiry, exceptions to them are not only methodologically significant, but may eventually reveal that we need to restrict the laws' scope.

There is also another, harder to reconcile tension in Newton's account of laws. In the passage above, he suggests that particles and forces are ontologically prior, and that laws are dependent on them: "it may also be allowed that God is able to create Particles . . . and forces . . . and thereby [eoque; pacto] to vary the Laws of Nature."[18] There are similar statements elsewhere in the *Opticks*, for example, "[particles have] a *vis inertiae*, accompanied with such passive laws of motion as naturally result from that force . . ." (Newton, 1952: 401). However, the notion of dependence is left unspecified. Moreover, Newton also suggests the opposite. In De Gravitatione (an early text, written before the discovery of universal gravitation), he speculates that "If we should suppose that . . . impenetrability is not always maintained in the same part of space but can be transferred here and there according to certain laws, . . . there will be no property of body which it does not possess" (Newton, 2004: 28). This earlier text suggests that the force of impenetrability and the laws of nature are fundamental, and these give rise to all other features of body—including the forces studied elsewhere in De Gravitatione—and the higher-order laws that govern them. Even later in life, Newton suggests the same. In the *Opticks*, Newton highlights the difference between his philosophy and its competitors:

"These principles I consider, not as occult qualities, supposed to result from the specific forms of things, but as general laws of nature, by which the things themselves are formed . . .". (Newton, 1952: 401)

It might be tempting to think that only nonuniversal features of body might result from the laws of nature, since Newton's target here is specific forms (i.e., the forms of water, wood, and gold). However, the context of the passage suggests otherwise. We quote more fully:

It seems to me farther, that these Particles have not only a Vis inertia, accompanied with such passive Laws of Motion as naturally result from

---

[18] The passage first appeared in Latin: "*illud insuper concedi necesse est, uniq, posse Deum creare ...,vario quoque; numero & quantitate pro ratione Spatii in quo insunt, forte etiam & diversis densitatibus diversisq, viribus, eoq, pacto variare Leges Naturae, mondosq; condere diversa Specie, in diversis Spatii universi particus*" (Newton, 1706: 347). In English, "*eoq, pacto*" was rendered as "thereby," but it has rather strong causal connotations, and can also be rendered as "in that manner," or replicating the Latin most closely, "thereby, in that manner."

that Force, but also that they are moved by certain active Principles, such as is that of Gravity, and that which causes Fermentation, and the Cohesion of Bodies. These Principles I consider, not as occult Qualities, supposed to result from the specific Forms of Things, but as general Laws of Nature, by which the Things themselves are form'd. (Newton, 1952: 401)

We read the passage as asserting that in Newtonian philosophy, active principles—even universal ones like gravity—are purely general laws of nature, and those laws form the things, which they govern. Of course, there are other ways to read the passage.

However, our point is that those are available because there is an inherent tension in Newton's account, one which we can now state explicitly: in certain passages, Newton asserts that laws are expressions of the natures of the entities they govern, and in other passages, he asserts that the natures of various entities (including forces) arise from the laws that govern them. As far as we know, Newton does not address this tension directly or provide means for solving it. The available evidence suggests that he saw his position as entirely coherent. And although it may not seem so from a contemporary perspective, it fits well with a form of causation Newton endorsed in other contexts: formal causation. For, as David Hume correctly discerned, for Newton laws are genuine, so-called secondary causes (EHU §7.25, n. 16; for this terminology see Hattab, 2000). However, they are not efficient. They are akin to formal causes in a traditional, Scholastic sense (Kuhn, 1977: 21–30). In one sense, a formal cause constitutes (and is thus prior) to the substance in which it inheres. Without a formal cause, a substance would not be what it is or do what it does. In another sense, a substance is prior to its form. Substance is the genuine existent and subject of predication, and form's existence is grounded in the existence of substance. It is thus possible to speak both of forms as forming substances and as substances as grounding forms. This tension—which is built directly into the notion of formal causation— mirrors the tension in Newton's account of laws. In one sense, bodies and forces are prior to the laws of nature, since from different forces and bodies arise different laws of nature. In another sense, laws of nature are prior to bodies and forces, since they are the principles by which the "things themselves" are formed (McGuire, 2007; see also chapter 5).

The second tension in Newton's account eliminates several other interpretations of his understanding of 'laws.' To begin with, it is clear that Newton is not a Humean *avant la letter*; he does not think of laws as supervening on states of affairs. Moreover, he is not a straightforward power-theorist. Laws do not (merely) arise for him from underlying powers and capacities, but are constitutive of the forms that delimit those powers and capacities. The tension also places Newton in an unusual, hybrid position vis-a-vis contemporary accounts of laws. Its proto-Kantianism is best brought out through comparison with Katherine Brading's account. Brading (2011, 2012) has argued that Newton's laws are partially constitutive of Newtonian metaphysics, particularly his account of body. To ask what body is, according to her, is to ask what the laws say about bodies. Thus, for Newton, the metaphysics of body is turned into a part of empirical inquiry: in order to determine what bodies are, determine what the laws of nature are. In subsequent work, Brading has explained how the reception of Newton's *Principia* and eighteenth-century developments in physics can be understood as an unfolding of such metaphysics. Brading's account dovetails nicely with the conception of laws as formal causes: they (at least partially) define what bodies are and how they behave. And although Brading eschews causal language in her law-constitutive account,[19] this sort of constitution was often understood (before Kant) as formal causation (Kuhn, 1977: 21–30). It is for this reasons that we call laws the 'formal' causes of bodies and forces, thereby also picking out something resembling Baconian forms (see chapter 5). We are not suggesting that Brading is Kantian. Rather, we are suggesting that Brading's account is useful for understanding Newton's allegiance to the law constitution central to Kantian accounts.

DiSalle (2002, 2006) offers such a Kantian account. He argues for the constitutive character of Newton's laws by showing that the fundamental concepts of Newton's system are required in order to give empirical meaning to the laws. For example, he holds that to find the meaning of absolute time and motion, we must ask how such concepts can be interpreted in a way that makes possible physical inquiry guided by the laws. DiSalle's case is compelling; however, we do not think that Kantian accounts like his do full justice to Newton's position. The crux of the issue is that Newton's accounts of body and the laws are embedded in a set of broader ontological commitments, which Kantian approaches (and Brading's law-constitutive approach) do not

---

[19] Brading does so for principled reasons. Her account relies on the conception of laws inherent in the *Principia*, and that conception leaves much (intentionally) unspecified. Our account here relies on Newton's more speculative statements in the *Opticks*. Although the overall significance of the Queries is a subject too detailed to address here, we take them to indicate Newton's deeply held beliefs, albeit ones that he could not sufficiently support on empirical grounds.

adequately represent. It is true that Newton's approach to natural philosophy opened the door for a more thoroughgoing law-constitutive approach to metaphysics. But that approach was not yet wholly in place for Newton himself. We can see this in two ways.

First, Brading and DiSalle's positions underdetermine the (unusual) modal status ascribed by Newton to the laws. The laws really could have been otherwise in our region of the universe and, for all we know, may be otherwise in other regions—that is (recall), they are true of all the known systems of bodies of the actual world and not for all the systems of bodies in all possible worlds within our universe (see also Miller [2009]). How are we to describe their modality on accounts that stress their constitutive character? Since laws become the guides to metaphysics on such accounts, questions regarding their own modality become difficult to answer.

The same holds of accounts that take laws as primitives (e.g., Maudlin's). Maudlin (2007) treats laws as our ultimate guides to physical necessity and possibility. This may very well be the case, of course, but it is not the way Newton treated his laws. For Newton, while the laws are epistemically useful guides to discovering physical possibility within our world (in his metaphysically rich sense), they are neither primitive *tout court* nor do they constitute metaphysical possibility. The laws could have been otherwise, and they could have been otherwise because they are dependent on—in the way in which formal causes are dependent on their bearers—the particles and forces God chose to create. Thus laws are crucial to Newton's epistemology, as commentators have rightly noted, but not to his metaphysics nor his views on modality (see also chapter 8). This connects to our second concern.

Second, accounts that stress the constitutive character of Newton's laws do not do full justice to the fact that Newton explicitly claims that the laws are grounded in (contingent) bodies and forces. As we saw, in the *Opticks* the laws arise from God's creation of particular sorts of matter and force. Constitutive approaches represent the deep connection between bodies and laws, but emphasize only half of the picture. Newton believes both that laws constitute bodies and forces ("from which the things themselves are form'd" (Newton, 1730: 377) and that laws are grounded in the kinds of bodies and forces that exist ("as naturally result from" them (Newton, 1730: 376). The two halves of the picture are compatible (if we keep formal causality in mind), but many of Newton's metaphysical commitments simply disappear if we only keep the former in mind. Newton is more of a neo-Aristotelian about the metaphysics of laws. When God creates, he creates bodies with properties. What those bodies are is given by laws, but the laws have no existence apart from their grounding in those bodies. Whether passive bodies can count as 'substances' for Newton is a separate question (see chapter 2).

6.3 Conclusion

In conclusion, we have argued that by using "Axioms, that is, laws of motion," Newton signals that he is developing his account from shared premises found in the successful development of seventeenth-century mechanics (e.g., as it culminates in Huygens' work). Newton's trust in their certainty (as described in Rules 3 and 4 of the *Regulae Philosophandi*) arises from their shared status, as well as Newton's growing confidence that they latch on to fundamental features of reality. This dual use of 'axioms' and 'laws' was not entirely novel in Newton's time. Even so, Newton's position is a striking rejection of the position endorsed by many mechanical philosophers, such as Huygens. The *Principia* does not offer mere mathematical models that provide an intelligible explanation of the phenomena. Rather, its laws characterize the exact and true relations of features of reality, in the spirit of fallibilism. Newton's confidence in the truths of his laws gives rise to two tensions. The first arises from the juxtaposition of his confidence in their truth and his commitment to their variability and contingency. We have argued that this tension is resolvable, since Newton's statements about variability and contingency are set in a metaphysical, speculative register, while his statements about certainty and fallibility are set in a more cautious register, one appropriate for natural inquiry. The second tension arises from Newton's various statements about the grounding and fundamentality of the laws of nature: he ascribes fundamental status to both the laws of nature and to the bodies and forces which they govern. We have argued that this tension is not resolvable, but that it is inherent in a neo-Aristotelian conception of formal causation.

Laws, we have suggested, are formal causes of bodies and forces in Newton's universe. Their existence is grounded in the existence of certain types of bodies and forces, but the natures of those bodies and forces are grounded in God's choice of laws. We have related this conception to two contemporary accounts of laws, highlighting in particular their relation to Kantian-inspired conceptions.[20]

[20] We thank Michela Massimi and Angela Breitenbach for their incisive comments. In addition, we are grateful to Andreas Hüttemann, Ruth Groff, Barry Loewer, and, especially, Katherine Brading, for extremely helpful comments. The usual caveats apply.

7

Newton's Philosophy of Time

For in him we live, and move, and have our being.

Acts 17:28

Introduction

In this chapter I explain what Newton means with the phrase "absolute, true, and mathematical time" (Newton, 1999: 408) in order to discuss some of the philosophic issues that it gives rise to[1]. I do so by contextualizing Newton's thought in light of a number of scientific, technological, and metaphysical issues that arose in seventeenth-century natural philosophy. In the first section, I discuss some of the relevant context from the history of Galilean, mathematical natural philosophy, especially as exhibited by the work of Huygens. I briefly discuss how time measurement was mathematized by way of the pendulum and explain the significance of the equation of time. In the second section, I offer a close reading of what Newton says about time in the *Principia*'s Scholium to the definitions. In particular, I argue that Newton allows us to conceptually distinguish between "true" and "absolute" time. I argue that from the vantage point of Newton's dynamics, Newton needs absolute, mathematical time in order to identify and assign accelerations to moving bodies in a consistent fashion within the solar system, but that what he calls "true" time is an unnecessary addition. In the third section, in the context of a brief account of Descartes' views on time, I discuss the material that Newton added to the second (1713) edition of the *Principia* in the "General Scholium" and I draw on some—but by no means all the available—manuscript evidence to illuminate it. These show that Newton's claims about the identity of "absolute" and "true" time have theological origins and significance.

[1] This chapter first appeared as Eric Schliesser, "Newton's philosophy of time." In Heather Dyke and Adrian Bardon (eds.). *A Companion to the Philosophy of Time*. London: Blackwell (2013): 87–101.

Newton's Metaphysics. Eric Schliesser, Oxford University Press. © Oxford University Press 2021.
DOI: 10.1093/oso/9780197567692.003.0009

7.1 The Inheritance of Galileo: Huygens[2]

Galileo bequeathed his successors two technical questions that when properly answered would provide a breakthrough in physical timekeeping: first, does an (ideal) pendulum really describe an isochronous curve, that is, is the period of a pendulum independent of its amplitude? An isochronous curve is one where from all possible starting points an object falls (along the curve) to the bottom of the curve in the same amount of time. Given that the physical (as opposed to, say, the mathematical) version of this question presupposes uniform gravity, in practice the issue is intermingled with gravity research. After considerable experimental work, it seems Galileo decided that an ordinary pendulum did do so.

Second, while a clock can be calibrated for local time at noon by the passage of the Sun through the meridian, can one also create a reliable, univocal measure of time available anywhere on Earth such that one can compare events viewed far apart (a crucial issue for astronomy and longitude research)? Soon after Galileo's telescope aided discovery (1610) of the so-called four Medicean planets or satellites of Jupiter, he realized that reliable tables of their eclipses would create a frequent, repeatable event visible in numerous places at the same time (on cloudless nights). Throughout his life Galileo worked on improving such tables, and he tried to interest various governments to sponsor this research as a means to the solution to the problem of finding longitude at sea—a crucial problem in navigation.

On the first problem, Christiaan Huygens subsequently discovered that the cycloid is isochronous as Newton credits (1999: 553–555). In fact, Galileo had initiated the mathematical study of this curve, which had usages outside of pendulums. For example, Newton relies on the cycloid to construct an approximate means of finding a body that is moving on a Keplerian ellipse [Newton, 1999: 513–514]. One nice feature is that effects of many resisting mediums on the movement of the bob do not undermine the isochronicity of the pendulum, something Newton demonstrated and exploited experimentally [Newton, 1999: 700–723]).

Huygens also discovered that a pendulum's swinging in between two equal evolutes of a cycloid might constrain the pendulum such that it would follow a cycloid (Yoder, 1988). Throughout the remainder of his life Huygens tried

---

[2] This section is indebted to discussion with Geoff Gorham, who recommended Ariotti (1968), and Maarten Van Dyck, who recommended Bedini (1991).

to design and build pendulum clocks based on his mathematical insights that would keep time reliably on land and on sea.[3] Leaving aside the complex engineering difficulties to keep a clock properly calibrated on a damp and rolling and rocking seafaring ship, it is not even an easy task on land, due to friction when the cord of the pendulum hits its cycloidal walls. Through trial and error, Huygens realized that when one keeps the arc of the swing of a long pendulum relatively small it approximates isochronousness (as Galileo had thought) (Schliesser and Smith, forthcoming).

From the middle of the seventeenth century onward, mechanical, relatively reliable pendulum clocks appeared throughout Europe, making possible an important systematic correction to (local) timekeeping. For since ancient times it had been known that the solar day (as measured by the passage of the sun through a local meridian at 'noon') is irregular. As Newton writes in the Scholium to the definitions of the Principia, "For Natural days, which are commonly considered equal for the purpose of measuring time, are actually unequal" (Newton, 1999: 410). Huygens created a table of daily corrections—known as the "equation of time"—that in effect smooth out the irregularities in any given day. The equation creates a mean day that can be used to calibrate and set local clocks and (when widely adopted) ensures that astronomers are using the same temporal framework with which to interpret astronomical data.

As an aside, in the wake of Galileo's and his students work on reliable tables of the eclipses of Jupiter's satellites, the foremost astronomer of the middle of the seventeenth century, Cassini, was able to calculate extremely accurate ephemerides. Together with Huygens, Cassini was brought to Paris to head the Royal Academy of Science. His ephemerides were used to calculate the longitude of different places on Earth in a systematic fashion. One such effort was undertaken by Cassini, who remained in Paris; his French colleague, Picard; and his Danish host, Ole Rømer, at Tycho Brahe's old observatory in Uraniborg (outside of Copenhagen) in 1671. There they noticed that there was a systematic discrepancy in the expected values and recorded values of the observed eclipses. This led to the discovery that light had a finite speed, which Huygens calculated.[4]

---

[3] See, for example, Huygens (1669), reprinted here: http://adcs.home.xs4all.nl/Huygens/06/kort-E.html, accessed October 5, 2012.

[4] See recent treatment by Kristensen and Pedersen (2012); for the English context, including the adoption of Rohmer's theory by Newton, see Willmoth (2012).

7.2 Time in Newton's Dynamics[5]

The main aim of the *Principia* is, according to Newton, "to determine true motions from their causes, effects, and apparent differences, and, conversely, of how to determine from motions, whether true or apparent, true causes and effects. For to this was the purpose for which I composed the following treatise" (Scholium to the Definitions; Newton, 1999: 413–414; see also Newton's "Preface to the Reader" of the first edition of the *Principia*). In particular, Newton infers forces, which he treats as such "true causes and effects," from the measurement(s) of accelerations. These (theory-mediated) measurements as well as the laws of motion on which they are predicated presuppose a conception of time.

Newton's most extensive explicit treatment of his conception of time in the *Principia* occurs in the Scholium to the definitions. He introduces the topic as follows,

> Although time, space, place, and motion are very familiar to everyone, it must be noted that these quantities are popularly conceived solely with reference to the objects of sense perception. And this is the source of certain preconceptions; to eliminate them it is useful to distinguish these quantities into absolute and relative, true and apparent, mathematical and common. (Newton, 1999: 408)

So, time in Newton's physics is a quantity.

In order to elucidate its nature Newton introduces a threefold distinction between a popular and a theoretical conception of time. The popular conception—relative, apparent, and common time—that Newton wishes to dispel is presumably the Aristotelian "notion of time depending on the motions or existence of the material world." (Samuel Clarke to an unknown correspondent, 1998: 114; Clarke goes on to cite Newton's Scholium to the definitions approvingly.) While much of the Scholium (and subsequent scholarly discussion) is devoted to space, place, and motion, Newton begins with time:

---

[5] My discussion in this section is very indebted to generous, private correspondence with Niccolo Guicciardini and Nick Huggett (who, together with Dan Kervick, commented on a blog post about these matters), as well as Smeenk and Schliesser (2013).

Absolute, true, and mathematical time, in and of itself and of its own nature, without reference to anything external, flows uniformly and by another name is called duration. Relative, apparent, and common time is any
sensible and external measure (exact or nonuniform) of duration by means
of motion; such a measure—for example, an hour, a day, a month, a year—
commonly used instead of true time. (Newton, 1999: 408)

As Richard Arthur has shown, Newton's conception of uniform flowing time
is deeply indebted to Christian Epicureans such as Gassendi and Charleton,
perhaps mediated by Newton's (more nominalistically inclined) teacher,
Barrow.[6] More recently Steffen Ducheyne (2008) has explored the significance of Van Helmont. These studies suggest that Newton's views were
developed out of existing discussions. It remains, however, not immediately obvious what Newton means by "Absolute, true, and mathematical"
time.[7] This only gets elucidated (after Newton has offered more informative definitions of space, place, and motion) a few paragraphs down in the
Scholium:

In Astronomy, absolute time is distinguished from relative time by the
equation of common time. For natural days, which are commonly considered equal for the purpose of measuring time, are actually unequal.
Astronomers correct this inequality in order to measure celestial motions
on the basis of a truer time. It is possible that there is no uniform motion by
which time may have an exact measure. All motions can be accelerated and
retarded, but the flow of absolute time cannot be changed. The duration or
perseverance of the existence of things is the same, whether their motions
are rapid or slow or null; accordingly, duration is rightly distinguished from
its sensible measures and is gathered from them by means of an astronomical equation. Moreover, the need for using this equation in determining
when phenomena occur is proved by experience with a pendulum clock
and also by eclipses of the satellites of Jupiter. (Newton, 1999: 410)

[6] See Arthur (1995). However, while Arthur s treatment is very penetrating he fails to note that
the fluxion of the temporal variable need not always be constant (see, e.g., the very important proposition of *Principia*, Book 2, Proposition 10, discussed in Guicciardini [1999: 42, 245–246]), so his
claims that fluxional and absolute time are identical cannot be accepted. I thank Niccolo Guicciardini
for discussion.

[7] Any modern discussion of Newtonian spacetime is deeply indebted to Stein (1967). For a good
introduction to the literature inspired by it, see Rynasiewicz (2011). However, what follows was
prompted by a more recent paper, Huggett (2012), which has shown the fruitfulness to attending to
the distinction between absolute and true motions in the Scholium.

I assume that according to Newton "true" time and "absolute" time are not necessarily identical concepts, although they can, in fact, coincide. Here I treat them as two ways of conceiving time that can be combined into a single conception, but need not be so combined. In what follows I treat "true" and "absolute" as instances of "mathematical" time. As a first approximation, we can say that "absolute" time is approximated by our clocks (or some other measure of relative time) corrected by the astronomical equation of time (as Newton was familiar with from the work of Huygens and later Flamsteed). As we have seen above, the equation of time is derived from, and simultaneously corrects, ordinary ("for example, an hour, a day, a month, a year") sensible measures. The equation of time corrects the solar time allowing thus a *measure* of true time, but is explicitly not identical to it. So while the equation of time is absolute and mathematical, it is not itself true time.

Moreover, even time corrected by the equation of time is not (absolutely) "absolute." For, while the equation of time is a theoretical construct, the corrected time still relies on sensible or mechanical measures, and these always leave room for improvement (except, of course, for the "perfect mechanic of all"—Newton, 1999: 382). So there is a sense in which even absolute time is itself a useful, regulative ideal.[8] No doubt this is all very confusing.

Let me explain. It is by now a familiar fact from scholarship on Newton that he recognized something akin to an inertial frame of reference (see, especially, his treatment of a system of bodies sharing a common acceleration in Corollary 5 and, especially, 6 to the laws of motion: Newton, 1999: 423). But it has been less remarked upon that Newton treats the equation of time as something akin to a shared "temporal frame." By this I do not mean that Newton thinks an "inertial frame" is independent of such a "temporal frame." This "temporal frame" has to be used as part of determining the "inertial frame." In the context of this chapter, we can, however, treat the characteristics of Newton's conception of this "temporal frame" in isolation.

In particular, the equation of time governs the temporal frame of the solar system. As he writes in the original, suppressed version of the final part of the *Principia*, the *Treatise*: "That the Planets, in respect of the fixed Stars, are revolved by equable motions about their proper aces. And that (perhaps) those motions are the most fit for the equation of time" (Newton, 1740: 58). Once the solar day is corrected by the (mathematical) equation of time, one obtains a shared temporal frame suitable for one's physics—this is what

---

[8] I thank Niccolo Guicciardini, Dan Kervick, and Nick Huggett for discussion on this point.

approximates absolute time. A mathematical equation that governs temporal relations among the *whole* system of fixed stars (including the solar system) would closely *approximate* true time.

So absolute (mathematical) time governs the shared temporal frame in the solar system (and nearby objects such as comets, etc.); it is what's presupposed in one's physics. But absolute time does not presuppose that a moment of time is spread beyond the local "frame." We can treat it is as a regulative ideal within one's physics because it pushes one to improve one's clocks, timekeeping, and the equation of time itself. By contrast, true time is a counterfactual (to humans) equation of time that obtains for the infinite universe. It requires the idea that a moment of time is identical at any spatial location. Now, absolute (mathematical) time maps on to true (mathematical) time if one assumes that a moment of time spreads to every place in the whole universe. But there is little empirical pressure to do so.

To put this point slightly differently:[9] from the vantage point of Newton's dynamics, Newton needs absolute, mathematical time in order to identify and assign accelerations to moving bodies in a consistent fashion. But at the same time Newton relies on inertial motion, that is, motion in the absence of force to measure absolute time. The best examples of these are the rectilinear motions of a body in empty space, which can "regulate" a good clock (which approximates the flow of absolute time).[10] Newton uses this clock in order to measure true accelerations. And when the acceleration is measured as zero, then the clock allows one to "deduce" that the motion is inertial. That is, "absolute time" can be identified with our closest thing to our inertial clock. (And this is why "the flow of absolute time cannot be changed.") One might think there is a contradiction here. But if one treats absolute time as akin to a regulative ideal, one can see Newton's strategy as a very useful approximation in the spirit of Newton's methodological "Preface to the Reader": "[T]he principles set down here will shed some light on their this mode of philosophizing or some truer one" (Newton, 1999: 383).

True time is an unnecessary addition to Newton's conceptual framework of absolute and mathematical time given the particular problems addressed in the *Principia*. One way to put this point is that in the first edition of the

---

[9] This paragraph quotes with minor changes from an email by Niccolo Guicciardini.

[10] In Law 1, Newton also mentions "spinning hoops" and the circular motions of "larger bodies—planets and comets"; as Guicciardini has noted, the Sun, the Earth, and Jupiter's satellites are such spinning hoops, and the principal bodies used by astronomers to regulated time until fairly recent technological innovations.

Principia (that is, without the addition of the "General Scholium"), true time could have been understood not as an ontological posit, but as a regulative principle that itself can be successively approximated by correctable astronomical equations (that is, absolute, mathematical time). This makes sense of Newton's use of the *measure* being both revisable and "truer."

But even in the Scholium to the definitions, Newton provides enough of a hint to suggest that he had other, theological uses for "true time" in sight:

> if the meanings of words are to be defined by usage, then it is these sensible measures which should properly be understood by the terms, "time," "space," "place," and "motion," and the manner of expression will be out of the ordinary and purely mathematical if the quantities being measured are understood here. Accordingly those who there interpret these words as referring to the quantities being measured do violence to the Scriptures. And they no less corrupt mathematics and philosophy who confuse true quantities with their relations and common measures. (Newton, 1999: 413–414)

Newton is responding to unnamed authors that argue from the truth of the Copernican hypothesis to the falsity of Scripture. (Cf. Spinoza's treatment in the *Theological Political Treatise* 6.55; III/92.) This is not the place to explore the full details of Newton's argument (Janiak, 2012) but the passage is a forceful reminder that time, space, place, and motion also have metaphysical and apologetic roles to play in Newton's theology. I turn to a discussion of some of these now.

7.3 Time in Newton's Metaphysics[11]

In the previous section I argued that Newtonian absolute time should not be conflated with Newtonian true time. Moreover, I argued that Newton introduces more conceptual distinctions than required by his physical

---

[11] This section has benefited from discussion with Emily Thomas. For historical background (with special attention to the antecedents within Gassendi and Barrow) to Newton's views, see Gorham (2012). See also Ducheyne (2012). The most thorough treatment of Newton's conception of time is also by Gorham (2011). Gorham and I differ slightly on the proper historical conceptual framework for understanding Newtonian emanation (he emphasizes Descartes, while I focus on Bacon), but we agree that time is something like an attribute of God. Moreover, we understand Newton's God as a kind of *causa sui* or formal cause of time (see chapter 5).

180   NEWTON'S METAPHYSICS

theory; his dynamics requires no more than absolute (mathematical) time as a contrast to "relative, apparent, and common time" without resort to "true" time. While, as we have seen, Newton offers considerable argument for the existence of, say, absolute space, he offers, as others have noted, no argument for the existence of true time (or "duration"), which, "in and of itself and of its own nature, without reference to anything external, flows uniformly." [12,13] Given that much of Newton's Scholium to his Definitions can be properly understood as an attack on Descartes' *Principia* (see Stein [1967] and Rynasiewicz [2011]), I digress, briefly, to discuss Descartes' position.[14]

In *Principia* 1.57 Descartes writes:

> Now some attributes or modes are in the very things of which they are said to be attributes or modes, while others are only in our thought (*in nostra tantum cogitatione*). For example, when time (*tempus*) is distinguished from duration taken in the general sense (*duration generaliter*) and called the number of movement (*numerum motus*), it is simply a mode of thought (*modus cogitandi*). For the duration which we find to be involved in movement is certainly no different from the duration involved in things which do not move. This is clear from the fact that if there are two bodies moving for one hour, one slowly and the other quickly, we do not reckon the time to be greater in the latter case than in the former, even though the amount of movement may be much greater. But in order to measure the duration of all things (*omnium durationem*), we compare their duration with the greatest and most regular motions, which give rise to years and days, and call this duration 'time' (*hancque durationem tempus vocamus*). Yet nothing is thereby added to duration, taken in its general sense, except a mode of thought.[15]

To modern readers it is tempting read Descartes as a subjective, idealistic, or even conventionalist (these are not the same, of course) position about time. But given that thought is a (created) substance ontologically on par

---

[12] See Gorham (2011) also for earlier sources.
[13] In private correspondence, Niccolo Guicciardini suggests that when it comes to the time reference system, geocentrists and heliocentrists agreed: they can use the very same "equations of time," so there was little reason for a thorough defense.
[14] My treatment is indebted to discussion and correspondence with Abe Stone, Jeff McDonough, Alan Nelson, and Noa Shein.
[15] I quote from the translation offered in Gorham (2007). My treatment here of Descartes is indebted to Geoffrey Gorham.

with extension for Descartes this inference is not automatically warranted. In Descartes, it does not follow from being a mode of thought that it is thereby merely subjective (or ideal). It is striking that *even* Descartes' measure, "the greatest and most regular motions" (that is, planetary orbits), is neither arbitrary or conventional nor subjective. This is simply the most suitable measure for Descartes' physics. For the most "regular" motions provide stability to the measure, while the greatest motions are the easiest to use as a measure. So the motion that combines both is simply the best measure on Descartes' account.[16]

This reading of the nature of Descartes' measure is compatible with the further fact that the measure is the product of the mental operation (and, hence, a mode of thought), abstraction, which traditionally (in Scholasticism and Platonism) is used to isolate a particular feature of nature and make it amenable to analysis (Gorham, 2007). Newton also uses this notion of "abstraction" in the Scholium to the definitions in the *Principia* ("instead of absolute places and motions we use relative ones . . . in [natural] philosophy abstraction from the senses is required" [Newton, 1999, 410; see also *De Gravitatione*, a manuscript unpublished in Newton's time], and the following: "we have an exceptionally clear idea of extension by abstracting the dispositions and properties of a body" [Newton, 2004: 22]; Domski, 2012, 2013 on the nature and significance of abstraction in Newton.)

Time understood as a measure is contrasted by Descartes with duration in its most general sense; the latter is not treated as a mode of thought. It is not entirely clear what the status of duration is in Descartes. A plausible line of interpretation suggested by Descartes' *Principia* (1.48) is that generic duration is a category of existence of any type of entity.[17] Be that as it may, Gorham (2007, 46) has shown nicely that according to Descartes duration always involves constant succession. As we will see, these features have some affinity with Newton's views.

So, let's now return to Newton.[18] True time has similar characteristics as Newtonian absolute space ("of its own nature without reference to anything external, always remains homogeneous and immovable") so it is very

---

[16] Here I differ from Gorham (2007), who treats the same passage about the measure of time as evidence that the measure is a convention.

[17] I am indebted to Emily Thomas and Abe Stone for this suggestion. I have benefitted from discussion with Alan Nelson, Lex Newman, Noa Shein, and Jeff McDonough on these matters.

[18] Newton's most devastating criticism of Descartes' conceptual apparatus does not center on time; rather according to Newton "Cartesian motion is not motion, for it is has no velocity, no determination, and there is no space or distance traversed by it" (Newton, 2004: 20).

tempting to think that the arguments for the existence of absolute space simply carry over by analogy to arguments for the uniform flowing true time. This analogy was fairly standard during the seventeenth century (Gorham, 2011a).

However, Newton does not always assert an analogy between space and time (Ducheyne, 2011). In the context of Newton's famous treatment of space and God's sensorium and the infamous missing *tanquam* passage of Query 31 of the *Opticks* (Koyré and Cohen, 1961), Newton only discusses space and makes no mention of time at all.

One may think that Newton encourages the thought that there is parity between time and space: "just as the order of parts of time is unchangeable, so, too, is the order of the parts of space . . . for times and spaces are, as it were, the places of themselves of themselves" (Newton, 1999: 410; in Newton's terminology "places" are occupied by things). But the analogy works in the other direction; Newton takes the fact of the unchangeability of the order of parts of time as *basic*. For Newton *all* moments of time have identical fixed relations. This is a nontrivial metaphysical claim—it rules out, first, the thought that God created time at creation of the universe for then the first moment of time would stand in a very different relation to other moments (as in lacking an earlier moment); second, it rules out thought-experiments in which God either makes time irregular or moves temporal places around. The lack of argument for either the existence claim or the nature of time's characteristics is thus all the more puzzling even if we allow that the whole *Principia* provides overwhelming evidence for the idea that Newton's approach is coherent and empirically adequate.

The "General Scholium" was added to the second edition of the *Principia*. In what follows, I will not be concerned with establishing to what degree the doctrines stated therein are developments of Newton's views or merely make explicit a preexisting position. While the General Scholium continues the polemic with vortex theorists and makes explicit Newton's (evolving) methodological stances, much of it is a public statement of Newton's metaphysical and theological views (see chapter 3 and 8). From the present vantage, the most crucial addition is the following treatment of the "supreme God," who is:

Eternal and infinite, omnipotent and omniscient, that is, he endures from eternity to eternity, and he is present from infinity to infinity . . . He is not eternity and infinity, but eternal and infinite; he is not duration and space, but he endures and is present. He endures always and is present everywhere, and by existing always and everywhere he constitutes duration and space. Since each and every particle of space is *always*, and each and every indivisible moment of duration is *everywhere*, certainly the maker and lord of all things will not be *never* and *nowhere*. (Newton, 1999: 941; emphasis in original)

Now this passage resolves one of the previously noted puzzles; time is not created by God. God and time (and space) coexist eternally—so there is no first creation of a moment of time.* God is always immanent within the order of nature (understood as existing in space and time; see Westfall, 1982 and chapters 2 and 5). Newton does not shrink back from claiming that time (and even nature) is eternal. In the fourth letter to Bentley, Newton allows "there might be other systems of worlds before the present ones, and others before those, and so on to all past eternity, and by consequence that gravity may be coeternal to matter, and have the same effect from all eternity as at present" (Newton, 2004: 102).

According to Newton God occupies all temporal places (forever). But while true time therefore clearly has a separate status from God, there is an ontological sense in which God's existence is more fundamental than the existence of time because in virtue of existing everywhere God "constitutes" duration (and in virtue of existing always God constitutes space). Clearly, time's existence is in some sense a necessary consequence of God's existence.[19] But in the "General Scholium," Newton leaves unclear in what way God's existence is the source of time's existence and why it is so significant to him that God is always everywhere. I return to this below.

Not unlike his treatment in the Scholium to the definitions, Newton treats time in the "General Scholium" as an entity with indivisible momentary "temporal places"; here he adds that these are *spread out* over infinite space. This allows the inference that two events spatially apart happen at the same "true time" regardless of the absolute "temporal frame(s)" in

---

* To what degree it makes impossible an end of time, I do not pursue here. But it would seem that Newton comes close to Spinozism.

[19] As Gorham notes, in a Preface (or Avertissement) Newton drafted for a 1720 edition of the Leibniz–Clarke correspondence, where he cautions that the "unavoidable narrowness of language" must not confuse readers: space and time are "properties" of God only in the sense that they are "unbounded consequences of the existence of a substance which is really necessarily and substantially omnipresent and eternal." For references, including to earlier work by Cohen and Koyré, see Gorham (2011). I explore the nature of this unbounded consequence in chapter 8.

which they occur. As he puts it in De Gravitatione: "The moment of duration is the same at Rome and at London, on the earth and on the stars, and throughout all the heavens . . . we understand any moment of duration to be diffused throughout all spaces" (Newton, 2004: 26).[20] The physical significance of this move is sufficiently known since Einstein. (Smeenk, 2015) The motives behind Newton's claim are clearly theological (see Stein, 1967, 2002); the diffused spatial identity of a moment of time grounds two of Newton's theological commitments: (i) "God" is "one and the same God always and everywhere" (Newton, 1999: 941); (ii) God is not "like a body, extended and made of divisible parts"—God is extended, but indivisible (De Gravitatione, Newton, 2004: 26; God and minds share this property).

Moreover, in these "General Scholium" passages Newton tends to treat time and space as strict analogies if only because there is a kind of parity in God's relationship to space and time: "the supreme God necessarily exists, and by the same necessity he is *always* and *everywhere*" (Newton, 1999: 941; emphasis in original). It is worth noting that the modal status of God's spatial existence and God's temporality is said to be identical. To put this somewhat informally: if one can say that existence is added to God's being then God's temporality is added in the same way; so if God is, time is. To put it in the language of De Gravitatione: duration "is an affection of a being just as a being. No being exists or can exist which is not related to time "in some way." (Newton, 2004: 25; in the quote I have replaced space with duration/time, following Newton's instruction: "the same may be asserted of duration.")

That is to say, time is (with space) a condition of possibility of all existing things. The Newtonian puzzle is how to have duration and space follow from God without God himself being spatial or durational and how to have God in time and space without making him a divisible body or subject to change.[21] Most commentators rely on De Gravitatione to understand Newton's answer to the puzzle. Now De Gravitatione is heavily studied in recent scholarship by historians of philosophy because it is Newton's most sustained effort to critically engage with Descartes' metaphysics and articulate his own view. There is little consensus either over its date of composition or over the proper historical-conceptual framework with which to interpret it.[22]

---

[20] I thank Geoffrey Gorham for calling my attention to the significance of this claim.
[21] I thank Daniel Schneider for this formulation.
[22] In addition to cited work by Gorham, some important papers are McGuire (1978), Stein (2002), Jalobeanu (2007), and Slowik (2009). See also the discussions in Janiak (2008) and Ducheyne (2012), chapter 4.

Because De Gravitatione has quite a bit to say about the nature of space and time and how their existence follows from God's existence, it offers considerable material for understanding the consequence relation between God and time in the "General Scholium" if we assume (i) that in De Gravitatione space and time are treated as strictly analogous, and (ii) Newton did not change his view between the composition of De Gravitatione and the "General Scholium." In fact, on the first point, in De Gravitatione Newton explains the "immobility of space" by the immobility of "duration." In particular,

> just as the parts of duration are individuated by their order . . . so the parts of space are individuated by their positions, so that if any two could change their positions, they would change their individuality at the same time and each would be converted numerically into the other. The parts of duration and space are understood to be the same they really are only because of mutual order and position; nor do they have any principle of individuation apart from that order and position. (Newton, 2004: 25)

Newton asserts here the same doctrine as we have discussed above in the context of the Scholium to the Definitions of the *Principia*.[23] So, again time is treated as the more basic concept.

Finally, in De Gravitatione Newton does articulate a so-called emanative account of the way God and space are related, and this is often taken to shed light on what it may mean that God constitutes duration. (This, too, has not generated any consensus; see Ducheyne (2012): chapter 5.) If we may replace space by time, these are the four passages that bear on this issue. The first passage reads:

> [N]ow it may be expected that I should define extension [space—ES] as substance, accident, or else nothing at all. But by no means, for it has its own manner of existing which is proper to it and fits neither substances nor accidents. It is not substance: on the one hand, because it is not absolute in itself, but it is as it were an emanative effect of God and an affection of every kind of being; on the other hand, because it is not among the proper affections that denote substance, namely actions, such as thoughts in the mind and motions in body. (Newton, 2004: 21)

---

[23] This has surely tempted Stein (2002) to read De Gravitatione as a source of illumination of the *Principia*.

The second reads:

> space is eternal in duration and immutable in nature because it is emanative effect of an eternal and immutable being. (Newton, 2004: 26)

The third passage reads:

> Space is an affection of a being just as a being. No being exists or can exist which is not related to space in some way. God is everywhere, created minds are somewhere, and body is in the space that it occupies; and whatever is neither everywhere nor anywhere does not exist. And hence it follows that space is an emanative effect of the first existing being, for if any being whatsoever is posited, space is posited. (Newton, 2004: 25)

The fourth passage reads:

> [l]est anyone should ... imagine God to be like body, extended and made of divisible parts, it should be known that spaces themselves are not actually divisible and furthermore, that any being has a manner proper to itself of being present in spaces. (Newton, 2004: 26)

We best understand the God-like emanative source of space and time as akin to a formal cause (chapters 5–6). For the second and third passages suggest that God, the first existing being, transfers some of his own qualities to space and time. The first and fourth passages suggest that space and time are akin (but not identical) to substance because they are indivisible (and, as we have seen, unchanging and unmoving). Space and time have places that all other entities occupy. So space and time are literally the things that are presupposed for the existence of all entities. This fits well with the treatment of time in the Scholium to the definitions, where Newton defined absolute, true, and mathematical time as being "in and of itself and of its own nature, without reference to anything external." But unlike substance, time is not the source of activity; it is *passive*. So, infinite, eternal, omnipresent (etc.) God is the substantial (as it were emanative) source of infinite, eternal, omnipresent (etc.) time. This helps explain what Newton means in the "General Scholium" that God "endures always and is present everywhere, and by existing always and everywhere he constitutes duration and space." That is to say, excepting God, time is freestanding. But time requires God's existence.

So the proper way to understand the consequence relationship between God and time is that they are related by way of self-causation. Newtonian time is an attribute-like aspect of a self-causing God. By "attribute" here I mean a property required for the very existence of the substance it is a property of. (See chapter 2) We know from Newton's early notebooks that he interpreted the ontological argument of Descartes' fifth Meditation in terms of self-causation: "A Necessary being is ye cause of it selfe or its existence after ye same manner yt a mountaine is ye cause of a valley . . . (wch [sic] is not from power or excellency, but ye peculiarity of theire natures" (Quaestiones, folio 83r, quoted in McGuire, 1978: 485).

It might appear unlikely that in De Gravitatione Newton would endorse this reading, because he insists that it is "repugnant to reason" that God created "his own ubiquity" (Newton, 2004: 26). One might be tempted to claim that this rules out any *causa sui*. But this conclusion cannot be established because emanation is a doctrine that avoids creation in time. Emanation as a form of divine causation is traditionally distinguished from conceptions that refer to God's will.[24] Newton is clearly signaling that his God does not stand outside nature; even God exists temporally and spatially. Thus, this passage offers a final insight into Newton's embrace of true time and why it has no beginning. Newton's *rational theology* requires that his philosophical God is always present somewhere such that creation of the (material) world takes place in space and time, and God can provide or maintain the being of the entities in it.[25]

---

[24] See Internet Encyclopedia of Philosophy; http://www.iep.utm.edu/e/emanatio.htm, accessed October 5, 2012.
[25] Newton's striking appeal to reason should also make us cautious about reading Newton's empiricism back into De Gravitatione (see Stein, 2002: chapter 2).

Postscript to Chapter 7

Since chapter 7 first appeared, Katherine Brading has illuminated Newton's philosophy of time in two important papers (Brading, 2017 and 2019). Readers may naturally wonder how I would respond to her criticism (Brading, 2017). Part of our disagreement is terminological and part is philosophical. Some of our differences are merely apparent, but a few are, perhaps, not. My interest here is to convey the significance of her approach and use it to develop my position; along the way I mark some of our possible disagreements over Newton's metaphysics with the aim to make more precise how I understand Newton's philosophy of time.

First, I use Brading's approach to call attention to and analyze four disanalogies in Newton's treatment of space and time. Second, I then explain our exegetical disagreement over Newton's terminology of absolute and true time. Third, I remove an apparent disagreement over the empirical nature of Newton's metaphysics. Fourth, I try to pinpoint our disagreement over the structure of Newton's metaphysics. Since this postscript is greatly indebted to discussion with her, it is possible there is no disagreement left.

In order to set this up, let's return to the Scholium of the Definitions: "Time, space, place, and motion" are quantities, Newton remarks, "it is useful to distinguish these quantities into absolute and relative, true and apparent, mathematical and common." Brading and I *agree* that Newton here offers us three fundamental distinctions to be applied to these four basic quantities.[1] Newton then writes:

> 1. Absolute, true, and mathematical time, in and of itself and of its own nature, without reference to anything external, flows uniformly and by another name is called duration. Relative, apparent, and common time is any

---

[1] I introduce 'fundamental' and 'basic' in order to keep track of these distinctions and quantities. The terminology is meant to convey significance, but not meant to do further philosophical work absent further argument. I do not mean to attribute this terminology to Brading.

Newton's Metaphysics. Eric Schliesser, Oxford University Press. © Oxford University Press 2021.
DOI: 10.1093/oso/9780197567692.003.0010

sensible and external measure (precise or imprecise) of duration by means of motion; such a measure—for example, an hour, a day, a month, a year—is commonly used instead of true time.

2. Absolute space, of its own nature without reference to anything external, always remains homogeneous and immovable. Relative space is any movable measure or dimension of this absolute space; such a measure or dimension is determined by our senses from the situation of the space with respect to bodies and is popularly used for immovable space, as in the case of space under the earth or in the air or in the heavens, where the dimension is determined from the situation of the space with respect to the earth. Absolute and relative space are the same in species and in magnitude, but they do not always remain the same numerically. For example, if the earth moves, the space of our air, which in a relative sense and with respect to the earth always remains the same, will now be one part of the absolute space into which the air passes, now another part of it, and thus will be changing continually in an absolute sense. (Newton, 1999: 408–409).

Now there are four dis-analogies in these two paragraphs. The first one is related to the first sentence of each paragraph: 'true, and mathematical' are present as modifications of (absolute) time and absent as modifications of (absolute) space. The second is related to the second sentence of each paragraph: that "apparent, and common" are present as modification of (relative) time and absent as modification of (relative) space. These two dis-analogies are pertinent for my terminological disagreement with Brading. But before I get to that, let's focus on the third dis-analogy.

For, third, there is an asymmetry in the measures Newton proposed. In order to measure time (a quantity) Newton proposes to use the motion of bodies. In order to measure space (another quantity) Newton proposes to use the "situation" of a part of space to another part. And, crucially, the measure of space is identical in "species and in magnitude" to the thing it is a measure of (space). This identity (in species and magnitude) is omitted in the measure of time. Motions of bodies, even the regular motions of the solar system or pendulum clocks, are not identical in species and magnitude to duration.

Brading (2019) explains what's going on in the third dis-analogy. In her terminology, rods (that is, measures of space) are geometrical just as space is. Whereas clocks (that is, a regular motion of bodies) are, for Newton, dynamical systems. And whatever time "in and of itself and of its own nature" might be it is not natural to think of it as a dynamical system. That is to say, there seems to be no possible gap between the measure(s) of space and space (Brading, 2019: 160–161); there does seem a possible gap between the measure(s) of time and time (Brading, 2019: 162).

The reason why clocks are a dynamical system on Newton's account is very nicely explained by Brading:

> It is central to the project of the *Principia*, that forces and the motions of bodies are inter-dependent. Newton emphasizes this in the Preface to the first edition:
>
> > For the basic problem of philosophy seems to be to discover the forces of nature from the phenomena of motions and then to demonstrate the other phenomena from these forces (Newton, 1999, p. 382)
>
> This means that all clocks in the *Principia* are dynamical systems, because they are systems of bodies in motion, and motions and forces are inter-related (Brading, 2019: 162).

That there is a possible gap between the measure of time and time in of itself is not just caused by the fact that the measure is a dynamical system and it seems odd to say this about time itself (where forces and bodies are absent), but also, and more important, because while time "flows uniformly" according to Newton's stipulation (Brading, 2019: 164), there is no guarantee that any measure, which is a natural process of bodies in motion, does so. And even when one constructs an artificial measure (a pendulum clock) or an abstract measure (a mathematical equation of time), there is no guarantee that its regularity and time's uniformity are identical (see also Brading, 2017: 34).

Now, Brading's use of 'rods' is anachronistic, which, given the ends of her paper, is legitimate. Newton himself uses "situations." Unfortunately, Newton does not explain what a situation [*per situm*] is, and whether he intends to use is it as technical jargon, or just to convey something about location. We can infer that a 'situation' is a kind of mental inspection or abstraction from other phenomenal details. As an aside, in his *Essay*, Locke, who I am treating here as an independent guide on contemporary use (not a source), treats 'situation' as an ingredient of complex ideas of (solid) bodies (2.23.9) and sometimes as a way to convey a relationship among sensible parts (2.4.4). It's the latter use that seems to be Newton's in the Scholium to the definitions.

Obviously, if you want to stabilize situations for intersubjective use, rods or other sticks are tempting instruments. Throughout the *Principia*, Newton uses 'situation' when he discusses bodies apart in motion. In context, in each case, some preexisting measurement or geometric practice is presupposed (as Brading pointed out to me in correspondence).[2]

The more important point, and here I am drawing on a note from Chris Smeenk, is that "situation of the space with respect to bodies" (August 28, 2020, personal correspondence) itself involves bodies apart in some sense; because unoccupied spaces are invisible and so difficult to use as markers of situations. Now, Smeenk worries, I think, this implied role of bodies, as the markers or relative locations of situations, reintroduces dynamics. If that is so, then the gap that Brading diagnoses on the measure of time vs. time side may also exist on the space vs. measure of space side. It looks as if a measure of space presupposes elements of a dynamic system—that is, bodies in motion.

A reflection on the epistemology of observation implied by Newton's *Principia* may partially diffuse Smeenk's worry. The reason why there need not be such a gap between the measure of space and space is that one can evaluate/inspect a situation at an instance if a situation is small enough. I say this for two reasons: (i) in the early modern period it is often thought that some ideas are secure, or adequate, if in their presentational aspect they can be inspected at once or instantaneously and be evident (without intermediate *relata*; see De Pierris [2015: 37–42] on Descartes and Locke). And perhaps Newton thinks that some relatively small situations can function as foundational measures in this way such that situations and spaces are the same in "species and in magnitude." And, as I note in chapter 7, (ii) keeping situations small echoes the manner in which a pendulum (a body in motion) can be a reliable measure of time by keeping the arc small, as was well known to Newton and Huygens (Schliesser and Smith forthcoming).[3]

The second ((ii)) also helps qualify the nature of the gap between the measure of time and time diagnosed by Brading. For once the ideal timekeeper has been mathematically articulated and shown to be possible, Newton can bind the amount of error associated with departures from this ideal case. In fact, from the mid-1680s Huygens shows how to start doing this for a pendulum in practice. And so that the right thing to say here, and this is very much in the spirit of George Smith's teaching on Newton (which Brading, Smeenk, and I share as common ground), is that Newton also creates a research program of successive approximation into theoretically and empirically establishing the error bounds and methods for correcting

---

[2] See, for example, Newton (1999: 567).
[3] This epistemology of observation goes beyond Brading's interests in (2017).

them among our measures; in Smeenk's terms of "systematically improving time measurements to approach the ideal of a truly periodic system." So Newton helps initiate a research project into measure theory. (Smith, unpublished; see also Bokulich 2021 which draws on Smith 2014.) And so, if there is a gap on the space vs. measure of space side, there is also a forward-looking attempt to learn how to recognize and close it over time.

The fourth dis-analogy is that the measure of time (bodies in motion) presupposes space or its measure while the measure of space (a rod) does not presuppose time. For in his criticism of Descartes' physics, Newton argues in De Gravitatione that in order to be able to conceptualize and analyze motion, one must make reference to some "motionless being such as extension alone or space in so far as it is seen to be truly distinct from bodies" (Newton, 2004: 19–20). One need not agree with Newton's metaphysics here, to see that some fixed or ordered coordinate system, independent of the bodies, is required for the analysis of motion (for the metaphysical significance of this see chapters 8–9).

Okay, with that in place, let's turn to Brading's (2017) criticism of my position. The terminological disagreement between Brading and myself is about Newton's use of "absolute time" and "true time." Remember that time is one of four basic quantities (space, place, motion, and time) and that each of these four basic quantities is in turn treated in light of three fundamental distinctions between absolute and relative, true and apparent, and mathematical and common. Brading, quite naturally, interprets Newton's use of each of these three distinctions as uniform over the four basic quantities (2017: 19). The effect of this is that "absolute time" has the most fundamental and widest scope in Brading's analysis (2017: 19); whereas according to Brading "true time" is the system relative time parameter of the "system of the world"—that is, our solar system (2017: 27 and 34).

Brading and I agree that "Newtonian absolute time should not be conflated with Newtonian true time" (Brading, 2017: 17–18), although they can coincide. (I return to this possible coinciding below.) But I have argued that "true time" has the most fundamental and widest scope, whereas "absolute time" is the system-relative time parameter of the "system of the world," our solar system. So whereas for Brading the three fundamental distinctions (between absolute and relative, true and apparent, and mathematical and common) are treated uniformly when applied to the four basic quantities (space, place, motion, and time), I claim, by contrast that Newton is not uniform in his use of the fundamental distinctions.

This lack of uniformity in the usage of the three fundamental distinctions surprised me about my own position and has been a source of unease about my exegesis. (It is so intuitive to assume that 'absolute' is the broadest category!) And clearly the burden is on me to prove it. I do not have much new to add to my argument of chapter 7. But let me note two distinct further arguments. First, the four dis-analogies between Newton's treatment of space and time suggest that the basic quantities are not fundamentally alike. They have nontrivial differences beyond the ones I mentioned in the chapter itself. So that makes a position like my own at least not implausible.

Second, there seems to be a passage that does not naturally fit Brading's exegesis: "In astronomy, absolute time is distinguished from relative time by the equation of common time." (Newton, 1999: 410). It seems to me that if Brading's hermeneutics were correct, then Newton should have written here, "in astronomy, *true* time is distinguished from relative time by the equation of common time." For, on her proposed usage, here true time ought to be the time parameter appropriate to the needs of mathematical astronomy, which concerns itself with the system of the world (and bodies that interact with it).[4] This does not settle the terminological disagreement between us, but I hope to have shifted the burden.[5]

I now turn to what I take to be a merely apparent disagreement over the empirical nature of Newton's metaphysics. Because I claim that Newton's introduction of true time is not required by Newton's physics—it is superfluous structure from the vantage point of Newton's dynamics—but seems motivated by what I called Newton's "rational theology," Brading mistakenly thinks that I treat true time as not empirical in Newton's philosophy (Brading, 2017: 17). She then argues, correctly, that Newton's third Rule of Reasoning "plays a crucial role in enabling Newton to extend results from terrestrial experiments to the celestial bodies of the solar system." While

[4] When confronted with the body of the text, Brading writes, "For Newton, there is one time that is both true and absolute, so he could equally well write "in astronomy, absolute and true time is distinguished from relative time," just as Newton writes "true and absolute motion," and frequently contrasts true motion with relative motion (Brading, personal correspondence). I read Newton as using different descriptions depending on what he is trying to draw our attention to, and he's pretty careful I think . . . So I find this passage completely unproblematic, from my point of view." (Brading, personal correspondence, October 2020)

[5] I do not want to overemphasize the difference between Brading and myself here. (What follows is due to correspondence with Niccolo Guicciardini.) If we assume our clocks can be improved by successively improved equations of time (and on the basis of the advancements of the sciences [physics, geology, astronomy, etc.])—as has happened—how can we read this improvement if not under the assumption that it is a series of approximations toward a "true" time? And we might equally think that what we approximate with "absolute" time is also "true" time. A true feature of the world. So, in the converging limit, when science is completed, my disagreement with Brading dissolves.

there are epistemic risks, the rule also encourages to project "to bodies be-yond the solar system" (Brading, 2017: 18).

The key part of the rule that Brading quotes is as follows: "Those qualities of bodies that cannot be intended and remitted and that belong to all bodies on which experiments can be made should be taken as qualities of all bodies universally" (Newton, 1999: 795). But notice that in the quote from Rule 3, Newton explicitly and repeatedly mentions bodies and their qualities, but not the basic quantities (time, space, place, motion). I suggest that's because there is an implied contrast between the two that Newton makes explicit in the passages from the General Scholium I quoted in my treatment of his phi-losophy of time in chapter 7 and in a passage in Query 31 of the *Opticks*, where Newton writes "it may be also allowed that God is able to create particles of matter of several sizes and figures, and in several proportions to space, and perhaps of different densities and forces, and thereby to vary the laws of nature, and make worlds of several sorts in several parts of the universe" (Newton, 1952: 403–404; also recall chapter 6 with Biener, which engages with Brading's influential views on laws of nature). Here just note an important contrast: for Newton it is quite conceivable that bodies and the laws they obey could be dif-ferent in distinct solar systems ("worlds of several sorts in several parts of the universe") whereas he does *not* think space *could* be different.

My claim is that in this respect time is more like that other basic quantity, space, than like bodies and its laws. And that's because crucially, as I argued in chapter 7, space and time cannot be created differently by God: "by ex-isting always and every where, [God] constitutes Duration and Space . . . 'Tis allowed by all that the supreme God exists necessarily; and by the same ne-cessity he exists always and every where" (Newton, 1999: 942).

Now Brading appears to think that when I use "rational theology" I mean "unempirical" (Brading, 2017: 17). Now it is true that I argue that unlike ab-solute time, true time is of no use in Newton's dynamics or the physics proper of the *Principia*. But by this I do not mean that true time is thereby unempir-ical. For Newton is quite explicit in the "General Scholium" that theology is, in part, an empirical enterprise: "to treat of God from phenomena is certainly a part of natural philosophy" (Newton, 1999: 943).

Rather, by "rational theology" I meant (by way of analogue to his use of "rational mechanics" [Newton, 1999: 382]) that there are features of Newton's theology that, while attentive to the phenomena, rely on conceptual moves that are not overdetermined by the empirical findings. So, for example, what grounds (true) time in God is a constitution relation governed by a species of necessity. In chapters 8–9, I explore the nature of such modal metaphysics.

Of course, this is not just a mutual misunderstanding. On my reading of Newton's project, Newton means to turn into an empirical science features of metaphysics and theology that fit awkwardly with our present conceptions about the nature of physics and the questions pertinent to it. As Brading puts it, "Newton transformed the methodology by which these questions should be addressed, providing empirical purchase on them and rendering them empirically tractable" (Newton, 2017: 38). But Brading thinks "these questions" are a subset of the questions that I think Newton is pursuing.

And so finally this gets met back to the nature of Newton's basic quantities. Brading and I agree that the key problem that Newton's physics intends to solve is "the Copernican dispute" which "concerns whether the system of the world is geocentric or heliocentric or whether there is no fact of the matter. Book III of Newton's *Principia* is called 'The System of the World,' and this is where Newton marshals the resources developed in Books I and II to give his answer to the Copernican question. Addressing this question is the overall purpose of the *Principia*" (Brading, 2017: 19). And this is why for Newton, "the basic problem of philosophy seems to be to discover the forces of nature from the phenomena of motions and then to demonstrate the other phenomena from these forces" (Newton, 1999: 382). So far Brading and I are in agreement.

But I claim that Newton adds more structure than he needs for the project(s) described in the previous paragraph. In particular, he stipulates that "the only places that are unmoving are those that all keep given positions in relation to one another from infinity to *infinity* and therefore always remain immovable and constitute the space that I call immovable." (Newton, 1999: 412; emphasis added.) As we have seen, in the "General Scholium," Newton also treats time, by implication, as infinite because infinite and eternal God is always and everywhere (Newton, 1999: 941) He also seems to stipulate that absolute and true time coincide ("Absolute, true, and mathematical time, in and of itself and of its own nature" [Newton, 1999: 408]).

In order to settle the Copernican dispute and to discover the forces of nature there is no reason that these basic quantities, space and time, are stipulated to be infinite.[6] Strikingly Brading agrees "that, for the purposes of the *Principia*," as Brading more narrowly than I conceives them, "Newton does not need his time parameter to extend from infinity to infinity" (Brading, 2017: 18).

As it happens, Newton thinks the infinite cosmos *may* be populated with "fixed stars" that "are the centers of similar systems" (Newton, 1999: 940; Newton uses 'if'). But he adds in the third edition, that these systems have negligible interaction, as "the systems of the fixed stars will not fall upon one another as a result of their gravity, he has placed them at immense distances from one another" (Newton, 1999: 940).[7] So this leaves open whether from within such systems it is even possible to construct temporal frames that would incorporate these other systems.

By contrast, I claim that the infinite (and eternal) nature of the basic quantities are due to Newton's interest in discovering the empirically features of the "most perfect mechanic" (Newton, 1999: 381). And, in particular, to supply the minimal conceptual structure for a world in which God "exists always and every where."

So let me try to summarize. On my view true time and absolute time are conceptually distinct with true time being the wider or more general concept. I argue that strictly speaking Newton does not need true time in the science of the *Principia* and that it is introduced for theological reasons. The characteristics of absolute time are discovered as we learn more about the relationships among the temporal frames of different solar systems. In the context of enquiry one assume that these temporal frames require the same absolute time until, as the Rules of Reasoning suggest, there is systematic empirical reason to revise one's conception of absolute time. By contrast the characteristics of true time are based on a kind of modal metaphysics of necessity (which I explain more at length in chapters 8–9). This model metaphysics is itself informed by what we discover about God and about empirical

---

[6] Newton uses "infinite time" in a corollary to Proposition 1, Theorem 1, in Book 2, (Newton, 1999: 633). But he is describing a counterfactual, "if a body, devoid of all gravity, moves in free spaces by its inherent force alone." There are many geometric constructions in the *Principia* where a line is extended to infinity in order to construct or illustrate a proof.

[7] I use 'negligible' because Newton (and Huygens) explicitly recognizes that "the light of the fixed stars is of the same nature as the light of the sun, and all the systems send light into all the others" (Newton, 1999: 940). Because for Newton, light rays are constituted by corpuscles, it looks like he thinks some bodies could move among solar systems despite the immensity among them. This might motivate an indeterminately large time and space parameter.

phenomena, but not obviously subject to empirical revision or improvement. Newton seems to assert by stipulation a coincidence between true time and its structure, and absolute time (and its mathematical structure) that I claim he is not really entitled to. But given his stipulation, it is no surprise that for Newton at the end of enquiry true and absolute time coincide.[8]

[8] I am very grateful to Niccolò Guicciardini, Chris Smeenk, and especially Katherine Brading for their generous comments on earlier versions of this postscript.

8

Newton's Modal Metaphysics and Polemics with Spinozism in the "General Scholium"

8.1 Introduction

It is clear that in the General Scholium, first published in the second (1713) edition of the *Principia*, Newton defends the legitimacy of knowing God by his design and final causes (the classic source is Hurlbutt (1965 [1985]). In fact, Newton claims it is the only way humans can know God (his phrasing is compatible with other kinds of minds—angels—knowing him through different sources):[1]

> We know him only by his most wise and excellent contrivances of things, and final causes; we admire him for his perfections; but we reverence and adore him on account of his dominion. For we adore him as his servants; and a God without dominion, providence, and final causes, is nothing else but Fate and Nature. (Newton, 1999: 942)

The implied contrast is here with *Spinozism*, which asserts (i) the doctrine in which God just is nature, *Deus sive Natura*, and which (ii) denies the reality of final causes (in the Appendix to *Ethics* 1, Spinoza had called these mere projections of the human mind). For Spinozism all there is, to quote the next line of the General Scholium, is (iii) "blind metaphysical necessity."[2] In what follows, I treat Spinozism as committed to (i–iii) and as distinct from Spinoza's texts (Moreau, 2014).

---

[1] I leave aside what justifies ruling out other (more rationalistic) sources as possible knowledge of God by humans.

[2] There are, in fact, other doctrines commonly associated with Spinoza—substance monism; the real infinitude of space—that Newton shares. See Blüh (1935) and also chapters 2, 4, and 5.

Newton's Metaphysics. Eric Schliesser, Oxford University Press. © Oxford University Press 2021.
DOI: 10.1093/oso/9780197567692.003.0011

In this chapter I remain agnostic to what degree Newton was familiar with Spinoza's writings and to what degree he also identified the historical Descartes, Leibniz, and some of the English Deists with elements of such Spinozism.[3] My argument does not require that these lines are only focused on Spinozism. In addition, there are other passages in the "General Scholium," which I ignore here, that are at odds with Spinoza's philosophy, including the rejection of the plenum and the denial that God is corporeal.

In the "General Scholium" in support of his claim about how we come to know God, Newton asserts that "All the diversity of created things, each in its place and time, could only have arisen from the ideas and the will of a necessarily existing being" (Newton, 1999: 942). The underlying argument inductive: the (apparent) functionality of natural things everywhere and always must be ascribed to the mind and volitions of a necessarily exiting designer.[4] But Newton does not explain why this designing Being must be "necessarily existing" [entis necessario existentis]. Even if one were to grant the argument that goes from apparent functionality of all things at all times and places[5] to the existence of a designer, the modal status of such a designer—that s/he is necessarily existing—is not evident. There are suppressed premises about the nature of modality here.

Newton, too, must have felt some weakness in his position because in the very same paragraph, he inserted the following sentence into the third (1726) edition version of the Principia's "General Scholium": "No variation of things arises from blind metaphysical necessity, which must be the same always

---

[3] For as Steffen Ducheyne pointed out to me, Newton attributes to Leibniz the claim that "nihil omnino in rerum natura dari præter materiam et motum, nullas esse causes finales, omnia fato regi, Deum esse intelligentiam supra-mundanam, totam philosophiam naturalem in eo versari ut per hypotheses explicemus quomodo omnia per materiam et motum absque providentia et causis finalibus produci potuerunt" (i.e., he ascribed [blind] fate and absence of providence to Leibniz's system (Cambridge University Library [hereafter CUL] Add. Ms. 3968, f. 469r-v. [Ducheyne, 2011: 258–260]) Here Newton clearly ascribes to Leibniz the denial of final causes and providence, but he does not attribute to Leibniz (i) the identification of God and Nature (rather he treats the Leibnizian God as supra-mundane), which I treat as characteristic of Spinozism.

[4] Not all of this is an argument. Newton clearly seems to take for granted that things are created: "Tota rerum conditarum" could also just be translated as an axiom: "all things are created."

[5] It is notable that for Newton, everything is adapted to its environment. In the "General Scholium", Newton does not offer evidence or cite authorities for the claim that "we find" such functionality always and everywhere. Presumably he treats it as a well-established phenomenon (in Newton's sense of a robust natural regularity agreed upon by the expert community of inquirers). Having said that, the Principia did offer hints in this direction. Newton, Huygens, and Kant all took Newton's findings about the felt gravity at other planets as offering some such evidence. In the only use of 'God' in the first edition of the Principia, Newton writes "Therefore God placed the planet at different distances from the sun so that each one might, according to the degree of density, enjoy a greater or smaller amount of heat from the sun" (Newton, 1999: 814; he removed the passage in later editions). For more on this, see chapter 3 above.

and everywhere [*A cæca, quæ utique eadem est semper & ubique, nulla oritur rerum variatio.*]" (Newton, 1999: 942). The sentence appeals to the character of metaphysical necessity (*necessitate metaphysica*). In context, the sentence exhibits something like the following argument:

A1: Metaphysical necessity implies homogeneity and homogeneity implies metaphysical necessity.

A2: Homogeneity and variety are disjunctive alternatives (suppressed).

P1: We observe a particular kind of variety.

C: Therefore, no metaphysical necessity.[6]

I have phrased P1 the way I did to do justice to Newton's claim in context that we find "diversity of created things . . . suited to different times and places." (One may rephrase P1 as 'we observe variety.')

I label this "the homogeneity argument." The argument pattern has independent interest. For with the help of some auxiliary premises (A1 and A2), it appeals to empirical evidence (P1) in order to settle (C) a metaphysical debate. In the wake of the success of the *Principia*, this argument pattern—to appeal to empirical facts in order to settle metaphysical debate—becomes increasingly influential throughout the eighteenth century. So, it is worth paying attention to not just for its intrinsic interest, but also its historical legacy in generating what I call "Newton's Challenge to philosophy" (Schliesser, 2011).

The homogeneity argument itself is meant to establish the existence of a providential God by ruling out the most plausible alternative to such a God.[7] For the previous sentence reads, "God without dominion, providence, and final causes, is nothing else but Fate and Nature" (Newton, 1999: 942). That is to say, Newton treats the denial of metaphysical necessity as *ipso facto* a denial of a Spinozism.

While I have been unable to find another use of "blind metaphysical necessity," very similar phrases in Henry More and Samuel Clarke nearly always pick out Spinozism: "A Blind and Unintelligent Necessity" (*Demonstration*, VIII; Clarke, 1998: 38); "A Blind and Eternal Fatality" ("Introduction" *Demonstration*, Clarke, 1998: 5); "blind mechanical necessity" (H. More,

---

[6] The reconstruction does not full justice to the temporal dimension of Newton's sentence (X cannot arise out of Y), but when the discussion turns to cosmogony below, this temporal feature will be partially captured.

[7] This is not to deny that in seventeenth-century practice showing design in nature was tantamount to offering an argument about God's nature. But as we will see, the focus of the argument is less about design and more about God's nature.

Confutation of Spinoza, 91); Spinoza is a "completely blind and stupid philosophaster." (H. More, 91). The use of 'blind' refers to Spinoza's famous denial of final causes in his *Ethics* (E1, Appendix),[8] such that formal and efficient causes are unguided and without an end (see also chapter 4).

In the passages quoted in the previous paragraph, 'Epicureans' are sometimes lumped together with Spinozism, but when the authors need to, they disambiguate the two systems by calling the Epicurean approach the system of 'chance' (a reference to the famous 'swerve') and Spinozism the 'system of necessity' or 'fate.' While it is worth exploring why in this passage of the "General Scholium" Newton does not allow the Epicurean system as a viable alternative to his own, we will not pursue this issue in this chapter (but see chapter 3). Sadly, in the other high profile use by Newton of a closely related phrase—"blind Fate could never make all the Planets move one and the same way in Orbs concentrick," (*Opticks*, Query 31; Newton, 1730: 402)—the exact target is left unspecified.[9] However, the quoted passage in the previous sentence exhibits an instance of P1: in nature we observe a highly specific kind of variety.

This chapter is primarily devoted to explaining the nature and use of such metaphysical necessity in Newton's "General Scholium". For the homogeneity argument presupposes nontrivial (and unusual to modern eyes) metaphysical commitments about (i) the nature of modality; (ii) the nature of formal causation; and (iii) God's existence. These metaphysical issues turn out to be interesting in their own right. In order to explain these, I draw on Clarke's (1705) *Demonstration* and Clarke's subsequent correspondence with Joseph Butler. While there are many differences between Clarke and Newton, here I assume without argument that Clarke and Newton have pretty much the same understanding of metaphysical necessity. Clarke's *A Demonstration* is itself, in part, a polemical critique of Hobbes, Spinoza and, especially, Toland's Spinozism. So at various points I elucidate Spinoza and Toland.

In order to clarify some philosophical distinctions, I treat Toland's Spinozism, in particular, as the target of Newton's homogeneity argument. I argue this by showing that (a) the homogeneity argument is not entirely convincing against Spinoza's *Ethics*; and (b) Newton's argument is anticipated

---

[8] It is undeniable that More was concerned with Descartes' Spinozism before he was familiar with Spinoza. And Maclaurin attributes Spinoza's Spinozism to Descartes' baleful influence (Schliesser, 2012).

[9] It's the denial of general final causes—associated with God's foresight—that turns a system into a 'blind' one. Apologies to those that object to able-ist language, but here we're dealing with recalcitrant sources.

in Clarke's (1705) polemic with Toland's *Letters to Serena* (1704). Toland turns out to be a complex target because he presents himself as largely agreeing with the Newton of the first edition of the *Principia* and sometimes presents himself as a fierce critic of Spinoza (Clarke was not fooled). I argue that Toland generated a dilemma for Newton and the Newtonians that, if left unchallenged, would lead straight to Spinozism. Toland's creative response to the *Principia* has been underexplored, even though a case can be made that it helped shape subsequent responses. Along the way, I'll provide suggestive evidence that Newton was in a decent position to distinguish the thought of Descartes from Spinozism.

8.2 Newton vs. Spinoza

It is well known that Spinoza is committed to blind metaphysical necessity. This is associated with four different doctrines. First, he is against final causes (E1, Appendix). Second, by Spinoza's lights the course of nature is deterministic (E1p29; E1p33; see also E1p21–22). Third, nature has no 'start': the causal sequence goes back into infinity (E1p28). The course of nature is also exception-less (E4, preface). But what about variety?

In recent scholarship there has been considerable attention to Spinoza's embrace of a version of the PSR (Della Rocca, 2008; Schneider, 2014). One might plausibly think that the PSR has low tolerance for arbitrariness or variation. For it would seem that there is no sufficient reason for arbitrary differences. It is often thought (Bertrand Russell is a prominent instance) that the PSR is incompatible with brute facts. And indeed Spinoza's system has well-known problems to explain mode (or finite entity) existence (in the scholarly literature this concern is often associated with Hegel's criticism of Spinoza; Melamed, 2010). In fact, accounting for variety may just be a special case of the main problem in Spinoza's system; as Clarke and the eighteenth-century Newtonian, Colin Maclaurin, emphasized, Spinoza has problems explaining the origin, nature, and particularity of motion. This is directly relevant because in Spinoza motion and entities seem interdefined (chapter 4).

To put the problematic of the previous paragraph in terms of a possible Spinozistic cosmogony: to get empirically observed variety, it would seem that there would have to be some variety from any given arbitrary point of the universe going back to infinity. For, if, by contrast, one assumes that in the infinite recess of time the universe is a homogeneous plenum (Spinoza, E1p15S), then no 'genuine' variation should be produced.[10,11] So it's quite natural to associate a Spinozistic system of necessity with a lack of variation.

This is not, in fact, Spinoza's position; *Spinoza* is not committed to homogeneity. For, according to Spinoza "from the necessity of the divine nature must follow an infinite number of things in infinite ways–that is, all things which can fall within the sphere of infinite intellect" (E1p16). He then adds, "However, I think I have shown sufficiently clearly (by Prop. xvi) that from God's supreme power, or infinite nature, an infinite number of things–that is, all things have necessarily flowed forth in an infinite number of ways" (E1p17S). Rather than being committed to some kind of homogeneity thesis, Spinoza is, in fact, committed to the idea that all possible variety is a necessary consequence of the divine nature![12]

It is not at all obvious Spinoza is entitled to the denial of homogeneity. So even if Newton's argument fails against Spinoza's actual position, it may still be taken to refute Spinozism. But even if Spinoza or a Spinozist were to have sufficient explanation for variety, he does not have a detailed mechanics that can explain (a) the *determinate* variety we observe; as Newton puts in another version of P1 of the homogeneity argument in the "General Scholium": "they could by no means have at first deriv'd the regular position of the orbits themselves from those laws." Moreover, Newton has considerable evidence against the idea that nature is a plenum, including (b) the experiments with "Mr. Boyle's vacuum," (recall chapter 3). So while here it is not my intention to show that Newton decisively refutes Spinoza or Spinozism, it helps explain partly why Spinozistic metaphysics (with its attack on final causes and embrace of necessetarianism) seemed not just undesirable, but also vulnerable intellectually throughout much of the eighteenth century to those familiar with Newtonian science (see chapters 3-4).

---

[10] The scare quotes around 'genuine' are designed to do justice to the fact that Spinoza makes a distinction between imagined and rational cognition; the former might provide evidence of apparent variation, while the latter would rule it out.

[11] The famous Newton-Bentley exchange may have been set off by Bentley asking Newton (repeatedly if we can judge by Newton's evasive responses) what the universe would look like if one assumed a Spinozistic homogeneous plenum and then introduced Newtonian forces (as innate qualities of matter) into them. Descartes denies such inhering qualities of matter, but Spinoza (and Leibniz) allow them.

[12] How to think about the nature of such possibility in Spinoza's system is not so easy (there may be an internal tension as Clarke strongly implies), but need not concern us here (see chapter 4).

One may doubt if Newton has any knowledge of Spinoza. For to the best of my knowledge, Newton never discusses Spinoza nor did he own Spinoza's works.[13] There is, however, a passage in which we can see Newton describing a *Spinozist* position in terms that clearly distinguishes it from Newton's perception of Descartes. Newton writes:

[A] Even arguments for a Being if not taken from Phænomena are slippery & serve only for ostentation. [B] An Atheist will allow that there is a Being absolutely perfect, necessarily existing & the author of mankind & call it Nature: & [B⁺] if you talk of infinite wisdom or of any perfection more then he allows to {say} in {natur} heel reccon at a chemæra & tell you that you have the notion of *finite* or *limited wisdom* from what you find in your self & are able of your self to {prefin} the word *no{t}* or *more then* to any *verb* or *adjective* & without the existence of *wisdome not limited* or [C] *wisdome more then finite* to understand the meaning of the phrase as easily as Mathematicians understand what is meant by an infinite line or an infinite area. [D] And heel may tell you further that the Author of mankind was destitute of wisdome & designe because there are no final causes & [E] and that matter <is space & therefore necessarily existing & having always the same quantity of motion, would> in infinite time would run through all variety of forms... Isaac Newton [Letters added to facilitate discussion][14]

This passage is frequently discussed in Newton scholarship. Scholars tend to claim that Newton is targeting Descartes or Cartesianism in it (Henry, 2013: 132ff.). While one cannot rule out concern with some elements of Cartesianism, this interpretation is a mistake; it is understandable because Newton does indeed criticize "the Cartesians" earlier in the page of the notebook, and Descartes had famously rejected final causes in physics (Osler, 1996: 391) for complications). While claim [A], which criticizes cosmological arguments, could indeed target both Cartesianism and Spinozism, claims [B, D-E], in particular, target Spinozism and not Cartesianism (and [C] may also do so, but is a bit more tricky to parse).

---

[13] It's striking how separately Newton and Spinoza are treated in the seminal volume by Force & Popkin, (1994).
[14] Cambridge University Library, Library Ms. Add. 3970 (B), f. 619r-v. http://www.newtonproject. sussex.ac.uk/view/texts/normalized/NATP00055. I am grateful to Steve Snobelen for sharing his transcription of the ms with me. I thank Alison Peterman for reminding me of the significance of this text. See also http://cudl.lib.cam.ac.uk/view/MS-ADD-03970/1257 and http://cudl.lib.cam.ac.uk/view/MS-ADD-03970/1258.

For in [B] Newton is characterizing a version of Spinoza's famous *Deus sive Natura* doctrine. As the eminent Dutch scholar, Piet Steenbakkers (2003: 45) has argued, unlike any other doctrines found in Spinoza, it is not just characteristic of Spinoza, but (nearly) unique to him. It's true that [B⁺] sounds vaguely Cartesian and echoes things we might read in the *Meditations* (a book Newton was very familiar with). But crucially, when in De Gravitatione Newton explicitly discusses Descartes' account of an indeterminately large as distinct from *infinite* space, Newton insists that Descartes avoids atheism (and he ascribes to Descartes a "fear" of it): "if he should consider space infinite, it would perhaps become God because of the perfection of infinity" (Newton, 2004: 25).[15] So B⁺ is not a nod to Descartes but an allusion to Spinoza's genealogy of error in which God is an anthropomorphic projection (E1, Appendix). Of course while some modern commentators would be disinclined to treat Spinoza's pantheism as a species of atheism, this was quite normal in the early modern period, especially among Spinoza's critics, including More (Nadler, 2010: 240; Nadler also treats Spinoza as an atheist).

Moreover, the denial of general final causes [D] is a clear allusion to Spinoza's infamous doctrine of Appendix 1 of the *Ethics* whereas Descartes always remained officially agnostic about this. Finally, [E] is a clear reference to Spinoza's doctrines of E1p16 and E1p28, which entail the necessary existence of an infinite time and matter's eternity with all variety of forms. By contrast, even in the creation myth of chapters 7–8 of the posthumously published *The World*, where Descartes is careful to distinguish nature from God, Descartes still requires God to get movement started (like a cosmic billiard ball player) and impose laws on matter. It's true that the idea that "quantity of motion" is preserved has a Cartn origin, but it is not incompatible with a Spinozist physics (Garrett, 1994: 79; Schliesser, 2017). It's not uncommon for Newtonians to think that Spinozism just arises out of Descartes in this respect (Schliesser, 2012: 314).

Finally, and more speculatively, I take [C] (in the context of the sentence starting at [B⁺]) to be a nod to Spinoza's famous criticism of the mathematical treatment of infinity in the "Letter on the Infinite." For there Spinoza insists that mathematicians and mathematical natural philosophers actually tend to be confused when they are talking about infinities. In addition, Spinoza offers a more fundamental criticism of the mathematization of nature (Melamed, 2000; Peterman, 2015: Schliesser, 2017).

---

15 I thank Andrew Janiak for discussion of this crucial point.

In particular, on my reading, Newton attributes two claims to the "Atheist": (i) that God is infinite; (ii) that the (ordinary, theist friendly) conception of infinity is likely to be misguided; it's on that the "Atheist" gives a so-called error theory about our anthropomorphic projection. Admittedly, my reading of [C] ignores further complications, and it is certainly possible that Newton is drawing a kind of composite portrait of the Spinozistic "Atheist" that includes both Spinoza's and Cartesian elements.[16]

So does this prove that Newton had read Spinoza? No, of course not. But it shows that Newton was familiar with some of the highly technical, textual, and philosophical criticism by some of Spinoza's fiercest critics, including (almost certainly) Clarke's *Demonstration*, which anticipates in great detail many of the terser design arguments of the "General Scholium"—something that is missed when we focus on Clarke's debts to the "General Scholium" in his exchange with Leibniz.[17]

Given that Spinoza does not, in fact, embrace the homogeneity of nature thesis and given that Newton does not explicitly mention Spinoza in the key passages under discussion, one may well wonder if Spinoza is relevant here beyond the circumstantial evidence I have offered. Moreover, we have made no progress in understanding the commitments to metaphysical modality in both the homogeneity argument and the larger passage in the "General Scholium". To discern the connections among these issues, it's helpful to turn to some of Clarke's less studied writings in recent scholarship (but see Yenter, 2014; Barry, 2016; Yenter and Vailati, 2020.)

8.3 Clarke on Metaphysical Necessity

Clarke's *Demonstration* is according to its subtitle officially aimed against Hobbes, Spinoza, and "their followers." This book is full of fascinating arguments, and I consider it a shame that Clarke is more known today for his later exchange with Leibniz than this book, which brings together ancient and modern arguments for the nature and existence of God and updates these in light of then recent scientific developments as well as important new

[16] I thank Ori Belkind, Steve Snobelen, and Alison Peterman for discussion. They are unlikely to be wholly convinced by my reconstruction.

[17] The claim in this paragraph is compatible with allowing genuine philosophical differences between Clarke and Newton (see Postscript to chapter 2).

seventeenth-century criticisms of a Christian conception of God (especially by Spinoza).

One reader of the *Demonstration*, (a then very youthful) Joseph Butler, was not satisfied with elements of Clarke's argument and pressed for clarification of the material I have been discussing above. In one of his responses, Clarke explains a key assumption (recall A1) in the homogeneity argument. He writes: "Necessity absolute and antecedent in the order of nature to the existence of any subject has nothing to limit it; but if it operates at all (as must needs do), it must operate (if I may so speak) everywhere and at all times alike...." (Clarke's "Answer to Butler's Third Letter," Dec 10, 1713.)

To put Clarke's point informally: absolute necessity has the same impact everywhere and all times and should have the same consequence everywhere and all the time. So, more precisely, I call *metaphysical modality* the idea that if *such* 'necessity' operates in some respect, Y, then we ought to expect Y to be homogenous in relevant ways. And if the most fundamental form of necessity operates in all (possible) respects, then we ought to expect general homogeneity (as in A1 in the homogeneity argument).

Recall that the homogeneity argument was inserted in the "General Scholium" in the context of an otherwise puzzling, inductive argument that goes from the apparent functionality of created things at *all* times and places to the *necessarily* existing designing Being. Newton's inference seems to rely on the idea that if some effect Z is universal temporally and spatially, then we can infer or posit as the ultimate cause something that itself is necessary in a way that accounts for Z.[18] So given that we find evidence of design or functionality everywhere and at all times (let's accept this for the sake of argument), we need to infer a necessary designer. This account of metaphysical modality has consequences for Newton's claims about the attribution of universal qualities of matter. One reason to distinguish between universal and essential qualities of matter is that while both end up leading back to a necessary cause, only universal qualities could have been otherwise.

Whether the ultimate necessary cause so posited acts by way of intermediary (or secondary) causes we can leave unexamined here. However, whether such an ultimate, necessary cause so posited could have acted otherwise will be pursued below. The view behind this idea of metaphysical

---

[18] Notice that this is not treated as a brute fact, but, rather, empirical patterns *require* further causal explanation. Newton's "General Scholium" is famous for admitting being unable to offer such an explanation in the case of (apparent) universal action at a distance. What follows aims to provide some of the grounds to explain some of the relevant dis-analogies.

modality can be made intuitive as follows: if one accepts something like the PSR (as Clarke explicitly does), the existence or presence of a universal quality or property, which is not intrinsic or essential to the entity that has it, needs to be explained (without reference to a miracle) by a single cause (or single chain of causes).[19] The most austere kind of explanation that can be posited is something that has universal causal reach: Leaving aside God, necessity is the only available 'cause.'

This view of necessity, thus, turns out to be a nontrivial principle (cf. Boehm, 2016). I emphasize two features of Clarke's position. First, according to Clarke necessity is the "formal cause" of God (Clarke's "answer to sixth Letter to another Gentleman" [published first in 1738], see Clarke, 1998: 113). That is to say, Clarke does not just allow the question, 'what causes God?' to be intelligible,[20] but it even has a proper response: one can say that, for Clarke, God itself is in a certain sense, the consequence of a species of necessity, which is "antecedent, though not in time yet in the order of nature, to the existence of the being itself" (Clarke, 1998: 113; for an explication of this notion of formal causation, see Hübner [2015]).

There is, in fact, an unmistakable (but generally overlooked) echo of this doctrine of metaphysical modality in the relationship between necessity and God in the "General Scholium": "'Tis allowed by all that the supreme God exists necessarily; and by the same necessity he exists always and every where" (Newton, 1999: 390–391). In the quotation, Newton insists (i) that God's necessary existence is common ground among all philosophers.[21] Newton, then, (ii) ascribes to necessity explanatory 'power' to help account for (at least some of) God's properties.[22] It's this type of explanatory power of metaphysical modality that is captured by Clarke's claim that necessity is a formal cause.

Second, Clarke uses the 'necessity as formal cause of God principle' to rule out the existence of multiple substances (Clarke's first response to Butler,

---

[19] Obviously, multiple causes that produce a single universal effect is a coherent position (something Hume alerts us to in his *Dialogues Concerning Natural Religion*). But as Newton's first two rules of reasoning suggest, especially the second one, parsimony inclines one to search for a single such cause.

[20] This points to the fact that Clarke, too, accepts a version of the PSR. For discussion, see http://plato.stanford.edu/entries/clarke/#2.2.

[21] This is not the place, alas, to explore why it would be obvious to Newton that one cannot take a genuine, philosophical atheist seriously (as opposed to the Spinozist 'atheist').

[22] In fact, on my reading, these are like emanative properties that are coextensive with essential features of God's nature in Newton. (Something like this is true for Spinoza's attribute of extension, but not the property duration in Spinoza.) See chapter 5.

November 10, 1713 [Clarke, 1998: 99ff], and see *Demonstration* VII; Clarke, 1998: 35). So, like Spinoza (E1D1, E1p7, E1p5, and E1p7) Clarke seems to treat God as the only true substance. In particular, in order to make credible substance monism, Clarke relies on the idea that absolute necessity does not just have the same impact everywhere and all the time, but also that in the most absolute sense these consequences 'displace' the (logically possible) 'instantiation' of other hypothetical (but not genuinely possible) effects derived from hypothetical (but not genuinely possible) causes. So A1 and A2 in Newton's homogeneity argument rely on profound metaphysical commitments about the nature of modality and sufficient reason that are embraced by the Spinozists and Clarke![23]

If we attend to some of the further details of Clarke's letter to Butler, we also learn something about the explicit target of Clarke's argument:

> Necessity absolute and antecedent in the order of nature to the existence of any subject has nothing to limit it; but if it operates at all (as must needs do), it must operate (if I may so speak) everywhere and at all times alike. . . .
>
> The argument is likewise the same in the question about the origin of motion. Motion cannot be necessarily existing because, it being evident that all determinations are equally possible in themselves, the original determination of the motion of any particular body this way rather than the contrary way could not be necessary in itself, but was either caused by the will of an intelligent and free agent, or else was an effect produced and determined without any cause at all, which is an express contradiction: as I have shown in my *Demonstration* [p. 19]. (Clarke's Answer to Butler's Third Letter, Dec 10, 1713; Clarke, 1998: 105).

Before I turn to Clarke's *Demonstration,* in this passage Clarke relies on a version of metaphysical modality—that is, the principle that if necessity operates in some respect, Y, then we ought to expect Y to be homogenous in relevant ways—and, then, he applies this (metaphysical) necessity principle to the case of motion in order to show that the origin of motion cannot be explained by necessity (in the way that God is so caused). Rather than pointing to the empirical existence of variety (as Newton does in the "General Scholium"),

---

[23] Leibniz seems to have sensed this which is why he succeeds at putting Clarke on the back foot in their exchange. One might wonder how Newton avoids being a full-blooded Spinozist. I return to this issue below. But note that Newton never explicitly embraces the second feature that I attribute to Clarke and Spinoza.

in the letter to Butler, Clarke argues that there is nothing about matter and/
or motion *as such* that inclines it in one way rather than other ways (recall
chapter 4 with Domski).

One way to understand Clarke's claim is that a particular directionality
is a contingent fact about matter; there are different metaphysical positions
compatible with Clarke's claim here, but we can put his main point a bit met-
aphorically as follows: at the origin of the universe (or at any other given
time and space [recall that necessity works everywhere and all times in the
same way]), a particle does not 'know' in what direction it has to be oriented,[24] so that the very 'first' motion and its determinate direction has to
be nonnecessary and this entails either (a) that there was a willing God or
(b) no cause at all. But (b) is impossible because it denies the so-called causal
principle (see chapter 9), so therefore, there must be a (a) a willing God. Clarke
tacitly rules out here the possibility of a brute, contingent fact (because it
conflicts with the PSR).

Page 14 (in the 1998 edition this is p. 19) of Clarke's *Demonstration* is
devoted to refuting "Mr. Toland's pernicious opinion of motion being essential to matter" with a reference to Toland's (1704) *Letters to Serena* in
the margin.[25] Recall that the subtitle of *A Demonstration* reads "More
Particularly in Answer to Mr. Hobbs, Spinoza and Their Followers." Toland
is the only "follower" named in A Demonstration. In particular, Clarke
attributes to Toland the idea that a 'willing' God is not required to explain
the origin of motion.

In this section, I have relied on Clarke's writings to explain some of the
likely metaphysical presuppositions behind the homogeneity argument[26]
and why there is good reason to believe that in a closely related version
of the argument, Clarke is targeting Toland. We have seen that nontrivial
features of the metaphysical claims about necessity, what I have dubbed
"metaphysical modality," are also found in the "General Scholium." I now
turn to Toland to explain more fully the rationale of Clarke's and Newton's
argument.

[24] On some fairly standard matter theories (e.g., Aristotelian and Epicurean), bits of matter do have
a particular directionality ('down' or 'toward the center of the universe') built into or added to their
natures. In his correspondence with Bentley, Newton is at pains to disassociate his theory from such
theories.

[25] Confusingly, in the (1705) edition of Clarke's *Demonstration* scanned into Google Books, the
reference is to Letter III, but it has to be to the fourth letter.

[26] A skeptical reader might worry that I have given no evidence yet that Clarke endorses the homogeneity argument (see Yenter [2014] and Barry [2016] for discussion).

8.4 Toland's Appropriation of Newton

Toland's *The Letters to Serena* (1704) is a rhetorically complex work in five letters.[27] The first three letters include a genealogy of the idea of immortality of
soul, a proto-feminist tract, and an account of justice amongst other themes
discussed. The fourth letter is a self-styled "confutation of Spinoza"[28]—like
other English critics of Spinoza, Toland finds Spinoza's account of motion
wanting—often using the authority of Newton's then recent *Principia* in the
process; the fifth letter, by contrast, advances Spinozist themes by rejecting
Newton's account of the vacuum, space, and God (among other doctrines;
see Daniel, 1984: 11). Among the Spinozist positions that Toland adopts as
his own is that God is immanent in nature (and the denial of the immateriality of the soul).[29]

Here I ignore all the very interesting complexities *of Letters to Serena* and
focus on a key feature of *Letter* Four. In order to do justice to Toland's position, I introduce some anachronistic terminology: by '*anti-mathematicism*'
I mean the expressed reservations about the authority and/or utility of the application of mathematics. Such anti-mathematicism can come in many guises
and strands. Here I focus only on what I call *The global anti-mathematicist
strategy* by which I mean to pick out those arguments and positions that challenge and de-privilege the epistemic authority and security of mathematical
applications as such. To avoid misunderstanding, this strategy is compatible with allowing some subservient uses (for bits) of mathematics in one's
physics (and praise for *pure* mathematics). While the term is anachronistic,
the position pre-dates Toland. The canonical late-seventeenth-century expression of the global antimathematicist strategy can be found in Spinoza's

---

[27] On Toland's philosophy, see Daniel (1984) and Dagron (2009). For a useful introduction to
Toland's views, especially as they relate to the reception of Newton and Newtonianism see Jacob,
1969. Jacob treats Toland as a sincere critic of Spinoza and as a follower of Bruno. As my argument
notes there are nontrivial differences between Toland and Spinoza (see Dagron, chapter vii), but I do
not doubt that in the fifth letter to Serena, Toland is what I have been calling a Spinozist. This is not
the place to explore the commonalities between Bruno and Spinoza or the ways in which Toland's
criticism of Spinoza is a mere smokescreen. On these points, see Leask (2012; Leask turns Toland
into a Leibnizian of sorts—a topic that transcends this chapter). Toland has been read as interpreting
Newton with a Lockean epistemology (Wigelsworth, 2003). For present purposes I can remain
agnostic about Wigelsworth's main thesis, for he, too, recognizes Spinozism in Toland (see, especially, 530).

[28] Letters, 5.1, 163. I am quoting by letter, paragraph, and page-number to the original 1704 edition. See also Toland (2013), which also provides original page-numbers.

[29] It is useful to note that Toland coined the term 'pantheism' shortly after the *Letters* in 1705, in the
title of his work *Socinianism Truly Stated, By a Pantheist*. (Jacob, 1981: 22).

so-called Letter on the Infinite addressed to Lodewijk Meyer and published in his *Opera Posthuma*.[30]

In order to understand what is at stake in Clarke's response to Toland it is indispensable that we have some features of Toland's *global anti-mathematicist strategy* in view. Toland expresses a key aspect of his position as follows: "The Mathematicians generally take the moving force for granted, and treat of local motion as they find it, without giving themselves much trouble about its original [cause]; but the practice of the philosophers is otherwise, or rather *ought* to be" (Letters 4.8 [141]; emphasis added; Daniel, 1984: 102 also notes the significance of this).

This passage presupposes a hierarchically organized intellectual division of labor between the "mathematicians" who, Toland explains, find "rules of motions" by "observations learnt from . . . experience"; they only deal with "local motion" (or a "change in situation") in order to generate the "ordinary rules of motion" by "probable calculations" (*Letters* 4.8, 140). That is to say, these "mathematicians" are primarily engaged in what we would call 'empirical induction' and 'instrumentalist description.'[31] Higher in status are the "philosophers," who assign causes and the (causal) "principles" of "true" motion (*Letters* 4.8, 140). The distinction is functional; it's of course possible that the very same person acts as a 'mathematician' and as a 'philosopher' (See also Schliesser, 2020).

There are three key features of the hierarchical intellectual division of labor:

   (i) it is normative ("ought to be");

  (ii) mathematicians, or we might say, mathematical natural philosophers do not have last word on their own analysis (*qua* mathematician);

 (iii) mathematicians are incapable of supplying what we really want—a causal understanding of how the world operates.

It is the combination of (i–iii) that makes Toland's position instantiate a global anti-mathematicist strategy. (That's compatible with him allowing a pragmatic or instrumental role for mathematics in physics.) Crucially,

---

[30] For a lot more details on Spinoza's position, see Schliesser (2017). Throughout the seventeenth century in response to the aspirations of what we may dub 'Galilean science,' there were informed criticisms of the utility and application of mathematics to natural philosophy. For an inventory of such anti-mathematicist arguments, see Demeter and Schliesser (2019) Nelson (2017); and chapter 13 of Schliesser (2017).

[31] Toland's position allows that "mathematicians" may not realize that they are doing no more than this.

Toland asserts that Newton agrees with (ii) by appealing to the scholium to Proposition XI of Newton's *Principia* (which is quoted in Latin at the bottom of the page):

> "The Mathematicians compute the Quantitys and Proportions of Motion, as they observe Bodys to act on one another, without troubling themselves about the physical Reasons of what every person allows, being a thing which does not always concern them, and which they leave the Philosophers to explain: tho the latter wou'd succeed better in their Reasons, if they did more acquaint themselves before hand with the Observations and Facts of the former, as Mr. NEWTON justly observes." (*Letter* 5.9, 177)

In context, and throughout the *Letters*, Toland implies that the methodological stance of (the first edition of) the *Principia* explicitly recognizes a distinction between the discovery of *local* forces (which, Toland grants, are discovered by Newton) and the "general or moving force of all matter" (see *Letter* 5.29, 233–234; Toland quotes a passage of Newton's "preface" to the *Principia* in support of his claim). Toland may thereby be the first to interpret the methodological stance of Newton's *Principia* in an instrumentalist fashion—something that only becomes fashionable after the addition of the "General Scholium" and Clarke's exchange with Leibniz. For this precedes the changes to the *Principia* in response to the controversy over the status over action at a distance.[32] Here I leave aside the question to what degree Toland's interpretation of Newton can be defended in light of the details of the *Principia*.[33]

Another key move by Toland is to reject what we would call the invocation of a "God of the gaps" (Ratzsch, 2014). He does so by appealing to the authority of Cicero. He then goes on to imply that one reason to favor the idea that motion is essential to matter is that it minimizes such invocation of gap-filling-God:

> [T]hey are forc'd at last to have recourse to God, and to maintain that as he communicated Motion to Matter at the beginning, so he still begets and

---

[32] There are modern interpreters that also claim that Newton was an instrumentalist (McMullin, 2001); for criticism see Smith (2001); Ducheyne and Weber (2008); and Schliesser (2011).

[33] There is often (Levitin, 2016) a conflation between Newton's stance toward action at a distance and his stance toward causal explanations as such (see Smeenk and Schliesser [2013] or chapters 1–2 above).

continues it whenever, and as long as there's occasion for it, and that he actually concurs to every Motion in the Universe. . . . . As Cicero observes when the philosophers are ignorant of the cause of anything, they presently betake themselves for refuge and sanctuary to God, which is not to explain things, but to cover their own negligence or short-sightedness . . .

I hold then motion is essential to matter . . . as inseparable from its nature as impenetrability or extension." (L4.15–6, 157–158)

In these passages, Toland is not merely criticizing occasionalist views, but all views that require any intervention of God. Toland contrasts two (coherent) positions: first, one that claims that matter is passive and requires God to be the first and concurring cause; second, one that aims to minimize God's role altogether and, thereby, opts for active matter and 'activity' is an essential quality of matter. The second view dispenses with God's role as the first cause of motion and has a tendency to insist that the universe must have existed forever.

In a nutshell: Toland adopts the second position: "I deny that matter is or ever was an inactive dead Lump in absolute Repose, a lazy and unwieldy thing . . . I hope to evince that this Notion alone accounts for the same Quantity of Motion in the Universe, that it alone proves there neither needs nor can be any Void, that Matter cannot be truly defin'd without it, that it solves all the Dificultys about the moving force, and all rest which we have mention'd before" (Letter 5.16, 159–160). By contrast, in a Demonstration Clarke adopts the former position (matter is passive, etc.), strongly implying that this is also Newton's position.[34]

So the way to understand Toland's challenge that Clarke faced is as follows: the formal structure of the Principia is ultimately neutral on matter being truly active or passive and what the general source of motion is. While Newton qua mathematician talks of forces as causes in order to help keep track of observed regularities, these forces are merely 'local' and do not pick out genuine explanatory causes in nature—that's the task of the philosopher and by Toland's lights Newton acknowledges this division of labor. Toland takes up the task to offer a philosophical conception of matter that coheres with the Principia; one in which matter is essentially active and, with a nod to Cicero's authority, that dispenses with a need for God

---

[34] There is a lively debate among Newton scholars on how to understand Newton's position on the activity and passivity of matter (Kochiras, 2009; Ducheyne, 2014; and my own chapters 1 and 2 and its postscript).

(beyond a vague immanent substance monism in which God gives being to matter).[35] If, by contrast, you insist that matter is passive then you introduce the god of the gaps.

8.5 Clarke Responds

The main response to Toland by Clarke is, as he writes to Butler, offered on p. 19 of the *Demonstration*. I have gone over elements of the argument above, but I quote it in full:

One late author indeed has ventured to assert, and pretended to prove that motion is essential to matter. [Footnote to Toland *Letters to Serena*] . . . The essential tendency to motion of every one or of any one particle of matter in this author's imaginary infinite plenum must be [A] either a tendency to move one determinate way at once, or [B] to move every way at once. [A⁺] A tendency to move some determinate way cannot be essential to any particle of matter, but must arise from some external cause because there is nothing in the pretended necessary nature of any particle to determine its motion necessarily and essentially one way rather than another. [B⁺] And a tendency or *conatus* equally to move every way at once is either an absolute contradiction, or at least could produce nothing in matter but an eternal rest of all and every one of its parts" (Clarke, *Demonstration*, III, 1998: 19).[36]

The argument here is more fleshed out than the version in the letter to Butler quoted above. The aim of the argument is to deny that motion is intrinsic to matter and that there is no sufficient reason to think that a particular directionality is built into matter (see [A⁺]). A key step in Clarke's argument relies on what I have called "metaphysical modality" (recall: if such necessity operates in some respect, Y, then we ought to expect Y to be homogenous in relevant ways).

---

[35] Despite Toland's criticism of Spinoza in *Letter* 4, the position is decidedly Spinozistic. See for very good work on Spinozistic active materialism Wolfe (2010).

[36] A helpful referee called my attention to p. 19 of Andrew Baxter's 1745 *An Enquiry Into The Nature of the Human Soul*, Volume 1. 3rd edition. London: Millar. (The first edition dates to 1733.) This is paragraph xiv of Section 1, where Baxter argues that vis inertia is essential to matter and denies that an active power could be essential to matter. Paragraph xiv indeed uses Clarke's reasoning in order to explain that vis inertia and an active power are incompatible. Baxter, who praises Clarke (p. 329), was also a critic of Spinoza (p. 80) and Spinozism. Paul Russell has shown that Baxter and Clarke are among Hume's targets in the *Treatise*; see Russell (2008).

As is well known, Newton distances himself from the idea that gravitational directionality is inherent to matter in his correspondence with Bentley—that's the "Epicurean" conception of motion Newton definitely does not want to be associated with.[37,38] In fact, I have argued that in Spinoza's own outline of his physics, the so-called physical interlude of Ethics II, Spinoza removes directionality entirely from his account of even 'inertial motion' (Schliesser, 2017).

Clarke also offers a further argument [B+]: if motion is intrinsic to matter then there would be no motion. In fact, Clarke repeats the argument later in extremely careful and fine-grained detail in section VIII of the *Demonstration* when explicitly discussing Spinoza's position (1998: 44–45). At first it is not obvious how Clarke's argument is supposed to work. But if Clarke can make the argument work, it's a nice result because it does not appeal to otherwise contested features of his metaphysics or physics (such as the plenum, etc.). I offer the following reconstruction:

Axiom 1 (with Toland & Spinoza): Any body will have tendencies (or determinations) to motion in every direction at once.

Premise 1: For any determination d1 in a body, there will be an opposing determination –d1 that cancels it out.

Therefore, the body does not move (i.e., it will be in perpetual rest).

But why assume Premise 1? One way to go is to claim that it is an implication of Axiom 1.[39] I agree this is so. But it is not entirely clear why all these tendencies must be manifested in such a way as to produce rest.

Clarke is relying here on the tacit assumption of metaphysical modality (recall again: if necessity operates in some respect, Y, then we ought to expect Y to be homogenous in relevant ways). So, absolute necessity requires that each bit of matter must behave identically everywhere and at all times, so all bodies (i) either remain at absolute rest (recall A in Clarke, 1998: 19)

---

[37] There has been a huge controversy in the secondary literature over this correspondence (and what it entails about Newton's views on attraction), but on this precise narrow point, that Newton distances himself from the Epicurean conception of gravity, John Henry is right (Henry, 1994; Henry, 1999); see my chapters 1 and 2 and its postscript.

[38] Both Descartes' and Newton's understanding of "inertial motion" presuppose yet another (thinner) form of directionality. One can understand Spinoza's *conatus* doctrine as taking aim at the metaphysical foundations of this species of directionality.

[39] This was a suggestion of a perceptive referee.

or (ii) have identical absolute uniform motion (recall B in Clarke, 1998: 19). Crucially, both (i) and (ii) entail that *particular* kind of *variety* ought not be possible.

That is to say, and crucially for the chapter's overall argument, this argument explains why one might adopt (A1) in 'the homogeneity argument' that we find in Newton's General Scholium and that Clarke also directs against Spinoza later in the Demonstration (Yenter, 2014). As Clarke puts it "Necessity Absolute in it self, Simple and Uniform, without any possible Difference or Variety: And all Variety or Difference of Existence, must needs arise from some External Cause, and be *dependent* upon it" (*Demonstration* VII, 1998: 35). Leaving aside what Newton did or did not know about Spinoza or Toland directly or from Clarke (who offers detailed criticism of Spinoza's texts), from Clarke's perspective we can say that Toland rejects (i) and disallows God's role, so begs the question; while Spinoza rejects (ii), and disallows God's role, and so also begs the question.

Given that Clarke explicitly recognizes the implications of Spinoza's E1p16–17 that we have noted above—"It might exactly as well be argued, that if God (according to Spinoza's Supposition) does Always necessarily produce all possible Variety of things." (*Demonstration* IX, Clarke, 1998: 49)— we know that Clarke explicitly recognizes that the homogeneity argument fails against the letter of Spinoza's text; even so he clearly thinks it works well against Spinozism, or the spirit of Spinoza's text:

> "I affirm, contradictory to Spinoza's Assertion . . . That there is not the least appearance of an Absolute Necessity of Nature, (so as that any Variation would imply a contradiction. . . . Motion it self, and all its Quantities and directions, with the Laws of Gravitation, are intirely Arbitrary; and might possibly have been altogether different from from what they now are. The Number and Motion of the Heavenly Bodies, have no manner of Necessity in the Nature of the Things themselves." (*Demonstration* IX, Clarke 1998: 49)

Before I conclude, I remark on one otherwise curious feature of Clarke's text. At various points in the *Demonstration* he invokes the authority of Cicero against Spinoza's argument.[40] This is a decidedly odd rhetorical strategy

---

[40] When responding to Spinoza's denial of final causes in the Appendix to Ethics 1, Clarke suggests that the reader can consult Cicero *On the Nature of the Gods*, Galen *De Usu Partium*, Boyle *Of Final*

if one is really focused on Hobbes or Spinoza. But it makes a lot of sense if Toland is your proximate target because, as we have seen, Toland treats Cicero as a key authority. (See chapter 9 for why this is so.)

8.6 Conclusion

In first edition of the *Principia*, Newton is agnostic on activity/passivity of matter. So, in that edition it is unclear if, given Toland's argument, he would have dispensed with a "god of the gaps."[41] But after the *Demonstration*, Clarke's criticism of Spinozism makes passivity of matter the theologically more attractive option. So does the General Scholium entail an embrace of passive matter (as Kochiras, 2011 has argued)? Not necessarily. Some further options available to Newton include (a) attraction is superadded (as Locke suggested; see Henry [1999]); (b) attraction is a relational, contingent quality of matter (see chapters 1–2); (c) remain agnostic (see Smith [2001]). All three alternative interpretations of Newton have defenders (see Ducheyne [2014] for a survey).

Another option available to Newton is to limit metaphysical modality and deny *absolute* necessity: "God is able to create Particles of Matter of several Sizes and Figures, and in several Proportions to Space, and perhaps of different Densities and Forces, and thereby to vary the Laws of Nature, and make Worlds of several sort in several Parts of the Universe" (Query 31; Newton, 1730: 379–380; see chapter 6 with Biener). One thing this passage from the *Opticks* reveals (in addition to a nominalist spirit about laws) is that Newton only applies metaphysical modality to infinite beings (God, space, duration, etc.), and applies a different modality to (local) laws and finite beings.

Causes, and Ray's *Of the Wisdom of God in the Creation* (Clarke, 1998: 51; see also 81 fn. 58). But of these only Cicero is quoted explicitly elsewhere in the *Demonstration*.

[41] There is, a role for a providential God in the first edition of Newton's *Principia*; God is mentioned only once, (Newton, 1999: 814). Bernard Cohen argues that in his *Discourse on Gravity* (1690), Huygens comments on Proposition 8 that it showed what kind of gravity "the inhabitants of Jupiter and Saturn would feel" (Newton, 1999: 219). We know that Huygens had empirical reasons to doubt Newton's argument for the inverse-square law as a universal quality of matter, but Huygens certainly accepted an inverse-square rule for celestial gravity that governed the planets. For more on this see Eric Schliesser, http://digressionsnimpressions.typepad.com/digressionsimpressions/2014/09/huygens-and-newton-on-beauty-and-design.html.

Thus while Newton allows the inference that goes: if some effect Y is universal temporally and spatially, then we can infer or posit as the ultimate cause something that itself is necessary in a way that accounts for, he does not allow the inference to go in the other direction (i.e., a necessary existing Being could have created an alternative reality when it comes to finite beings and the laws that govern particular worlds).[42,43]

However, Newton and Clarke (tacitly) assume that Spinozism applies absolute necessity to infinite beings *and* finite beings (as one would expect from PSR). Even if this assumption is true of Toland (who is consistent), it turns out that it is not true of Spinoza, who bifurcates the realm of 'eternal truths' and the realm of expressed (finite) determinations (E1p28) without ever really explaining why this is so.[44] So Toland's version of Spinozism is more coherent than Spinoza's—and *thereby* a cleaner target for the homogeneity argument than Spinoza would be.

---

[42] A note on Newton terminology: the universe can (but need not) be composed of different worlds; a world is constituted by different kinds of "Particles of Matter" (and forces) that are to be found in it. Newton explicitly allows that the universe could be composed of different worlds that coexist. (See chapter 6 with Biener.)

[43] Hume offers an argument that Anscombe has called "obscure and dealt with rather sketchily" (Anscombe, 1974). She has in mind *Treatise* 1.3.3.4, and that argument shows Hume is aware of the issues discussed in this chapter. In context Hume is denying that causation is a necessary relation (so that for Hume cause and effect *can* be separated). While the direct target of Hume's argument is Hobbes (as it happens, in the very next paragraph, Hume footnotes Clarke) the larger context is Hume taking aim at the principle that whatever begins to exist, must have a cause of existence (*Treatise* 1.3.3.1)—a key premise in the cosmological argument(s) for God's existence. Hume wants to challenge the epistemic status (certainty) of the (cosmological) principle in order to undermine the claim that causation is a necessary relation. So Anscombe's suggestion that Hume is focused on creation here is not without merit. We are in the realm of *ex nihilo nihil fit*, or a weak-ish version of the PSR (akin to the principle of causality). Hume's argument does justice to the principle, which basically says there is no reason why God's existence should start at any given time or at any given place—no place or time has any distinctive quality or property which could single it out unless, and this is the whole point of Clarke's use of "absolute metaphysical necessity," there is a particular cause that does so. But as Hume discerns (and Clarke had exploited) this principle makes it hard to see why the universe is not a pure, homogeneous plenum. So, something special is required to ground any (non-Godlike) particular, nonnecessary existence (which breaks the pure homogeneity of a universe governed by necessity alone). For Clarke it's obvious this means you need something like God's (completely free) will. For Hume, however, it entails there is no necessary connection between creation at a given time and place and the object caused then and there. There is, thus, a sense that from the perspective of absolute necessity, a theory that emphasizes God's will about finite entities is akin to a theory that sees us governed by brute facts. Leibniz would invoke the PSR and identity of indiscernibles in order to explain God's choice and so avoid the unwelcome result (as Anscombe discerns). But the nice thing from Hume's perspective is that from an embrace of (i) absolute metaphysical necessity, and (ii) the existence of particular variety, he can infer that (iii) not all causal relations are necessary.

[44] See also Spinoza's distinction between *natura naturans* and *natura naturata*.

In so far as Newton and Clarke also wish to refute Spinoza they can continue to claim that they have the superior physics and so can tell a far more fine-grained story about cosmogony and the ways the appearances reveal themselves to us (see chapter 3 on Kant's UNH). In so doing they can shift the burden of evidence against Spinoza and Spinozism.[45]

[45] I thank the audience members at the General Scholium held at the University of King's College, 24–26 October 2013, at University of Rochester, the PSA, and UCSD, and special thanks to the generous comments by Steffen Ducheyne, Steve Snobelen, and Scott Mandelbrote. The usual caveats apply.

9

The Posidonian Argument

The Presupposition of Design in Natural Philosophy

9.1 Introduction

The foremost popular defender of Newton in France, Voltaire, published his *Dictionnaire philosophique* or *Philosophical Dictionary* in 1764. In the entry on atheism, Voltaire writes, "Unphilosophical geometricians have rejected final causes, but true philosophers admit them; and, as it is elsewhere observed, a catechist announces God to children, and Newton demonstrates them to the wise" (Voltaire, 1843: 162).

Voltaire defends the legitimacy of final causes.[1] The original entry is very brief. But after the publication of Holbach's (1770) *System of Nature* (published under an assumed name), Voltaire expands it greatly. Voltaire's stance in this post-1770 period is central to Jonathan Israel's interpretation of a split between a radical and moderate Enlightenment, in which Voltaire is treated as a leading figure of moderate Enlightenment (Israel, 2009: 210–219).[2]

Much of section I of the entry is an excerpt of *System of Nature*, "which is in some respects far superior to Spinoza" (Voltaire, 1843: 502). When in Section II, Voltaire, in turn, attempts to refute the Spinozism of the author of the *System of Nature*, Voltaire offers a cryptic claim, "[A] If a clock is not made in order to tell the time of the day, [B] I will then admit that final causes are nothing but chimeras, and be content to go by the name of a final-cause-finder—in plain language, fool—to the end of my life. [C] All the parts, however, of that machine the world, seem made for each other. [D] Some

---

[1] I have used an anonymous translation (Voltaire, 1843) with minor modifications because it is the only unabridged version of this entry in translation that I have been able to find. I have consulted Voltaire (1822) which has a version of the entry which seems to be the basis of the 1843 translation.
[2] Israel does not mention the *Philosophical Dictionary*.

Newton's Metaphysics. Eric Schliesser, Oxford University Press. © Oxford University Press 2021.
DOI: 10.1093/oso/9780197567692.003.0012

philosophers affect to deride final causes, which were rejected, they tell us, by Epicurus and Lucretius" (Voltaire, 1843: 505; capital letters added to facilitate discussion).

To those of us trained on William Paley's watch, and debates over it, Voltaire's argument seems familiar and yet somehow off.[3] It is by no means obvious that denying that clocks are designed to tell time entails that (all) final causes are illusions. It is even less obvious that acknowledging that the clock is designed to tell time entails that (a) final cause/ causes exist. And in particular that the universe is ordered. So why would [A] entail [B]?[4] Or that this is somehow synonymous to the thought that all the universe's parts are mutually useful. So, why think that ([A]->[B]) = [C]? In addition why report that the critics of final causes trace their own lineage back to Epicurus and Lucretius? A few lines down Voltaire argues from authority that "Cicero, who doubted everything else, had no doubt about final causes" (Voltaire, 1843: 505).

Even if one grants that Voltaire was, perhaps, not the most sophisticated philosopher, there seem a lot of suppressed premises before Voltaire's argument can be seen to be remotely plausible.[5] In addition, one may wish to know why Cicero is still treated as an authority this late into the eighteenth century (recall chapter 8)?

As it happens, in Section I of the entry, while Voltaire is quoting the *System of Nature*, there is a passage that helps explain Voltaire's argument. This passage is treated as an objection by the author of the *System of Nature*, and it is an argument endorsed, I think, by Voltaire:

---

[3] As Andrew Bailey suggested to me, one way to represent him looks like he is simply denying the antecedent:

1. If clocks aren't made to tell time, then there are no final causes
2. But clocks are made to tell the time.
3. So there are final causes (from 1 and 2).

[4] In the context of explaining and criticizing "the system of Spinoza," in his entry on God/Gods, Voltaire writes, "For my part, I see in nature, as in the arts, only final causes and I believe that an apple tree is made to bear apples, as I believe that a watch is made to tell the hour" (Voltaire, 1843: 562).

[5] Here is another way to think about Voltaire's argument:

4. If clocks are made to tell the time, then there are final causes in machines with parts that are made for each other
5. Clocks are made to tell the time
6. The world is a machine with parts that are made for each other
7. So there are final causes in the world (from V4, V5, and V6)

But it is unclear where Premise 6 comes from. And whether the final causes in Premise 7 are of the right sort for Voltaire's purposes. (Almost nobody denies humans have purposes.) In a way, the rest of the chapter helps explain and fleshes out this argument.

It will be observed and insisted upon by some, that if a statue or a watch were shown to a savage who had never seen them, he would inevitably acknowledge that they were the productions of some intelligent agent, more powerful and ingenious than himself; and hence it will be inferred, that we are equally bound to acknowledge that the machine of the universe, that man, that the phenomena of nature, are the productions of an agent whose intelligence and power are far superior to our own. (Quoting *System of Nature* in Voltaire [1843: 504])

In this passage, we can see the source of Voltaire's [A] → [B]. But [C] is absent. It is by no means clear why Voltaire thinks this argument, rejected by the author of the *System of Nature*, is so splendid. It is also not obvious why Voltaire would associate this argument with Cicero or Newton, and think it powerful against the Lucretians. In addition, why treat the Spinozist rejection (the system of necessity) as on par with the Lucretian one (the system of chance)?

In this chapter I argue that Holbach's argument, quoted by Voltaire, goes back to Cicero, perhaps even Posidonius. This argument was directed against both the system of chance and the system of necessity. I distinguish three interpretations of this argument: (1) a prima facie interpretation; (2) a 'neglected' interpretation (inspired by Hunter [2009]) and what I call (3) a 'transcendental interpretation.' I show that in the early modern period Cicero's argument was very widely discussed and its significance was not merely as a design argument; it connected scientific practice, even progress in science, to providential final causes. To show this I focus on Boyle and Locke before turning to Newton. In the final section, I return to Voltaire's response to Holbach and show how Voltaire adapts the argument and uses Newton.

9.2 The Posidonian Argument[6]

I introduce an once famous design argument transmitted, and perhaps invented by Cicero's *On the Nature of the Gods* (composed *c* 45 BC).[7] In this section I offer two interpretations of this argument: (1) a prima facie interpretation; and (2) a 'neglected' one.

---

[6] Material in the next five sections has previously appeared in Schliesser (2020).
[7] On the history of design arguments more generally, see, especially, Hurlbutt (1965, [1985]); Manson (2003); and Ratzsch and Koperski (2020).

After discussing other kinds of design arguments, Cicero's Stoic character, Quintius Lucilius Balbus, says:

> But if all the parts of the universe have been so appointed that they could neither be better adapted for use nor be made more beautiful in appearance, we must investigate whether this is chance, or whether the condition of the world is such that it certainly could not cohere unless it were controlled by intelligence of divine providence. If, then, nature's attainments transcend those achieved by human design, and if human skill achieves nothing without the application of reason, we must grant that nature too is not devoid of reason. It can surely not be right to acknowledge as a work of art a statue or a painted picture, or to be convinced from distant observation of a ship's course that its progress is controlled by reason and human skill, or upon examination of the design of a sundial or a water-clock to appreciate that calculation of the time of day is made by skill and not by chance, yet none the less to consider that the universe is devoid of purpose and reason, though it embraces those very skills, and the craftsmen who wield them, and all else beside?
>
> Our friend Posidonius has recently fashioned a planetarium; each time it revolves, it makes the sun, moon, and planets reproduce the movements which they make over a day and a night in the heavens. Suppose someone carried this to Scythia or to Britain. Surely no one in those barbarous regions would doubt that that planetarium had been constructed by a rational process. Yet our opponents [the Epicureans] here profess uncertainty whether the universe, from which all things take their origin, has come into existence by chance or some necessity, or by divine reason and intelligence. Thus, they believe Archimedes more successful in his model of the heavenly revolutions than nature's production of these, even though nature's role is considerably more ingenious than such representations. (*On the Nature of the Gods*, 2.87–88)[8]

There are many arguments from design. Let's dub the main one articulated in the quoted passage by Balbus, the "Posidonian argument." It deploys—in David Sedley's felicitous phrase—the "structural resemblance of state-of-the-art-planetary mechanism to the celestial globe." (Sedley, 2007: 207) For, it relies, to simplify, on the supposition that everybody (even barbarians) will grant that if a sophisticated complex machine, which is a

[8] (Cicero, 1978: 78); I have made some minor modifications. For the Latin Cicero (1958: 763–769)

scientific representation of nature, is the product of intelligent design, then (once granted) it turns to suggest that the represented complex (beautiful, well-adapted, etc.) machine must itself also have an intelligent author. The represented world need not be itself a mechanism (Sedley, 2007: 207).

I stipulate that despite the presence of minor variants, we are dealing with the Posidonian argument when (i) a presentation of a design argument is (ii) accompanied by a reference or allusion to Archimedes' planetary sphere; (iii) a reference to Posidonius' portable planetarium (iv) and some uneducated foreigner (ignorant barbarian, savage, etc.).[9] If all four of these are present in later texts we can be sure that the (original) source is Cicero, especially (v) if Cicero is mentioned.

However, there are variants of this design argument in which either (ii) or (iii) is dropped or different machines are used as examples. In these cases we may still be dealing with versions of the Posidonian argument (especially if Cicero is mentioned or knowledge of him can be presupposed). So, for example, in Voltaire's entry on final causes, in the context of responding to the *System of Nature*, (i), (iv), (v) are explicitly mentioned.

Of course, sometimes technical works on planetariums and astronomy (and their history) may note Cicero's mention of Posidonius' or Archimedes' planetarium without intending to offer a design argument (which is why [i] is necessary condition).[10]

Cicero seems to have been greatly fascinated by Archimedes' sphere because he mentions it in a number of works. The sphere (a kind of planetarium) seems to have been part of the Roman war loot of the conquest of Syracuse in 212 BC. It was in possession of the consul Marcellus (the grandson of the Marcellus who had sacked Syracuse), when he (the consul) figures as a character in Cicero's *De Re Republica*. In Cicero's presentation, Archimedes' sphere is said to be able to represent accurately eclipses of the sun (Jones, 2017: 130).

Posidonius was a famous Stoic philosopher active on Rhodes, where he was almost certainly met by Cicero. It is very likely that Cicero encountered

---

[9] According to Sedley (2007, 207, n.6), Archimedes' sphere is "likely to be the original Stoic example," and the naming of Posidonius' a "localizing touch" (Posidonius was one of Cicero's teachers). In addition, see Kidd, 1988: 74–75 and Berryman, 2010: 150–155. Details on Archimedes' sphere can also be gleaned from (among others) Cicero (e.g., *Republic* 1.21–22; *Tusculan Disputations* 1.63), *Sextus Empiricus*, M. 9.115, and Proclus (*A Commentary on the First Book of Euclid's Elements*, Book I, Chapter XIII). (All available to early modern readers.) For more such references see http://www.math.nyu.edu/~crorres/Archimedes/Sphere/SphereSources.html.

[10] See, for example, this entry of Beeckman's diary in 1629: http://adcs.home.xs4all.nl/beeckman/IIIv/1629v.html#105, or Huygens' description of his planetarium: http://dbnl.nl/tekst/huyg003oeuv21_01/huyg003oeuv21_01_0110.php?q=Posidonius. (Huygens, 1944: 588).

a portable planetarium designed by Posidonius. But until the discovery of the Antikythera mechanism,[11] it was not entirely clear what Cicero may have had in mind with Posidonius' portable planetarium. But thanks to the discovery and reconstruction of the Antikythera mechanism, we now know that the craftsmanship and the scientific ingenuity of these portable planetariums were true marvels.[12] What is especially noteworthy for present purposes, is that they represented many calendric and astronomical features in addition to orbits and eclipses.

Before I offer a modest reconstruction and evaluation of the argument,[13] I note a few features related to (i-iv). Balbus suggests that there are three ordering principles of the universe: chance, necessity, or intelligence.[14] The chance option is historically associated with the Epicurean position, an identification that continued in subsequent history through the eighteenth century (see chapters 3 and 8 above). Sometimes I treat, by stipulation, this option as synonymous with "brute fact."[15] Sometimes this position is treated as "blind chance" ([Berkeley, PHK 93]; it's called "blind" because of the denial of providence and final causes).

The necessity option (which also denies final causes), I'll associate with Spinozism.[16] Often it's called "fatalism" or "fatal necessity" (e.g., Berkeley, PHK 93). Thanks to Cudworth and Bayle, Strato became the ancient figure associated with the system of necessity in the early modern philosophy.[17] It is not altogether unlikely that the system of necessity was associated with Stoicism itself in Cicero's day.[18] Finally, Balbus asserts that a divine mind is

---

[11] The literature on it is huge. In addition to Jones (2017) and Solla Price (1974) see de Freeth (2006). I first learned about the connection between the Antikythera mechanism and Cicero's discussion of Posidonius from Marchant (2009).

[12] The Antikythera mechanism dates from a slightly later period than the mechanism presumably known to Cicero.

[13] For a very careful and illuminating rational reconstruction of Cicero's argument, see Hunter (2009): 235–245. Hunter's aim is to show how Cicero's argument is an instance of a "nontrivial valid argument leading from the admission that certain artifacts require a designer to the conclusion that certain natural entities, or the natural world as a whole, also require one" (236).

[14] I speak of "ordering principle," because for a Stoic it would be unintelligible to allow that the universe could be caused by nothing. I return to this in the text when I discuss the role of the (PSR) in the argument.

[15] I treat this as stipulate because this association of chance with brute fact is not self-evident to all contemporary metaphysicians. Traditionally, the Epicurean system of chance involves arbitrary variation, and this is why it is self-evident to most of the philosophers I describe in the chapter. By contrast, contemporary non-Humeans often treat necessity as brute. I thank Andrew Bailey for discussion about the material in this note and the following one.

[16] Not all contemporary positions that embrace necessity deny final causes. But in the early modern period the system of necessity always refers to Spinozistic denial of providence.

[17] The systems of Hylozoistic atheism (Cudworth's polemical description of Strato) and divine fate (Stoics) are run together in the wake of the Spinoza controversy (Brooke, 2006: 391ff).

[18] Divine fate works through an open-ended series of causes. While most Stoics insisted they believed in a providential order, it is no surprise that "According to Epicurus, Letter to Menoeceus,

the final option. He clearly associates the products of a divine mind with a providential order. Often I'll simply use 'God' to refer to this option. Without explanation, Balbus seems to think that chance, divine mind, and necessity exhaust the genuine possibilities.

If one takes Balbus' exposition at face value,[19] the argument relies on the analogy between the apparent beauty of the well-adapted, manufactured (etc.) machine and the beauty and well-adapted nature of the heavens to infer an intelligence behind the universe.[20] Moreover this argument may be read to rely on a further aesthetic premise, "nature's attainments transcend those achieved by human design,"[21] and in conjunction with the empirical assertion that human craft produces nothing without reason and art to argue for the existence of a higher excellent intelligence behind nature's order.[22] Here, I do not foreground the significance of this analogy, but it is not irrelevant (Berkeley PHK, 106–109; Hume EHU XI; Schliesser, 2020).

With that in place, I offer a rational reconstruction of Balbus' argument. My reconstruction is not meant to be exhaustive (again I drop the role of analogy), but I intend for it to capture the gist of the implied argument and for it to be a valid argument.

(1) All of nature's parts are ordered; they exhibit apparent design and beauty.

(2) Artificial, complex machines are the product of rational design.

---

preserved in Diogenes Laertius at 9.133 (LS 20A), the Stoic doctrine of fate would involve an 'inexorable necessity." Quoted from Brouwer (2019: 36). See also Cicero, *De Divinatione* 1.125–126.

[19] For example, Hicks, 1883: 64, quotes the same passage at length but without comment as an illustration of how Cicero anticipates modern arguments. Without analysis, he treats the argument as relying fundamentally on "analogy."
[20] Hunter nicely describes the "logical motor" of "standard" design arguments as follows: it "motivates its conclusion with a 'how much more so' question" (Hunter, 2009: 236). Hunter does not attribute the standard version to the Ciceronian passage under discussion because he wishes to interpret the passage as exemplifying a 'new' argument. But the Ciceronian passage can plausibly be interpreted as articulating both the 'standard' as well as the 'new' arguments (as Hunter admits "even the 'how much more so' comparison of the standard version is not entirely absent" [Hunter, 2009: 240]). Nothing hinges on my disagreement with Hunter here.
[21] One need not understand "excellence" (*perfecta*) in strictly aesthetic terms. One can interpret it in terms of magnitude, size, or power (etc.). But this is not to deny the presence of aesthetic elements in Cicero's argument ("beautiful"). These aesthetic issues, especially in terms of the inhabitants of planets of other solar systems, matter a lot to Clarke and Newton (see the "General Scholium" and chapters 3 and 8).
[22] Jantzen (2014: 36–40) treats Balbus' argument primarily as an argument by analogy. In the argument by analogy, nature (represented by the machine) is also a machine (or machine-like). Jantzen offers two more interpretations of Balbus' argument: in one it is assimilated to another argument for the improbability of order, in the other it is assimilated to a Socratic argument from purpose. Given the structure of Cicero's text it is indeed likely that the argument for the improbability of order is in the background of Balbus' exposition.

(3) A planetarium is a complex machine or concrete representation of the heavens.

(4) Nature's complexity is greater than the complexity of a planetarium.

(5) Posidonius' planetarium is a successful representation of the heavens.

(6) Even a barbarian will acknowledge that Posidonius' planetarium is a complex machine (when she is confronted by it)

(7) So, even a barbarian, who correctly accepts premise (2), will acknowledge that Posidonius' planetarium is produced by rational design.

(8) Nature's order is caused either by chance, or by necessity, or by a divine mind.

(9) Nature's order is not caused by necessity or chance because it is impossible that something less complex can be the product of rational design while the more complex thing (i.e., nature) it represents is not.

(10) For if you thought otherwise, then the designer of the less complex concrete model would be superior to the cause of the thing represented by the model (nature). That is, if nature's order has greater complexity than the clock and the clock is a product of design then so is nature's order.

(11) Therefore, nature's order is caused by the divine mind's rational design.

Strictly speaking the conclusion does not require premises (2) through (7). But the particular appeal of this argument rests on these premises. The obvious weak spot in the argument is (8): Balbus assumes without argument that (8) is exhaustive. If it is not exhaustive, then the argument is obviously not sound. In addition, I have formulated (8) so as to take the *origin of nature* off the table. That's because due to the embrace of a version of the PSR—*ex nihilo nil fit*—[23] for a Stoic, and most ancients,[24] nature must be caused. That is, there is a suppressed premise in the argument:

---

[23] Some readers may wish to distinguish between causal principles and the Principle of Sufficient Reason (PSR). In particular, if they associate the PSR with Leibniz's version then they are inclined to assume that the PSR involves commitment to final causes or what is best in the way (other) causal principles (such as "ex nihilo nil fit") do not and also involves commitment to some kind of rational connection between the item that explains and the thing explained. I thank Janiak for discussion. By contrast, I follow Melamed and Lin (2016), in treating the PSR as a "family of principles" some weaker or thinner than others.

[24] Strictly speaking, the Stoics may not have embraced *ex nihilo nihil fit*, but they did embrace various causal principles that clearly rule out uncaused motion and, in some instances, un-caused existence. For very helpful discussion, see section 1.3.3 in Bobzien, 1998. Given their theology, it seems

(8⁺) Given the PSR, Nature must be caused.

Obviously when we return to the early modern, Christian context, the status of (8)/(8⁺) and the acceptance of PSR which it presupposes needs to be investigated.

Even if one were to grant that all of nature is beautiful, not everybody will naturally agree with (1)—as Diderot would argue, defective animals (so-called monsters) are born not infrequently (Wolfe, 2005). This suggests that not all individual parts of nature are best adapted for use. How to think about (1) in light of such naturally occurring imperfections is no easy matter. There are ways to account for nearby versions of (1) in which some apparent imperfections turn out to be very beneficial in light of the overall beauty and aptness of the universe.[25] Even so, it is no surprise that modern presentations of so-called deductive, abductive, and inference to the best explanation design arguments tend to require only that "some things in nature . . . exhibit exquisite complexity" (Ratsch).[26]

So the universal quantifier—"all of nature's parts are ordered"—in premise (1) also seems rather strong. But as I noted in chapter 8, Newton endorses a version of it in the "General Scholium" (a text very familiar to Voltaire), "All that diversity of created things which we find, suited to all different times and places."[27] As we have seen in chapter 8, Newton goes on to say that *such* diversity "could arise from nothing but the ideas and will of a Being necessarily existing" (Newton, 1999: 942). So in the "General Scholium" something like premise (1) is used to stipulate a designing God.

In addition (1) anticipates, and has at least a strong family resemblance to Voltaire's [C], "All the parts, however, of that machine the world, seem made for each other." I leave aside for the moment to what degree Newton embraces the idea that the world is a machine or accepts Cicero's argument. What is

the question of the origin of cosmic existence does not quite arise in the crisp way it does for Lucretius or Aristotelian thinkers, and later Christians (see Bobzien, 1998: 412). Cicero's De *Divinatione* 1.125–126 is worth reading on fate. See also Lucretius on his principle at *De Rerum Natura*, 1.149–156. I thank Eric Brown for helpful discussion.

[25] Leibniz is fond of such arguments.
[26] See Ratzsch and Koperski (2019). Of course, if one is in the grip of the PSR, one may well wonder why it's only some parts that appear as designed.
[27] By drawing on Motte's (1729) translation, I have adapted Cohen & Whitmann's translation of the latin [*Tota rerum conditarum pro locis ac temporibus diversitas*], <https://newtonprojectca.files.wordpress.com/2013/06/newton-general-scholium-1729-english-text-by-motte-letter-size.pdf>, accessed March 26, 2021. Motte obscures that all things are explicitly said to be created, but better than the modern translation captures the functionality of everything.

clear is that Voltaire embraces something like (1) and, if we go back to the argument quoted in the *System of Nature*, he accepts nearly all of Balbus' other premises, although some are suppressed.

While (1) is characteristic of arguments *from* design, one may well wonder if (1) is really required in Balbus' version of the argument. For one can derive the conclusion (1) without it. That is to say, the real work in this reconstruction of the argument is not being done by the existence of apparent design (1), but by (a) the (partial) morphism between the concrete model and reality, and (b) the relational complexity of model and reality (that is, (3), (4), and (10)—see Hunter [2009]). In modern terms, what (a–b) capture is that the machine is a simulation.[28]

Versions of the argument that include (1) I'll treat as "*prima facie* versions of the Posidonian argument." Versions of the Posidonian argument that drop (1) I'll call the "neglected Posidonian argument."[29] As should be clear now, in the entry on final causes, Voltaire explicitly quotes the neglected argument (in the objection discussed by the *System of Nature*), but also seems to endorse the prima facie version of the argument. I return to this in the final section of this chapter (and also sort out some other features of Voltaire's presentation).

Of course, some modern readers may also think that there is something fishy about premises (9–10). Surely some artifices are better than their natural counterparts? One can grant that (perhaps easier now in age of precision tools), and still think that (9–10) can survive scrutiny. In fact, I would argue that Balbus' point here is a more subtle (see also Hunter [2009]). Even the very best simulations or of reality are imperfect because they must leave out or abstract away from some of the intricacy of nature. This point is basically stipulated in (4). This is not to deny that one can imagine successful concrete representations of nature where the simulation or is (say) unnecessarily more complex than nature (e.g., by adding an extra gear); but, leaving aside questions about to what degree such complexity must be functional and efficient, even *that* concrete model will leave out other bits of the machinery of nature or the phenomena it tracks. So the stipulation (4) can be defended and survive scrutiny.

But that it is stipulated suggests that (9–10) may be dispensable. Here's a thought: premise (3) relies on the idea that the model inherits its features,

---

[28] David Chalmers makes the same point (forthcoming) about the Antikythera mechanism.
[29] In deference to the spirit of Hunter (2009). The details of his reconstruction different from mine.

or at least many of its significant ones, from reality *in virtue* of the effort in representing reality. That is to say, the particular complexity it has may be built by humans, but it is meant to track nature. The particular machine is built by humans, but—and this captures the intuition behind premise (10)— the morphism it exhibits with reality is not original in the human designer, but derived from nature (Hunter, 2009).

Note two features of the reconstruction of the prima facie version of the argument. First, the association between a divine mind and providence has been silently dropped. Of course, if one keeps (1) in the argument then one gets such providence for free; the world is product of design and full of 'fit' and beautiful. If one drops (1) from the argument, then the conclusion can, at least in principle, be separated from the existence of providence (because the divine mind need not be benevolent, etc.). Of course, for most folk interested in the argument the very idea of a world caused by a divine mind naturally entails providence. For such folk, a divine mind without such providence slides into Spinozism (which posits a God without concern for 'creation').

Second, note that due to technological developments (2) can increasingly seem less plausible, if we insist that it is only humans can build artificial machines. Artificial machines that design and create other artificial machines (can) exist now.[30] It is conceivable, even likely, that in the fullness of time such artificially designed machines by other machines will prompt people, or even the machines themselves, to assert that "such exquisitely crafted machines could never have been designed by feeble creatures like humans!"[31] Of course, as stated (2) makes no mention of humans (even if the implied referent may well be humans), and, in fact, the argument would work just as well if one were to think, as Paley suggests, of machines or robots engineering other machines. (I think that's nifty.)

I offer one final observation on the reconstruction of the prima facie argument. Premise (6) is a bit redundant if you accept premise (3). And premise (7) is redundant if you simply accept premise (2). In both cases a rhetorically arresting thought experiment is used to provide evidence for something one is likely to accept anyway. So, one may consider premises (6–7) dispensable.

Having said that, one role for the 'outsider' (e.g., foreigner, barbarian, savage, uneducated) in (6) seems to be to represent the judgment of a kind of untutored human nature uncorrupted by previous exposure to, if not philosophy,

[30] William Paley considers the possibility of an infinite chains of watches producing watches (Paley, 1808: 12–13).
[31] The idea was inspired by Dennett (2017).

then certain strands of sophisticated theism.[32] So her judgment about a feature of the machine looks like genuine independent evidence that the feature is really there. It shows that the judgment could be independent of any previous commitment to the (intermediate or ultimate) conclusions of the argument, and so deflects the charge that the premise in question is question begging.[33]

So to sum up the situation so far: the prima facie version of the Posidonian argument, which is an argument from design (and so contains [1]), contains within it—if we drop (1), (6), and (7)—a very clever argument for the existence of God. The 'neglected' Posidonian argument goes like this:

A. Artificial, complex machines are the product of rational design.

B. A planetarium is a complex machine or concrete representation or simulation of the heavens (and the complexity of the planetarium that it has in virtue of being a model of nature is derived from nature).

C. Nature's complexity is greater than the complexity of a planetarium.

D. Posidonius' planetarium is a successful representation of the heavens.

E. Nature's order is caused either by chance, or by necessity, or by a divine mind.

F. Nature's order is not caused by necessity or chance because it is impossible that something less complex can be the product of rational design while the more complex thing (i.e., nature) it represents is not.

G. Therefore, nature's order is caused by the divine mind's rational design.

As Hunter notes, this 'neglected' version of the Posidonian argument (so without [1]) does not become obsolete through the rise of Darwinism or, as I have suggested, even robots.[34] That's because it does not make any explicit claim about the appearance of design in nature. It's not an argument *from* design, but an argument *to* design. To anticipate: something like this neglected argument is the argument that Voltaire presupposes in Section II of his entry on final causes, and that he had quoted in the passage cited from Holbach's *System of Nature*.

---

[32] I doubt we're to assume that the "barbarians" are naturally atheist altogether.

[33] In this sense the argument would be an improvement over some of the question-begging features in Voltaire's argument at the start of this chapter. I thank Andrew Bailey for discussion.

[34] One might think that natural selection is a fourth source of order in E. I think that is Darwin's view. Huxley, by contrast, suggests that natural selection might itself be a consequence of necessity. For some informal discussion see Schliesser (2018a and 2018b). I thank Marij van Strien and David Haig for discussion.

Of course, there is a reason why this 'neglected' argument is neglected. A–D are dispensable. It's also not obviously sound because (E) may be incomplete—that is, there may be alternative ways to explain the origin of nature's complexity or order without a deist God, or the argument for necessity and/or chance can be made more robust and seem more explanatory than appeal to a divine mind. And (F) may be thought begging the question. In the next two sections I show that there was widespread familiarity with the (prima facie version) Posidonian argument in the early modern period.

9.3   Early Modern Posidonian Arguments

Most early modern learned readers would have been familiar with the Posidonian argument, because Cicero and *On the Nature of the Gods* continued to be read and quoted approvingly throughout the early modern period.[35] In his *Philosophical Dictionary*, in his entry on Cicero, Voltaire includes it in the "two noblest works that ever were written by mere human wisdom" (Voltaire, 1843: 305; the other is the *Tusculan Disputations*). Voltaire clearly admires Cicero as a (skeptical) critic of superstition (Voltaire, 1843: 170–171 in the entry on augury).

Even when he is critical, in *Evidences of Natural and Revealed Religion*, Samuel Clarke calls Cicero "that great Master" (Clarke, 1732: 209) and "the greatest and best philosopher, that Rome, or perhaps any other nation has ever produced" (Clarke, 1732: 292–293). It's not just Cicero's moral philosophy that is read; in drawing on Newton, as we have seen in chapter 8, Clarke quotes approvingly from *On the Nature of the Gods* in *A Demonstration* (e.g., Clarke, 1705: 110, where Toland is being mocked, and 229, where, as I discuss later, the Posidonian argument is explicitly discussed).

There were, in fact, lively, high-profile debates in the early modern period over Cicero's true philosophical views in *On the Nature of the Gods* (which anticipate the debates over Hume's views in the Dialogues—itself a work clearly modeled on Cicero's *On the Nature of the Gods*. [Sessions, 2002: 30–31 and Battersby, 1979]). For example, in his response to Collins' notorious (1713) *Discourse on Free Thinking* (which includes an epigraph from *On the Nature of the Gods*), Richard Bentley, a very serious classicist, does not merely criticize Collins' arguments and positions, but has a lengthy analysis of how to

---

[35] For the origin of this in Renaissance thought, see Glacken (1967: 54ff and 376).

interpret Cicero properly (Collins had treated Cicero as a fellow free-thinker; see Bentley [1734] [1713]: 246; for broader context Stuart-Buttle, [2019]).

Indeed, the prima facie version of Posidonian argument explicitly shows up, for example, in a number of philosophical works, such as the second Appendix (written by John Maxwell) to Cumberland's *A Treatise of the Laws of Nature*, Section 6. "If any Man" (says Cicero *On the Nature of the Gods* 2.13) "should carry to Scythia such a Sphere as Posidonius made, that doth but represent the Motion of the Planets, who amongst these Barbarians could doubt but that such a Sphere was made by Reason?" In Maxwell's hands the argument is directed against the Spinozistic doctrine that "things were not left to the blind Agitation of Matter, (which cannot Model, Distinguish, Proportionate, nor do things in Number, Weight, and Measure, nor do them so well as the greatest Reason can do no better)."[36] In Maxwell, the Posidonian argument is introduced by way of lengthy extracts about design found in nature from Newton's "General Scholium" and the Queries to the *Opticks* (especially in Maxwell's Section 2) as well as citations from Shaftesbury and Wollaston.[37] Rather than multiplying examples, I turn to Boyle and Locke, who are crucial for the material I want to discuss in Newton's "General Scholium" and Voltaire.

9.4 Boyle and Locke

For the most famous version of the prima facie Posidonian argument during the early modern period is probably to be found in Boyle, who frequently "compares the world as a whole with a clock." (Durland, 2015) This argument does more than just defend the "corpuscular hypothesis." Here's one of Boyle's celebrated instances of the analogy:

> '[T]is like a rare Clock, such as may be that at Strasbourg, where all things are so skilfully contriv'd, that the Engine being once set a Moving, all things proceed according to the Artificers first design, and the Motions of the little Statues, that at such hours perform these or those things, do not require, like those of Puppets, the peculiar interposing of the Artificer, or any Intelligent Agent imployed by him, but perform their functions upon

---

[36] Cumberland, 1727: *Appendix II: A Treatise concerning the Obligation, Promulgation, and Observance of the Law of Nature, CHAPTER II*. The Promulgation of the Law of Nature, §6.

[37] In the second appendix Maxwell basically recapitulates Clarke's arguments against Spinoza and Hobbes on providence in *A Demonstration*. (Maxwell's first Appendix reproduces Samuel Clarke's defense of the immateriality of a thinking substance against Toland.)

particular occasions, by vertue of the General and Primitive Contrivance of the whole Engine." (*A Free Enquiry*; Boyle, 2010: 448)

Boyle uses the world-clock analogy in order to drive home the idea that God's *general* providence works by general and original (this captures the sense of Boyle's "primitive" in light of the "first design") causes.[38] Something like this analogy is presupposed in Voltaire's comment in his entry on final causes.

It presupposes Boyle's voluntarist treatment of God's agency (Henry, 2009; cf. Harrison [2002]). Boyle's argument seems familiar to us, educated as we are to see Darwinism, in part, as a response to Paley's watch (e.g., editor's comment at Boyle, 1991: xvi). For my present purposes I want to focus on the Strasbourg clock because Boyle uses it elsewhere in ways that reveal his debt to Cicero (or some intermediary source).

For, on the *Usefulness of Natural Philosophy*, Boyle adds two claims to his treatment of the Strasbourg clock:[39] (i) "the various motions of the wheels and other parts concur to exhibit the phenomena designed by the artificer in the engine . . . "; (ii) "and might to a rude Indian seem to be more intelligent than Cunradus Dasypodius himself." (Essay IV quoted from Boyle, 1991: 160; Conrad Dasypodius was the designer of the famous Strasbourg clock.) Here I ignore Boyle's low regard, alas, for the intellectual achievements of native Americans. The phenomena exhibited by the (second) Strasbourg clock were primarily astronomical—that is, it was a gigantic, massive planetarium in which heavenly motions and phenomena were faithfully represented.[40] In context, Boyle is explicitly rejecting local final causes (and action at a distance). That Cicero is probably his inspiration is confirmed not just by the great similarities in tropes, but also by the fact a few pages later he explicitly cites On the Nature of the Gods (for different purposes, see Boyle [1991: 166]).

Okay, with that in place let's turn to Locke's *Essay*. Locke's high regard for Cicero is signaled from the start because he puts an epigraph from *On the Nature of the Gods* 1.84 on the frontispiece of his *Essay Concerning Human Understanding* (Stuart-Buttle, 2019: 22) And this fits a much wider engagement by Locke with Cicero's moral philosophy (chapter 1 of Stuart-Buttle, 2019).

---

[38] To modern eyes it is tempting to read the 'general' causes as laws of nature. But in this passage Boyle could also be relying on the traditional idea that the clock has a real essence (the hidden from sight "contrivance") from which effects follow in exception-less fashion (such a world would also be amenable to description by laws of nature, of course).

[39] In his works Boyle uses the Strasbourg clock to offer many different kinds of arguments to design. Many of these are logically distinct from the Posidonian Argument.

[40] Check out this wonderful image: https://en.wikipedia.org/wiki/Strasbourg_astronomical_clock#Second_clock

Boyle's version of the Posidonian argument (astronomical clock, ignorant foreigner) was familiar enough such that Locke would offer his own variant of a discussion of the Strasburg clock (without mention of Boyle) at *Essay* 3.6.3 and 3.6.9. but with "a gazing countryman" and no Indian. I quote from the version at 3.6.9:

Not the real essence, or texture of parts, which we know not. Nor indeed can we rank and sort things, and consequently (which is the end of sorting) denominate them, by their real essences; because we know them not. Our faculties carry us no further towards the knowledge and distinction of substances, than a collection of those sensible ideas which we observe in them; which, however made with the greatest diligence and exactness we are capable of, yet is more remote from the true internal constitution from which those qualities flow, than, as I said, a countryman's idea is from the inward contrivance of that famous clock at Strasburg, whereof he only sees the outward figure and motions. There is not so contemptible a plant or animal, that does not confound the most enlarged understanding. Though the familiar use of things about us take off our wonder, yet it cures not our ignorance. When we come to examine the stones we tread on, or the iron we daily handle, we presently find we know not their make; and can give no reason of the different qualities we find in them. It is evident the internal constitution, whereon their properties depend, is unknown to us: for to go no further than the grossest and most obvious we can imagine amongst them . . . The workmanship of the all-wise and powerful God in the great fabric of the universe, and every part thereof, further exceeds the capacity and comprehension of the most inquisitive and intelligent man, than the best contrivance of the most ingenious man doth the conceptions of the most ignorant of rational creatures. Therefore we in vain pretend to range things into sorts, and dispose them into certain classes under names, by their real essences, that are so far from our discovery or comprehension. A blind man may as soon sort things by their colours, and he that has lost his smell as well distinguish a lily and a rose by their odours, as by those internal constitutions which he knows not. (Locke, *Essay* 3.6.9)

We know that the epistemic modesty about hidden, real essences appealed to Voltaire. And that Voltaire interpreted Newton in the earlier *Lettres Philosophiques* in Lockean fashion (Fowler, 2017). Recent scholarship has since Stein (1990) quite rightly emphasized the nontrivial differences between Locke and Newton (but see Ducheyne, 2011: 257).

Even so, it is no surprise that Voltaire is inclined to interpret Newton along Lockean lines. The "General Scholium" clearly echoes Locke's 3.6.9:

> As a blind man has no idea of colors, so we have no idea of the ways in which the most wise God senses and understands all things. He totally lacks any body and corporeal shape, and so he cannot be seen or heard or touched, nor ought he to be worshiped in the form of something corporeal. We have ideas of his attributes, but we certainly do not know what is the substance of any thing. We see only the shapes and colors of bodies, we hear only their sounds, we touch only their external surfaces, we smell only their odors, and we taste their flavors. But there is no direct sense and there are no indirect reflected actions by which we know innermost substances; much less do we have an idea of the substance of God." (Newton, 1999: 942)

Even though Newton eschews Locke's language of "real essences," preferring "innermost substances," Newton compares our lack of knowledge of innermost substance with our lack of understanding of God's understanding (and his substance) with the blind man's lack of knowledge of color just as Locke does.

Okay with that in place, let's turn to some distinctive features of *Essay* 3.6.9. Echoing Plato, Aristotle, and Descartes (Daston and Park, 1998; Schliesser, 2005b), Locke allows that science begins in wonder. Familiarity with the objects of inquiry and their relations to other objects of inquiry removes our wonder but does not produce, infamously, knowledge of hidden, inner real essences. Here Locke anticipates the Berkeleyan and Kantian move in which empirical science becomes knowledge of (the law-governed) relations among the phenomena (Ducheyne, 2006b). This is familiar enough.

My reason to return to 3.6.9 is that Locke also offers a (rather skeptical) version of the prima facie version of the Posidonian argument. There is (i) the "famous clock at Strasbourg," which was also a planetarium; there is (ii) a "countryman" who stands in for the possible knowledge of "the most ignorant of rational creatures"; and (iii) "The workmanship of the all-wise and powerful God." What is unusual about Locke is that here he deploys it not to argue for the existence of God. Rather he *assumes* the validity of the prima facie version of argument *in order* to suggest that our best science's ignorance of hidden workings of nature *exceeds* the ignorance of the least capable person of the internal mechanism of a complex clock/planetarium ("the workmanship of the all-wise and powerful God in the great fabric of the universe, and every part thereof, further exceeds the capacity and comprehension of the most inquisitive and intelligent man, than the best contrivance of the most ingenious man doth the conceptions of the most ignorant of rational creatures").

As an aside, on the frontispiece of Locke's *Essay*, alongside the epigraph from Cicero's *On the Nature of the Gods* is the following passage: "As thou knowest not what is the way of the Spirit, nor how the bones do grow in the womb of her that is with child: even so thou knowest not the works of God, who maketh all things"—Eccles. 11. 5. So we can say that Essay 3.6.9 is central to the claims that Locke wishes to convey in the *Essay*.

So rather than using the prima facie version of the Posidonian argument to argue for final causes or the existence of Intelligence, Locke assumes widespread knowledge of (Boyle's version of) the Posidonian argument in order to argue for a species of epistemic humility. And while Newton transcends the limits of knowledge thought possible by Locke (Stein, 1990), Newton echoes this very passage (Essay 3.6.9) to illustrate his own views on the limits of knowledge in the "General Scholium." But in the material I quoted from the "General Scholium" the Posidonian argument itself seems absent. Before I complete my treatment of Newton (and Voltaire), I turn to yet a different version of the Posidonian argument which can help explain, I think, why Locke took the Posidonian argument for granted.

9.5 The Transcendental Version of the Posidonian Argument (Clarke)

Recall that the 'neglected' version of the Posidonian argument goes like this:

A. Artificial, complex machines are the product of rational design.
B. A planetarium is a complex machine or concrete representation of the heavens (and the complexity of the planetarium that it has in virtue of being a model of nature is derived from nature).
C. Nature's complexity is greater than the complexity of a planetarium.
D. Posidonius' planetarium is a successful representation of the heavens.
E. Nature's order is caused either by chance, or by necessity, or by a divine mind.

F. Nature's order is not caused by necessity or chance because it is impossible that something less complex can be the product of rational design while the more complex thing (i.e., nature) it represents is not.

G. Therefore, nature's order is caused by the divine mind's rational design.

Recall that (F) seems to beg the question. It primarily seems to be present to clarify or make explicit (E). Recall also that the PSR seems to be presupposed in E. I made no effort to suggest that there is an important link between the PSR and F. But in the early modern period, they get linked together in a very famous passage:

> It follows from this both that (a) something cannot arise from nothing, and also (b) that what is more perfect—that is, contains in itself more reality— cannot arise from what is less perfect. And this is transparently true not only in the case of effects which possess (what the philosophers call) actual or formal reality, but also in the case of ideas, where one is considering only (what they call) objective reality. (Descartes, *Meditations on First Philosophy*, AT VII 40–41; CSM II 28–29; quoted from Descartes, 1988: 91; letters added to facilitate discussion.)

In the context in which this passage is taught and discussed in the scholarly literature, the main interest is in connecting the two causal principles (a and b) to Descartes' theory of ideas; the relationships actual, formal, and regarding objective reality; and understanding how these connect to his argument for the existence of God. In the *Meditations*, Descartes treats (a) and (b) as clearly linked.[41]

Descartes claims to derive or infer (a and b) from another principle: "Now it is manifest by the natural light that (c) there must be at least as much (reality) in the efficient and total cause as in the effect of that cause." This (c) is pretty much treated as axiomatic by Descartes. It has its roots in neo-Platonic conceptions of emanation which merged with Aristotelian ideas about efficient causation. The underlying idea is that an effect receives its qualities from a cause and these (cause and effect) are conceived hierarchically (Lin, 2004: 32; see also chapter 5).

---

[41] It seems one can derive (b) from (a). If something with more reality arose from something with less reality, then the quantity (more reality or less reality) would have arisen from nothing. I thank David Gordon for this discussion.

I'll call (a–c) "Descartes' causal principles." The relationship between (a) and (c) is easy to discern (although one may wonder which one is truly the more fundamental principle). In addition, it's clear that with his embrace of (a)/(c) that Descartes is clearly committed to something akin to the PSR. That (b) follows from (c) is also pretty clear.

Obviously (b) is not identical to (F) in the neglected Posidonian argument. But if we are allowed to treat complexity as a species of perfection, as Descartes does in the First Replies (Dennett, 2008: 337–338), then (b) underwrites (F). For (b) and ([C], from which [F] is derived) rely on the same intuition that the model or copy derives its key, effective properties, the ones that track or are morphic with reality, from reality. And reality is more perfect or nobler than a copy.[42]

Of course, these reflections entail that (A–B) are misleading if we think of (to use Descartes' terminology) the total cause of a complex machine that tracks nature (say a portable planetarium) as limited to the human artificer that designed and built the concrete model. *Crucially, on the interpretation pursued here, the total cause of a planetarium includes something of, or presupposes, the natural order which it tracks.*

Now how much of nature's order is presupposed in a concrete model or simulation of it is worth careful consideration. The relationship between models and reality is now a burgeoning literature in the philosophy of science (Frigg and Hartmann, 2020; Winsberg, 2010). But if we import these Cartesian causal principles into the Posidonian arguments we have been considering, we get a new premise:

(I) A condition of the possibility of (an intended) successful (concrete) scientific representation of nature is that nature is orderly.

With this premise (I) we can drop (A–C) and doing so will be the basis of what I call a "transcendental Posidonian argument." Before I articulate the full version of it I motivate it by looking at Samuel Clarke. I argue that this—possibly even more anachronistic—transcendental interpretation of the Posidonian argument is a natural outgrowth of reflection on the scientific revolution in progress. The transcendental interpretation focuses on the fact that a scientific model of the world plays a crucial role in the Ciceronian argument.

---

[42] I assume, is that these causal principles are more widely shared in the early modern period until we see them explicitly challenged. See chapter 5.

An important aspect of the enduring attraction of the prima facie version of the Posidonian argument in the early modern period depends on the fact that *if* one is committed to the idea that science can reveal the design of nature then progress in the sciences aided by, say, microscopic and telescopic technology keeps revealing new and ever more sophisticated evidence of design. I suspect the first post-Copernican person to make *this* move was the unusual Leuven Jesuit, Leonard Lessius (Leys) in his 1612 book, *De Providentia Numenis*. Without taking a stance on the Copernican hypothesis (his language is compatible with it, but also the Tychonic one), Lessius summarizes the key findings of Galileo's *Starry Messenger* (1610), including the discovery of whole new celestial bodies. Lessius does so to argue that these discoveries reinforce the idea of the design of nature by an incorporeal mind (Lessius, 1612: 25). In context he explicitly mentions book 2 of Cicero's *De Natura Deorum*, but not the Posidonian argument.

That, despite his impressive knowledge of Cicero, Lessius does not connect the discoveries to the Posidoniun argument is not an accident because Lessius wants to argue that the orbits of these (infinitely) new bodies are unknowable to our finite minds: God "hath disposed the course of the starrs with that stupendious art and skill, as that they are in no sort subiect to the apprehension of man's understânding" (Lessius, 1612: 25 [slightly modernized spelling of Edward Knott's 1631 translation.]) That is, while Lessius does not seem a general skeptic, when it comes to knowledge of these newly discovered bodies he is profoundly pessimistic. We see here the idea that the world is a stupendous artifice whose principles are unknowable to the agents within it even if they get better at mapping it.

Prior to the breakthroughs that led to the *Principia*, Newton was also very pessimistic about the possibility of exact knowledge of celestial orbits (e.g., the "Copernican Scholium" discussed by George Smith [2008]). But that pessimism disappears in the *Principia* and in the people it inspires. While I return to Newton later, here I discuss Samuel Clarke (for context see Force [2008]); I focus on his *Demonstration*, which started out as a Boyle lecture (for more on context see chapter 8).

In the context of his polemic with Toland, who in the *Letters to Serena* with his own appeal to Cicero's authority, rejects what we would call the invocation of a "God of the gaps," (recall chapter 8). Clarke points out that when in ancient times, "Epicurus and his follower Lucretius" imagined "finding fault in the frame and constitution of the Earth" this was somewhat plausible due to the "infancy of natural philosophy" (although Clarke notes with satisfaction that even then the "generality of men" were not persuaded). Clarke can point to recent discoveries in anatomy and physiology such as "the circulation of the blood, the exact structure of the heart and brain" as well as the discovery of a number of veins and other vessels neither known nor imagined in ancient times. (*A Demonstration*: 227–228; see also Hume's *Dialogues*: 11.11). Then Clarke writes:

> If Tully, from the partial and very imperfect knowledge in astronomy, which his times afforded, could be so confident of the heavenly bodies being disposed and moved by a wise and understanding mind, as to declare that in his opinion, whoever asserted the contrary, was himself void of all understanding: What would he have said, if he had known the modern discoveries in astronomy? The immense greatness of the world (I mean of that part of it, which falls under our observation), which is now known to be as much greater than what in his time they imagined it to be, as the world itself, according to their system, was greater than Archimedes' Sphere?" (*A Demonstration*: 228–229)[43]

The history of scientific progress becomes an added argument for the plausibility of the Posidonian argument.[44] In the *narrow* sense this is so because on Clarke's account, science discovers more evidence of apparent design where previously there just had been mystery or lack of knowledge and, thus, is capable of ever more refined representations of nature covering an increasing domain of nature. This fact feeds into a broader argument that Clarke makes in context: the history of scientific progress *itself* becomes a further argument for the existence of a designing and benevolent God. In fact, after listing a large number of Newton's then recent discoveries, Clarke interprets the history of scientific progress as an unfolding Biblical prophecy:

> We now see with how great reason the author of the Book of *Ecclesiasticus* after he had described the beauty of the Sun and Stars, and all then visible works of God in heaven and earth, concluded ch. 43, v 32 (as we after all the

---

[43] In the middle of this passage, Clarke quotes book 2 of *The Nature of the Gods* in Latin. The passage is a few paragraphs removed from the Posidonian argument. The reference to "Archimedes' sphere" reminds us that we are still in the ambit of the Posidonian argument.

[44] Gibbon seems to have had this point in mind in his "Address" collected in *Miscellaneous Works of Edward Gibbon* (1797: Dublin), volume 3:469, and the footnote that calls explicit attention to Cicero. See https://books.google.nl/books?id=17E8AAAAYAAJ&pg=PA469&dq=Newton+Posidonius&hl=nl&sa=X&redir_esc=y#v=onepage&q&f=false

discoveries of later ages, may no doubt still truly say,) "There yet hid greater things than these, we have seen but a few of his Works." (*A Demonstration* XI: 232–233).[45]

Responding to Locke's pessimism, in Clarke's hands Newton's then recent discoveries become an important evidentiary signpost for understanding scientific progress as an open-ended unfolding and confirmation of confident Biblical prophecy.

Now we can return to the Posidonian argument in order to explore the transcendental interpretation of it. As I have emphasized the Posidonium planetarium is more than a timekeeper; it is capable of predicting other heavenly phenomena, especially eclipses. So, rather than viewing the heavenly phenomena as portents of danger (revolution, omens, etc.) the scientific representation fits them into an ordered universe. (This point is made in Adam Smith's "The History of Astronomy" 3.1, [Smith. 1982: 48; for the Posidonian argument in Adam Smith, see Schliesser, 2017: 280–285].) That is, the empirical success of the planetarium points to the significance of predictable order; the designer of a planetarium presupposes an orderly celestial globe.[46] For example, in Newton's *Principia* this orderliness is crowned by the original closing pages of the book: Newton's ability to predict the orbits of comets (see chapter 3 for discussion and references).

So we can put the significance and true intuitive force lurking within all elements of the Posidonian argument in anachronistic fashion: a condition of the possibility of (an intended) successful scientific representation of nature is that nature is orderly or ordered. Thus *a history* of successful scientific representations, and technologies of simulation relying on these, becomes a distinct and over time increasingly compelling argument for an orderly nature.[47]

---

[45] This is a rare occasion where Clarke appeals (against his official purposes) to a Biblical text at all in a *Demonstration*. Given the noncanonical (or deuteron-canonical) status of *Ecclesiasticus*, Clarke's choice is worth exploring, but I cannot pursue the question here. I thank Peter Anstey for this discussion.

[46] This points stands even if one insists that the design of the planetarium is based on empirical data.

[47] This is so, even if the representations do not track the causal order of nature, but merely simulate or predict the phenomena. Hunter (2009) puts the point very precisely: "It is the complexity internal to both the model and its original. Let us call it the pattern (P) which both the original and its model instantiate. By virtue of P the original and the copy are partially isomorphic" (244). To anticipate my treatment of the Posidonian passage in the *System of Nature*: This is also true of statues and their model.

This leads me to the promised transcendental interpretation of the Posidonian argument.[48] The form of the transcendental interpretation is as follows:

    I. A condition of the possibility of (an intended) successful scientific representation of nature is that nature is ordered.

    II. Science has a history of success.

    III. The world's order is not produced by nothing.

    IV. Order is produced by necessity, or by chance, or by designing mind.[49]

    V. The particular order science finds is not the subject of chance or necessity, so it must be produced by a designing mind.

    VI. There is a designer of nature's order (or nature is designed).

Leaving aside the specific premises, I make three claims about this version of the argument. First, most early moderns who were interested in such arguments were interested in reaffirming that God was the cause of the world. By framing the argument in terms of 'nature's order,' I am simplifying the argument. Early moderns tend to assume there is a tight link because the cause of the existence of nature and the cause of its order.[50] So it would be historically more accurate to represent their commitments as follows:

    I. A condition of the possibility of (an intended) successful scientific representation of nature is that nature is ordered.

    II. Science has a history of success.

    III. The world's order is not produced by nothing.

    III+. The same cause or principle is response for the existence of the world and it's order.

    IV. Order is produced by necessity, or by chance, or by designing mind.

    V. The particular order science finds is not the subject of chance or necessity, so it must be produced by a designing mind.

    VI. There is a designer of nature's order (or nature is designed).

[48] Clerk Shaw was the first to remind me of the significance of Kant's *Critique of Judgment*, where the logical purposiveness of nature is presupposed as a "regulative principle," but not constitutive principle, in one's science (see the antinomy of the Judgment).

[49] For some readers to say that 'X is caused by chance' just is to say that 'X is caused by nothing.' And so part of (IV) would be contradicted by (III). But chance as used in this context here does not mean 'nothing' (for it can refer to the swerve, etc.) Of course, some early moderns did think that chance was tantamount to saying nothing. I thank Marij Van Strien for this discussion.

[50] There are important debates about the relationship between God and so-called secondary causes in historical context and in the secondary literature, but for simplicity's sake I ignore this issue.

Second, the transcendental version of the Posidonian argument changes the significance of the history of science and the character of science more generally in two (related) respects. First, on the prima facie approach to the Posidonian argument when it comes to the question of intelligent design in nature, science is, in principle, perceived to be a neutral means in order to establish the nature and existence, if any, of the God(s). Let's call this the "neutrality requirement."

By "neutrality requirement," I mean to capture the normative and epistemic idea that when it comes to establishing substantive facts about the world or its domain of inquiry, science ought to be a neutral means. So, for example, if one wishes to establish scientifically that there are no jumps or gaps in a part of nature one is studying (because one is in the grip of the law of continuity), then (say) the choice of mathematical machinery one deploys in one's science ought not to settle this matter one way or another.[51] (Of course, if formulations with competing mathematical techniques—discontinuous or continuous functions—have differing empirical success then this might be taken as evidence one way or another.) The neutrality requirement is also compatible with science playing no role in establishing the properties of the God(s). Living up to the neutrality requirement helps give science epistemic authority because it means that when it makes claims about the world the dice are not loaded one way or another, so to speak.[52]

However, on the transcendental version of the Posidonian argument, there is a nontrivial sense that science itself presupposes for its very possibility and intelligibility that there is order and a source of order. On this approach science foregoes the neutrality requirement about at least one important topic. Such nonneutrality about a privileged feature is an important characteristic of transcendental and so-called indispensability arguments more generally.[53]

Third, on the transcendental version, the force of the Posidonian argument does not rely anymore on the (perhaps dubious to post-Humeans or post-Darwinians) explicit assumptions of the original argument in Cicero

---

[51] There was a lively eighteenth-century debate over this between Euler and D'Alembert; Iulia Mihai has called my attention to this. For some of the history, see van Strien 2014 and 2015.

[52] Of course, there is a lot more that goes into generating epistemic authority.

[53] I owe this last point to Ursula Renz. Violating neutrality requirements just is what transcendental arguments and indispensability arguments do. (This is one way of capturing why transcendental arguments tend to feel circular.) This is worth emphasizing because we live in a philosophical culture that is attracted to indispensability arguments which appeal to the authority of science (and so instantiate what I call "Newton's Challenge to philosophy").

(such as "nature's attainments transcend those achieved by human design") and the role of Descartes' causal principles in the neglected version of the argument. Moreover, on the transcendental approach, the Posidonian argument stresses the structural resemblance between state-of-the-art-planetary mechanism and the celestial globe, but does not appeal to aesthetics more generally nor claims about analogies between the designer of the planetarium and heavens.[54] As in the neglected version of the Posidonian argument, this argument does not appeal to apparent design to motivate the argument nor mentions it in its premises.

Of course, the premises of the transcendental version of the argument are not without controversy. Premise (I) may be thought too strong in two ways. First, it may be the case that all this is required is to presuppose in a scientific practice that a relevant region of nature is ordered. So one can imagine rewriting premise (I) as follows:

I[+]   A condition of the possibility of (an intended) successful scientific representation of a region of nature is that a region nature is ordered.

While this would require reformulating other premises (e.g., III), it does not undermine the conclusion of the argument. In fact, as it turns out that ever more bits of nature are orderly (comets, geology, etc.), it may well strengthen the overall argument. Even so, to what extent all regions of nature share the same order is also not obvious. (Newton himself seems to have allowed the possibility they do not; recall my discussion with Biener in chapter 6 of Query 31, *Opticks* 403–404; "thereby to vary the laws of nature, and made worlds of several sorts in several parts of the universe.")

Premise (II) is hard to evaluate—a lot depends on one's baseline and one's expectations. But it is no surprise that as the scientific revolution unfolds this premise seems very secure, despite the development of pessimistic metainduction arguments (say, by Jonathan Swift; see Schliesser, 2005b: 705–706).

Premise (III) relies, in its most general sense on a version of the causal principle or on the PSR. Not all thinkers embraced the PSR in the early modern

---

[54] Hunter puts the insight very nicely: "The new argument does not make the mistake of comparing unconnected instances of complexity. Its very different strategy is to exploit the fact that one and the same instance of complexity is found simultaneously in two places" (242). To be clear: Hunter's "new" argument is not my "transcendental" argument, but akin to what I have called the "neglected" version; Hunter's "new" argument preserves the neutrality requirement.

period, but I am unfamiliar with anybody that denies the causal principle. Of course, one can argue for restrictions on both (as Hume and Kant would). As I noted above (III⁺) is in one sense controversial—there were lots of debates over the relationship between God and secondary causes—but in other sense not so controversial (because God's omnipotence was simply assumed). In addition, there are hidden simplicity and parsimony assumptions packed into (III⁺).

Premise (IV) is inherited from the ancients. Much philosophical development in the early modern period struggles with it. In particular, new sources of apparent natural order (e.g., the mind's imposition, emergence, and even the development of a science that deals with the nature of order, entropy, etc.) start to challenge the naturalness of this premise. For example, in the Appendix to *Ethics* 1, Spinoza "makes," to quote his critic Voltaire, "a jest of final causes" (Voltaire, 1843: 652 in the entry on God/gods). But lurking in Spinoza's treatment is more than a genealogy of intellectual error to unmask final causes (Schliesser, 2017, 2018). It is a bit surprising Clarke and Voltaire do not address this. For one of Spinoza's key moves is not just to insist in proto-Lockean fashion that "the order of causes is hidden from us" (E1p33S1), but that order we assume "in nature [is no] more than a relation to our imagination" (E1, Appendix; in context Spinoza goes on to mock the philosophers "who have persuaded themselves that the motions of the heavens produce a harmony"). That is to say, in criticizing final causes, Spinoza anticipates here Hume (and Kant) in thinking that in some sense we are constitutive of nature's order (Schliesser, 2018). And so Premise (IV) starts to look doubtful.

Premise (V) requires argument. There are some philosophical proponents of chance (most notably Hume) as a source of order. And once probability theory is developed, the argument for chance is made in sophisticated fashion (by D'Alembert and Laplace) in the context of Enlightenment debates over cosmology and cosmogony (recall chapter 3).[55] But most of the philosophical debates circle around the challenges of Spinozism to this

---

[55] One further complication: by the middle of the nineteenth century, in Whewell's *Bridgewater Treatise*, the nebular hypothesis (explicitly associated not with Kant anymore, but Laplace) is treated as an instance of modern Epicureanism (so system of chance). In the revised preface of the 1864 edition, which has a lengthy quote of the passage from Cicero under discussion here in this chapter, Whewell makes a point of associating Darwin with the system of Epicurean chance, which at the cosmogenic level is to be associated with the nebular hypothesis (Whewell, 1864: 229–230). I thank David Haig for calling my attention to this. But Laplace's own argument presupposes necessity and his use of probability is primarily epistemic (lack of knowledge) not metaphysical. I thank Marij Van Strien for this discussion.

premise. In a nutshell critics of Spinozism (Clarke, Newton, MacLaurin) argue that while it predicts (by stipulation) necessary variety, it fails to predict the *particular* variety we observe. But as cosmogony starts developing (from Kant onward), the Spinozist position gets recast and becomes more plausible again (chapter 3). But in so doing another premise (III⁺) starts to become less likely.

That is, even if one is distinctly reserved about final causes or God's providence, the transcendental version of the argument tightly links scientific progress to God's or nature's order. If you accept this argument then, with scientific progress, you get deism for free. It also motivates commitment to the possibility of scientific progress. In the next section I offer a more speculative argument that Descartes inscribed the transcendental version of the Posidonian argument into his metaphysical natural philosophy (Garber, 1992).

9.6 The Transcendental Version of the Posidonian Argument (Descartes)

What follows is inspired by Dennett (2008). For he has discerned a version of what I call the transcendental version of the Posidonian argument in an unexpected place: Descartes' response to Caterus in order to make sense of Descartes' *a posteriori* argument for the existence of God (in *Meditations* 3). Dennett has many astute comments on Descartes' argument (including, not unexpectedly, some serious criticism), but is unaware of the wider trajectory of the Posidonian argument.

The key passage, which is repeated in *Principles of Philosophy* is this: "if someone possesses the idea of a machine, and contained in the idea is every imaginable intricacy of design, then the correct inference is plainly that this idea originally came from some cause in which every imaginable intricacy really did exist, even though the intricacy now has only objective existence in the idea" (Descartes, 1984: 76; see also 75; Dennett, 2008: 338). Descartes then goes on to claim "By the same token, since we have within us the idea of God, and contained in the idea is every perfection that can be thought of, the absolutely evident inference is that this idea depends on some cause in which all this perfection is indeed to be found, namely a really existing God." My interest here is in the incredibly intricate idea of a machine.

Descartes' argument has a family resemblance to the Posidonian argument, but I cannot prove that he was inspired by Cicero.[56] Descartes innovates on the traditional argument by insisting that an *intricate blueprint* ("the idea of a machine") can do the relevant work in the argument. It is one thing to take a working (or empirically successful) planetarium "to Scythia or to Britain" where "no one in those barbarous regions would doubt that that planetarium had been constructed by a rational process"; it's quite another to point to a blueprint of an intricate machine and, say, 'see.' Of course, an intricate blueprint of an intricate machine is not itself evidence for design. It needs to be a working machine.

The true novelty of Dennett's reading of Descartes' argument, and of present interest, is that he ultimately interprets Descartes as invoking a *working* machine in the argument to design:

> [Descartes] had used that very idea [of God] and no other as the sole foundation for his theory of . . . *Le Monde*. . . . The fact remains that it was a huge theory, full of intricacy, remarkably self consistent, often fiendishly persuasive even in today's hindsight. Any idea that could generate such a stunning intellectual edifice would be a prodigiously fecund idea . . . because his idea is not just made of lots of good parts (ideas available to everybody), and not because his idea is of a wonderful thing? God, it is because his idea is (he thought) a stunningly well-designed engine of scientific discovery. (Dennett, 2008: 342–343)

That is to say, on Dennett's reading, Descartes thinks he can point to his whole scientific theory as the working machine that represents and predicts the machinery of the universe. This is not the place to discuss to what degree pointing to *Le Monde* or the *Principles of Natural Philosophy* will really be more persuasive (in such a design argument context) than pointing to a working planetarium. Rather, my point here is that on Dennett's interpretation of Descartes, Descartes' scientific theory of the universe presupposes, eminently in Descartes' sense, if not God's existence, then at least the idea of God's existence. That's true even if Descartes' position fails to convince as an a posteriori argument for the existence of God (as Dennett argues).[57]

---

[56] Descartes has a healthy interest in Archimedes, see especially the Fourth Reply (Descartes, 1984: 168–171).

[57] Of course, the degree to which Descartes' physics or metaphysics is teleological is a matter for debate.

My interest here is not to argue that Dennett's reading of Descartes is persuasive.[58] But if Dennett is right then the transcendental version of the argument was inscribed into the new science in a very prominent place well before Clarke articulated his version of the argument. This would help explain why the new science, which drew on plenty of Epicurean commitments, was, while threatening to certain (flat-footed) readings of revelation not taken to undermine commitment to Theism and Deism (Funkenstein, 1986; Harrison, 2007), but repeatedly invoked by thinkers who were extremely ambivalent about revelation, the Trinity, and other Christian dogmas.[59] In effect, once one accepts something like the transcendental version of the Posidonian argument, every scientific breakthrough strengthens the Posidonian argument and also generates belief that further breakthroughs are not impossible—that is, it helps explain why the idea of permanent progress became a live option during the eighteenth century in the thoughts of scientists like Priestley and Condorcet.

Newton and the Posidonian Argument.

Despite seeming to echo Locke's *Essay* 3.6.9 in the "General Scholium," and his evident familiarity with Boyle's writings, Newton does not mention the characteristic features of the Posidonian argument in the *Principia*. He does, however, explicitly appeal to Cicero's *On the Nature of the Gods*, to claim that the Ancients agree with his claims about God's omnipresence, that "in him all things are contained and move" in a note to the "General Scholium" in the very sentences before he echoes Locke's denial of knowledge of real essences (Newton, 1999: 941 note c). My present concern is not to show that Newton embraces the Posidonian argument (although I show it is likely). Rather, I use the Posidonian argument to illuminate features of Newton's metaphysics and to explain why Voltaire inscribes Newton's natural philosophy into the Posidonian argument.

---

[58] It has been largely ignored in the scholarship on Descartes. But Dicker (2013: 140) cites Dennett's criticism of the argument in the *Meditations* approvingly (but it does not follow, of course, that Dicker endorses Dennett's reading of the Posidonian argument). The eminent Descartes scholar, Alan Nelson, expressed reservations about Dennett's argument when I presented it on my blog: http://digressionsnimpressions.typepad.com/digressionsimpressions/2015/05/on-descartes-dennett-and-design-ciceros-posidonian-argument-reconsidered.html?cid=6a00e54ee247e3883401bb082d652d970d#comment-6a00e54ee247e3883401bb082d652d970d

[59] At this level of abstraction, this describes Leibniz, Newton, Clarke, Wolfe, and pretty much the whole French and Scottish Enlightenment.

The transcendental version of the Posidonian argument requires the existence of natural order in order to conclude the existence of the God of order. That Newton thinks there is natural order is already explicit in the first edition of the *Principia*, from the Scholium to the Definition, where he writes "Just as the order of the parts of time is unchangeable, so, too, is the order of the parts of space" (Newton, 1999: 410; the Latin reads ["*Ut partium temporis ordo est immutabilis, sic etiam ordo partium spatii.*"]).

Once he adds the "General Scholium" in the second edition of the *Principia*,[60] Newton is explicit that his God is the "Lord of Lords" (Newton, 1999: 941)—that being a nod to *Deuteronomy* 10:17. That is, his God has dominion. What does this have to do with order? Newton goes on to make clear that God's dominion involves, in particular, his temporal and spatialized dominion:

> [H]is duration reaches from Eternity to Eternity; his presence from Infinity to Infinity . . . He is not Eternity and Infinity, but Eternal and Infinite; he is not Duration and Space, but he endures and is present. He endures for ever, and is every where present; and by existing always and every where, he constitutes Duration and Space. Since every particle of Space is always, and every indivisible moment of Duration is every where, certainly the Maker and Lord of all things cannot be never and no where. . . . There are given successive parts in duration, co-existent parts in space. (Newton, 1999: 941)

That is to say, Newton links by way of constitution [*constituit*] God's *ordered* existence with *nature's* order, in particular the eternal and fixed order of space and time (Force, 2008: 99).

Now one might argue that the clear evidence that Newton embraced special-temporal structure is not sufficient to claim that nature is ordered.[61] This objection comes naturally to those like Janiak (2008), who distinguish between (a) God, space, and time, which provide the structure to, or condition of possibility of, infinite nature, and (b) nature's order (i.e., finite matter in motion). So, if you keep (a) and (b) metaphysically separate more evidence is needed that nature is ordered for Newton.

Such evidence exists because the first few paragraphs of the "General Scholium" emphasize the *regularity* of the motions of the solar system.

---

[60] What follows is indebted to Yoram Hazony.
[61] I thank Janiak for the objection.

(Newton uses '*regulares*' three times.)[62] Newton argues from this regularity to God as its source ("This most elegant system of the sun, planets, and comets could not have arisen without the design and dominion of an intelligent and powerful being..." [Newton, 1999: 940; see also chapter 8]). So while the ordered structure of (a) is necessary in the way that the observed regularity and diversity of (b) is not, (b) itself shows that there is some ordering principle.[63]

We can, thus, summarize Newton's position as follows (I add, in brackets after each premise, in bold the premises of the transcendental version of the Posidonian argument):

(i)   There is a God of order ["It is agreed that ... God ... exists" Newton, 1999: 942] **[VI]**

(ii) A condition of the possibility of (an intended) successful scientific representation or concrete model of (a region of) nature is that nature is orderly (that is, "the order of the parts of time ... the order of the parts of space" [Newton, 1999: 410]). **[I]**

(iii) Nature's hidden order (could not be the product of chance [as suggested by Epicureanism] or necessity [as suggested by Spinozism]), but can only be *constituted* by God [as argued in "General Scholium"; see chapters 4 and 8 above]; **[III⁺-IV-V]**

(iv) Science is successful [The *Principia* as a whole] **[II]**

(v)   Science gives us some (indirect) access to the constitution of God's order, which is why "to treat of God from phenomena is certainly a part of 'natural' philosophy." (Newton, 1999: 943)

So, what I take to have shown is that it is no surprise that Clarke and Voltaire think Newton's natural philosophy is a natural fit with the Posidonian argument, for it really is.

Let me address three possible objections. First, on premise (iii), one may accept that Newton explicitly rules out Spinozism, or "blind metaphysical necessity" (Newton, 1999: 942) as the source of "laws of nature" or nature's order—that is, to quote Cotes' preface, "the fantasy that all things are governed by fate" (Newton, 1999: 397). But one may argue that he does not

---

[62] As I show in chapter 7, these celestial bodies in motion help constitute, with the astronomers' equation of time, nature's temporal frame for us. So a well-designed time-keeper/planetarium is appropriately partially isomorphic to it.

[63] Recall from chapter 8: "No variation in things arises from blind metaphysical necessity, which must be the same always and everywhere" (Newton, 1999: 942).

explicitly argue against Epicureanism and this is why some readers mistakenly interpreted him as a kind of Epicurean (see chapter 3). This objection is plausible enough. But Newton does signal he is ruling out the Epicurean doctrine, when he writes that "a god without dominion, providence, and final causes is nothing other than fate and nature" (Newton, 1999: 942) Epicureanism is a system of nature without final causes (which is why Voltaire criticizes it).

Second, one may think that Premise (iv) is something less than Clarke's idea that science has a *history* of success. Yet this idea too can be found in Newton's *Principia*. In joining together the study of projectile motion and collision theory, he is adamant that he is developing a conception and content of natural philosophy inherited from Galileo and that he is building on an existing consensus: "From the same laws . . ., Sir Christopher Wren, Dr. John Wallis, and Mr. Christiaan Huygens, easily the foremost geometers of the previous generations, independently found the rules of the collisions and reflections of hard bodies . . ." (Newton, 199: 424). If anything, as Biener and I argue in chapter 6, Newton invents a consensus of virtuosi as a kind of baseline. For later readers this naturally creates the sense of real progress. This strategy of creating a baseline is, as Thomas Kuhn emphasized, constitutive of progressive science (Schliesser, 2011).

Third, one may wonder whether Newton really explicitly embraces (ii). I am inclined to treat it as self-evident. But in what follows I offer evidence from two pre-*Principia* sources for Newton's commitment to (ii). First, there is an attenuated sense in which Newton's pessimistic "Copernican Scholium," also presupposes this commitment (for dating and context see Smith, 2020 and my chapter with Biener above). The second paragraph of this Scholium reads:

The planets neither move exactly in ellipses nor revolve twice in the same orbit. Each time a planet revolves it traces a fresh orbit, as happens also with the motion of the Moon, and each orbit depends upon the combined motions of all the planets, not to mention their actions upon each other. Unless I am much mistaken, it would exceed the force of human wit to consider so many causes of motion at the same time, and to define the motions by exact laws which would allow of an easy calculation. (Smith, 2008)

So, if we understand by 'order' something like 'repeating in sufficiently regular fashion' then Newton is clearly saying here that law-governed, causal astronomical science is only possible when the motions are orderly.[64] The shift from the De Motu drafts to the *Principia* is precisely the confidence that causal laws of motion are possible (and knowable).

Admittedly,[65] this (let's call it a 'phenomenal order') of the "Copernican Scholium" is a different notion of order than the fixed structure of space and time, or (say) 'meta-order' of the Scholium to the definitions and the "General Scholium".[66] What this reveals is that all versions of the Posidonian argument depend on something like the idea that there is an internal relation between the meta-order of space and time and the law-like phenomenal order discovered by science. This internal relation is secured by the positing of a God of order. One can understand Hume's and Kant's philosophy as a critical exploration of not just the legitimacy of this posit, but of the very idea that there is such an internal relation between the meta-order and phenomenal order.

Second, there is another pertinent passage that speaks to Newton's conception of order, as well as revealing he is committed to something in the neighborhood of this internal relation between the meta-order and phenomenal order. It's from "an early treatise on the apocalypse" (Force, 2008: 79):

9. To choose those constructions which without straining reduce things to the greatest simplicity. The reason of this is manifest by the precedent Rule. Truth is ever to be found in simplicity, & not in the multiplicity & confusion of things. As the world, which to the naked eye exhibits the greatest variety of objects, appears very simple in its internall constitution when surveyed by a philosophic understanding, & so much the simpler by how much the better it is understood, so it is in these visions. It is the perfection of God's works that they are all done with the greatest simplicity. He is the God of order & not of confusion. And therefore as they that would understand the frame of the world must indeavour to reduce their knowledg to all possible simplicity, so it must be in seeking to understand these visions. And they that shall do otherwise do not onely make sure never

---

[64] That is, I read the pessimism of the "Copernicium Scholium" not in terms of the complexity and multiplicity of the causes—although Newton is clearly worried about that, too—but in terms of the uniqueness of each apparent motion.

[65] Recall Janiak's objection.

[66] I use 'phenomenal' order to capture Newton's sense that behind the confusing data there are stable regularities, phenomena, visible to the philosopher-expert.

to understand them, but derogate from the perfection of the prophesy; & make it suspicious also that their designe is not to understand it but to shuffle it of & confound the understandings of men by making it intricate & confused. (Newton, *c.* 1670s–1680s, Rules for Methodising | Construing the Apocalyps. in Untitled Treatise on Revelation [section 1.1—quoted in Force, 2008: 60–61])[67]

In the passage, Newton draws a distinction between confusing appearances (which exhibit "multiplicity") and an orderly and simple hidden (internal) constitution. Newton's language is rather rationalist (a "philosophic understanding" grasps the hidden constitution of things), but in context it's clear that the interpretation of Scriptures (or prophetic visions) and of nature (they are treated as analogous enterprises) is also an empirical matter (e.g., "observation").

In the passage order is resolutely coupled with simplicity. There is a subtle distinction between order and simplicity. Let me first explain simplicity. For Newton there is a difference between (a) simplicity being a feature of God's works, especially the internal constitution of things, and (b) simplicity being a cognitive feature accompanying a philosophical understanding of that internal constitution of things. Truth just is when (a) and (b) match. For present purposes let's ignore skeptical scenarios that may undermine this notion of truth.

It also turns out that for Newton the simplicity in (b) also involves (c) a property or virtue of a successful epistemic reduction to a more limited number of elements. Newton remains committed to such epistemic reduction to limited elements throughout the *Principia* and *Opticks* (Hazony, 2014).To be sure, a Newtonian 'construction' is what we would call an 'interpretation,' but, given my interest in the Posidonian argument, I accept the equivocation in which 'construction' also evokes a concrete model.

In the passage order is, for Newton, a feature, or a perfection, of God: "He is the God of order." A natural reading of Newton is, then, that God's order explains the absence of confusion in the hidden constitution of nature. So, God secures the phenomenal order visible to the philosopher.

---

[67] Newton, Yahuda MS 1.1, f. 14r. See the Newton Project, http://www.newtonproject.ox.ac.uk/view/texts/normalized/THEM00135. I thank Laurie Paul for calling my attention to this passage.

9.7 Voltaire and the Posidonian argument

At the start of this chapter, I quoted Voltaire's enigmatic claim that "[A] If a clock is not made in order to tell the time of the day, [B] I will then admit that final causes are nothing but chimeras. . . [C] All the parts, however, of that machine the world, seem made for each other." (Voltaire, 1843: 505) I have now shown that [A–B] is an explicit response to a critic of the neglected version of the Posidonian argument, and that the natural implicature is that Voltaire is simply endorsing it. In addition, with [C] Voltaire both echoes, as we have seen, the "General Scholium" as well as, when we combine it with the neglected version of the Posidonian argument, thereby endorsing (the first premise of) the prima facie version of the Posidonan argument. Rhetorically, Voltaire follows Clarke's example by thereby inserting Newton into discussions of the Posidonian argument (something not uncommon in eighteenth-century natural theology).

I do not mean to be uncritical of Voltaire. As should also be clear by now, in the *Philosophical Dictionary*, Voltaire misses some of Spinoza's concerns and, in particular, seems unaware that Hume's philosophy, in particular, has made some of the key premises of even the transcendental version of the Posidonian argument unstable. I do not mean to suggest the Humean criticisms are in all respects unsurmountable; Paley is responsive to them. Teasing out the relative merits of Hume and Paley is for another occasion.

Here I call attention to one feature of Voltaire's response to the criticism of final causes. Because while much of what he says recycles familiar tropes he does say something distinctly modern and, in so doing, reinterprets the doctrine of final causes and its connection to the first premise of the prima facie version of Posidonian argument (recall "all of nature's parts are ordered"). A representative claim that I have in mind is:

> [E]verything is the result, nearer or more remote, of a general final cause; that everything is the consequence of eternal laws. When the effects are invariably the same in all times and places, and when these uniform effects are independent of the beings to which they attach, then there is visibly a final cause. (Voltaire, 1843: 506)

Voltaire here echoes and develops a terse argument by Newton in the "General Scholium." For just after asserting that it is legitimate to infer final causes, Newton adds, as we have seen, that "All that diversity of created things, suited to different times and places, could arise from nothing but the ideas and will of a Being necessarily existing" (Newton, 1999: 942). I have explored this claim and the metaphysical principles it relies on, in chapter 8.

In context, Voltaire is responding to the critic who derides the final-cause lover of producing endless just-so stories about local apparent adaptation. This critic is what Voltaire calls the "modern atheist" who argues in Epicurean fashion that:

> Can it be said that the conformation of animals is according to their necessities? What are those necessities self-preservation and propagation. Now, is it astonishing that, of the infinite combinations produced by chance, those only have subsisted which had organs adapted for their nourishment and the continuation of their species? Must not all others necessarily have perished?" (Voltaire, 1843: 157; see also 164, "in the infinity of ages, any one of the infinite number of combinations, as that of the present arrangement of the universe, is not impossible"; this argument is also found in Hume's *Dialogues*.) In the entry on final causes, this argument is repeated with embellishment in the long quotes from the *System of Nature* in the which the Posidonian argument is lodged as an objection (Voltaire, 1843: 503–505).

Voltaire pleads guilty as charged and sees adaptation almost everywhere:

> All animals have eyes and see; all have ears and hear; all have a mouth with which they eat; a stomach, or something similar, by which they digest their food; all have suitable means for expelling the feces; all have the organs requisite for the continuation of their species; and these natural gifts perform their regular course and process without any application or intermixture of art. Here are final causes clearly established; and to deny a truth so universal would be a perversion of the faculty of reason. (Voltaire, 1843: 506)

What he denies is that this requires special providence (which swerves uncomfortably close to endless miracles). Rather adaptation is the *effect* of general final cause(s), which are thereby subsumed under general and eternal laws of nature.

That is, there are three claims in Voltaire worth noting: first, nature is governed by natural laws; second, when one discerns such invariable laws of nature, one is ipse facto uncovering the effects or, to stick closer to Voltaire ("visibly"), manifestations of final causes. This point echoes Cotes' preface to the second edition of the *Principia*, that "this world—so beautifully diversified in its forms and motions—could not have arisen except from the perfectly free will of God, who provides and governs all things. From this source, then, have all the laws that are called laws of nature come, in which many traces of the highest wisdom and counsel certainly appear, but no traces of necessity" (Newton, 1999: 397).

The significance of the preceding paragraphs for the present argument is this: In the early modern period astronomy and biology, and any other science, provide, in virtue of the Posidonian arguments, mutually reinforcing, abductive evidence for the reality of final causes. From that vantage point, the Epicurean, proto-Darwinian argument looks like special pleading.[68]

Third, according to Voltaire the laws are eternal. This last point is worth pondering because it is a denial that they are created—they coexist with God just like Newtonian space and Newtonian time are constituted by God eternally. My own view is that this attitude toward the laws is not Newton's own position (see chapter 4, 5, and 8 above), but that is compatible with Voltaire thinking it is natural reading of Newton, say based on Cotes' preface ("it is the province of true philosophy to derive the natures of things from causes that truly exist, and to seek those laws by which the supreme artificer willed to establish this most beautiful order of the world, not those laws by which he could have, had it so pleased him" [Newton, 1999: 393]).[69]

I close with a remark on a detail. Recall that the version of the Posidonian argument that the author of *System of Nature* acknowledges as an objection (and Voltaire seemingly endorses) starts as follows "It will be observed and insisted upon by some, that if a statue or a watch were shown to a savage who had never seen them, he would inevitably acknowledge that they were the productions of some intelligent agent, more powerful and ingenious than himself" (Voltaire, 1843: 504).

This echoes a feature in Cicero that I had not remarked upon yet. Recall, "It can surely not be right to acknowledge as a work of art a statue or a painted picture, or to be convinced from distant observation of a ship's course that its

---

[68] To what degree Darwinism rules out natural teleology is itself a much more complex question than ordinarily allowed (see Haig [2020]).

[69] Cotes then goes on to assert what Locke denies is possible, "In mechanical clocks one and the same motion of the hour hand can arise from the action of a suspended weight or an internal spring. But if the clock under discussion is really activated by a weight, then anyone will be laughed at if he imagines a spring and on such a premature hypothesis undertakes to explain the motion of the hour hand; for *he ought to have examined the internal workings of the machine more thoroughly, in order to ascertain the true principle of the motion in question*" (Newton, 1999: 393 emphasis added).

progress is controlled by reason and human skill, or upon examination of the design of a sundial or a water-clock to appreciate that calculation of the time of day is made by skill and not by chance, yet none the less to consider that the universe is devoid of purpose and reason, though it embraces those very skills, and the craftsmen who wield them, and all else beside?"

What statues are models of is sometimes unclear. In the quoted passage from Voltaire, the 'savage' is only shown the statue not the person who sat for the artist. To me this suggests that in the example, the statue stands as a model for humanity. For, we are never told how good the statue represents a particular individual. Rather we are told that upon seeing it the proverbially hapless savage, or outsider, will automatically infer that it is the production "of some intelligent agent, more powerful and ingenious than himself."

One may doubt that this is what is thought by the savage. But it does point to a feature Hunter has emphasized. Hunter had noted that the Posidonian argument relies on a key step: "It is the complexity internal to both the model and its original. Let us call it the pattern (P) which both the original and its model instantiate. By virtue of P the original and the copy are partially iso-morphic" (Hunter, 2009: 244). This pattern P also holds for the statue and *its* model. Of course, in the logic of the Posidonian argument, the statue *is* the concrete model *of* reality and this is discernable by anybody. And so lurking here is the possibility that representation and reality are so symmetrical that they can be models for or simulations of each other.[70]

---

[70] This chapter has benefited from comments by Andrew Janiak, Marij Van Strien, Andrew Bailey, and incredible support by David Haig.

Bibliography

Albury W. R. 1978. Halley's Ode on the *Principia* of Newton and the Epicurean revival in England. *Journal of the History of Ideas*, 39(1): 24–43.

Anscombe, G. E. M. 1974. "Whatever has a beginning of existence must have a cause": Hume's argument exposed. *Analysis*, 34(5): 145–151.

Ariotti, Piero. 1968. Galileo on the isochrony of the pendulum. *Isis*, 59(4): 414–426.

Arthur, Richard T. W. 1995. Newton's fluxions and equably flowing time. *Studies in History and Philosophy of Science Part A*, 26(2): 323–351.

Bacon, F. 1870. *New Organon*. Translated by J. Spedding, R. L. Ellis, and D. D. Heath. In *The works of Francis Bacon* (Vol. IV). London: Longmans.

Barry, Galen. 2016. Reply to Yenter: Spinoza, Number, and Diversity. *British Journal for the History of Philosophy*, 24(2): 365–374.

Battersby, C. 1979. The *Dialogues* as original imitation: Cicero and the nature of Hume's scepticism. In *McGill Hume Studies*. D. F. Norton, N. Capaldi, and W. L. Robison (eds.), 239–252. Austin Hill Press.

Bedini, Silvio A. 1991. *The Pulse of Time: Galileo Galilei, the Determination of Longitude, and the Pendulum Clock*. Florence: Olschki.

Belkind, O. 2007. Newton's conceptual argument for absolute space. *International Studies in the Philosophy of Science*, 21: 271–293.

Belkind, Ori. 2012. Newton's scientific method and the Universal Law of Gravitation. In *Interpreting Newton: Critical essays*. A. Janiak and E. Schliesser (eds.), 138–168.

Belkind, Ori. 2017. On Newtonian induction. *Philosophy of Science*, 84(4): 677–697.

Bentley, Richard. 1692. Sermons preached at Boyle's lecture: Remarks upon a discourse of free-thinking. Edited by A. Dyce. London: Macpherson, 1838.

Berryman, Sylvia. 2010. *The Mechanical Hypothesis in Ancient Greek Natural Philosophy*. Cambridge, UK: Cambridge University Press, 2010.

Bertoloni Meli, Domenico. 2006a. Inherent and centrifugal forces in Newton. *Archive for History of Exact Sciences*, 60: 319–335.

Bertoloni Meli, Domenico. 2006b. *Thinking with Objects: The Transformation of Mechanics in the Seventeenth Century*, Baltimore: Johns Hopkins University Press.

Bertoloni Meli, Domenico. 2010. The axiomatic tradition in seventeenth-century mechanics. In *Discourse on a New Method: Reinvigorating the Marriage of History and Philosophy of Science*. M. Domski and M. Dickson (eds.), 23–42. Chicago: Open Court.

Biener, Zvi. 2017. De Gravitatione reconsidered: The changing significance of experimental evidence for Newton's metaphysics of space. *Journal of the History of Philosophy*, 55(4): 583–608.

Biener, Zvi, and Eric Schliesser (eds.). 2014. *Newton and Empiricism*. Oxford: Oxford University Press.

Biener, Z., and C. Smeenk. 2012. Cotes's queries: Newton's empiricism and conceptions of matter. In *Interpreting Newton: Critical Essays*. A. Janiak and E. Schliesser (eds.), 105–137. Cambridge University Press, UK: Cambridge.

Blüh, O. 1935. Newton and Spinoza. *Nature*, 135: 658–659.

Bobzien, Susanne. 1998. *Determinism and Freedom in Stoic Philosophy*. Oxford: Oxford University Press.

Boehm, Omri. 2014. *Kant's Critique of Spinoza*. Oxford: Oxford University Press.

Boehm, Omri. 2016. Spinoza must reject primitive necessity and deny that reason can set ends: A response to Eric Schliesser's review of Kant's critique of Spinoza. *Graduate Faculty Philosophy Journal*, 37(1): 173–186.

Bokulich, Alisa. 2020. Calibration, coherence, and consilience in radiometric measures of geologic time. *Philosophy of Science*, 87(3): 425–456.

Boole, G. 1854. *An investigation of the laws of thought, on which are founded the mathematical theories of logic and probabilities*. London: Walton and Maberly.

Brackenridge, J. B. 1995. *The Key to Newton's Dynamics: The Kepler Problem and the Principia*. Berkeley: University of California Press.

Brading, Katherine. 2011. On composite systems: Descartes, Newton, and the law-constitutive approach. In *Vanishing Matter*. D. Jalobeanu and P. R. Anstey (eds.), 130–152. London: Routledge.

Brading, Katherine. 2012. Newton's law-constitutive approach to bodies: A response to Descartes. In *Interpreting Newton*. A. Janiak and E. Schliesser (eds.), 13–32. Cambridge, UK: Cambridge University Press.

Brading, Katherine. 2017. Time for empiricist metaphysics. In *Metaphysics and the Philosophy of Science: New Essays*. Matthew Slater and Zanja Yudell (eds.), 13–40. Oxford: Oxford University Press.

Brading, Katherine. 2018. Newton on body. *The Oxford Handbook of Newton*. Edited by Eric Schliesser and Chris Smeenk. Oxford: Oxford University Press. Reprint forthcoming. https://www.oxfordhandbooks.com/view/10.1093/oxfordhb/9780199930418.001.0001/oxfordhb-9780199930418-e-10.

Brading, Katherine. 2019. A note on rods and clocks in Newton's *Principia*. *Studies in History and Philosophy of Modern Physics*, 67: 160–166.

Brading, Katherine, and Marius Stan. under review. How physics flew the philosophers' nest.

Brading, Katherine, and Marius Stan. Forthcoming. *Philosophical Mechanics in the Age of Reason*.

Brooke, Christopher. 2006. How the Stoics became atheists. *The Historical Journal*, 49(2): 387–402.

Brouwer, René. 2019. The Stoics on luck. In *The Routledge Handbook of the Philosophy and Psychology of Luck*. Ian M. Church and Robert J. Hartman (eds.), London: Routledge.

Brown, Gregory. 2016. Did Samuel Clarke really disavow action at a distance in his correspondence with Leibniz? Newton, Clarke, and Bentley on gravitation and action at a distance. *Studies in History and Philosophy of Science Part A*, 60: 38–47.

Buchdahl, G. 1992. *Kant and the Dynamics of Reason*. London: Blackwell.

Buchwald, Jed, and Robert Fox (eds.). 2013. *The Oxford Handbook of the History of Physics*. Oxford: Oxford University Press.

Burtt, E. A. 1927. *The Metaphysical Foundations of Modern Science*. London: K. Paul, Trench, Trubner & Co.

Cassirer, Ernst. 1951. *The Philosophy of the Enlightenment*. Translated by J. P. Pettegrove and F. C. A. Koelin. Princeton, NJ: Princeton University Press.

Cassirer, Ernst. 1953 (1923). *Substance and Function, and Einstein's Theory of Relativity*. Translated by W. C. Swabey and M. C. Swabey. New York: Dover.

Chalmers, Alan. 2009. *The Scientist's Atom and the Philosopher's Stone of Science*. New York: Springer.

Chalmers, David. 2021. *Reality+: A Philosophical Journey Through Virtual Worlds*. London: Penguin.

Cicero. 1958. *De natura deorum* 2 and 3. Edited by A. S. Pease. Cambridge, MA: Harvard University Press.

Cicero. 1978. *The Nature of the Gods*. Translated by P. G. Walsh. Oxford: Oxford University Press.

Clarke, Samuel. 1705. *A demonstration of the being and attributes of God: more particularly in answer to Mr. Hobbs, Spinoza and their Followers*. London: Botham.

Clarke, S. 1998 [1705]. *A Demonstration of the Being and Attributes of God*. Edited by E. Vailati. Cambridge, UK: Cambridge University Press.

Clericuzio, Antonio. 2001. Gassendi, Charleton and Boyle on matter and motion. *Late Medieval and Early Modern Corpuscular Matter Theories*. Edited by Christoph Lüthy. 467–482. Dordrecht, The Netherlands: Brill.

Cohen, I. B. 1964. "Quantum in Se Est": Newton's concept of inertia in relation to Descartes and Lucretius. *Notes and Records of the Royal Society of London*, 19(2): 131–155.

Cohen, I. Bernard. 1971. *Introduction to Newton's Principia*. Cambridge, MA: Harvard University Press.

Cohen, I. B., and G. E. Smith (eds.). 2002. *The Cambridge Companion to Newton*. Cambridge, UK: Cambridge University Press.

Colie, Rosalie L. 1959. Spinoza and the early English deists. *Journal of the History of Ideas*, 20: 23–46.

Colie, Rosalie L. 1963. Spinoza in England, 1665–1730. *Proceedings of the American Philosophical Society*, 107: 183–219.

Cunningham, A. 1988. Getting the game right: Some plain words on the identity and invention of science. *Studies in History and Philosophy of Science: Part A*, 19(3): 365–389.

Cunningham, A. 1991. How the *Principia* got its name: Or, taking natural philosophy seriously. *History of Science*, 29: 377–392.

Curley, Edwin. 1988. *Behind the Geometric Method: A Reading of Spinoza's Ethics*. Princeton, NJ: Princeton University Press.

Dagron, Tristan. 2009. *Toland et Leibniz: L'Invention du Néo-Spinozisme* [Toland and Leibniz: The invention of neo-Spinozism]. Paris: Vrin.

Daniel, Stephen H. 1984. *John Toland: His Methods, Manners, and Mind*. Vol. 7. Montreal, QC: McGill-Queen's Press-MQUP.

Daston, Lorraine, and Katharine Park. 1998. *Wonders and the Order of Nature, 1150–1750*. New York: Zone Books.

De Smet, Rudolf, and Karin Verelst. 2001. Newton's Scholium Generale: The Platonic and Stoic legacy: Philo, Justus Lipsius and the Cambridge Platonists. *History of Science*, 39: 1–30.

Della Rocca, Michael. 2008. *Spinoza*. New York: Routledge.

Demeter, Tamás, and Eric Schliesser. 2019. The uses and abuses of mathematics in early modern philosophy: Introduction. *Synthese*, 196(9): 3461–3464.

Dennett, Daniel C. 2008. Descartes's argument from design. *The Journal of Philosophy*, 105(7): 333–345.

Dennett, Daniel C. 2017. *From Bacteria to Bach and Back: The Evolution of Minds*. New York: WW Norton & Company.

Dempsey, Liam. 2006. Written in the flesh: Isaac Newton on the mind–body relation. *Studies in History and Philosophy of Science Part A*, 37(3): 420–441.

De Pierris, Graciela. 2015. *Ideas, Evidence, and Method: Hume's Skepticism and Naturalism Concerning Knowledge and Causation.* Oxford: Oxford University Press.

Descartes, René. 1984. *Principles of Philosophy.* Translated by Valentine Rodger Miller and Reese P. Miller. Dordrecht, The Netherlands: Kluwer.

Dicker, Georges. 2013. *Descartes: An Analytic and Historical Introduction.* Oxford: Oxford University Press.

Diderot, D. "Spinoza, Philosophy of". 2007 (1765). In *The Encyclopedia of Diderot & d'Alembert Collaborative Translation Project* (M. Eden, Trans.). http://quod.lib.umich. edu/cgi/t/text/text-idx?c=did;view=text;rgn=main;idno=did2222.0000.762.

DiSalle, Robert. 2002. Newton's philosophical analysis of space and time. In *The Cambridge Companion to Newton.* I. B. Cohen and G. E. Smith (eds.), 33–56. Cambridge, UK: Cambridge University Press.

Di Salle, Robert. 2006. *Understanding Space-Time: The Philosophical Development of Physics from Newton to Einstein.* Cambridge, UK: Cambridge University Press.

Dobbs, B. J. T. 1991. *The Janus Faces of Genius: The Role of Alchemy in Newton's Thought.* Cambridge, UK: Cambridge University Press.

Doepke, P. 1986. In defense of Locke's Principle: A reply to P. Simons. *Mind*, (378): 238–241.

Domski, Mary. 2012. Newton and Proclus: Geometry, imagination, and knowing space. *The Southern Journal of Philosophy,* 50(3): 389–413.

Domski, Mary. 2013. Kant and Newton on the a priori necessity of geometry. *Studies in History and Philosophy of Science,* 44(3): 438–447.

Downing, L. 2007. Locke's ontology. In *The Cambridge Companion to Locke's Essay Concerning Human Understanding.* L. Newman (ed.), 352–380. Cambridge, UK: Cambridge University Press.

Ducheyne, Steffen. 2001. Isaac Newton on space and time: Metaphysician or not? *Philosophica,* 67(1): 87–88.

Ducheyne, Steffen. 2006. The "General Scholium": Some notes on Newton's published and unpublished endeavours. *Lias: Sources and Documents Relating to the Early Modern History of Ideas,* 33: 223–274.

Ducheyne, Steffen. 2008. J. B. Van Helmont's *De Tempore* as an influence on Isaac Newton's doctrine of absolute time. *Archiv für Geschichte der Philosophie,* 90(2): 216–228.

Ducheyne, Steffen. 2009. Understanding (in) Newton's argument for universal gravitation. *Journal for General Philosophy of Science,* 40: 227–258.

Ducheyne, Steffen. 2011. *The Main Business of Natural Philosophy: Isaac Newton's Natural-Philosophical Methodology.* Dordrecht, The Netherlands: Springer.

Ducheyne, Steffen. 2014. Newton on action at a distance. *Journal of the History of Philosophy,* 52(4): 675–702.

Ducheyne, S., and E. Weber. 2008. The concept of causation in Newton's mechanical and optical work. *Logic and Logical Philosophy,* 16(4): 265–288.

Epicurus. 1972 (1925). "Epicurus letter to Herodotus." In *Lives of Eminent Philosophers.* Written by Diogenes Laertius and Edited by R. D. Hicks. Cambridge, UK. Harvard University Press. http://www.perseus.tufts.edu/hopper/text?doc=D.+L.+10.1&fromdo c=Perseus%3Atext%3A1999.01.0258.

Feingold, M. 2001. Mathematicians and naturalists: Sir Isaac Newton and the royal society. In *Isaac Newton's Natural Philosophy.* I. Bernard Cohen and J. Z. Buchwald (eds.), Cambridge, MA: MIT Press.

Feingold, M. 2004. *The Newtonian Moment: Isaac Newton and the Making of Modern Culture.* New York: New York Public Library.

Force, J. E., and R. H. Popkin. 1994. *The Books of Nature and Scripture: Recent Essays on Natural Philosophy, Theology, and Biblical Criticism in the Netherlands of Spinoza's Time and the British Isles of Newton's Time*. Dordrecht, The Netherlands: Kluwer Academic.

Fowler, James. 2017. 'Procedes Huc': Voltaire, Newton, and Locke in Lettres Philosophiques. *Neophilologus*, 101(1): 15–28.

Freeth, Tony et al. 2006. Decoding the ancient Greek astronomical calculator known as the Antikythera Mechanism. *Nature* 444(7119): 587.

Friedman, Michael. 1992. *Kant and the Exact Sciences*. Cambridge, MA: Harvard University Press.

Friedman, Michael. 2001. *Dynamics of Reason*. Stanford, CA: CSLI Publications.

Friedman, M. 2009. Newton and Kant on absolute space: From theology to transcendental philosophy. In *Constituting Objectivity: Transcendental Perspectives on Modern Physics*. The University of Western Ontario Series in Philosophy of Science. Michel Bitbol, Jean Petitot, and Pierre Kerszberg (eds.), 35–50. Dordrecht, The Netherlands: Springer.

Friedman, M. 2011. Newton and Kant on Absolute Space: From Theology to Transcendental Philosophy. In *Interpreting Newton: Critical Essays*. A. Janiak and E. Schliesser (eds.), 35–50. Cambridge, UK: Cambridge University Press.

Friedman, Michael. 2020. Newtonian methodological abstraction. *Studies in History and Philosophy of Science Part B: Studies in History and Philosophy of Modern Physics*, 72: 162–178. https://doi.org/10.1016/j.shpsb.2020.04.006.

Frigg, R. Y. Hartmann, and Stephan Hartmann. 2006. Models in science. In *Stanford Encyclopedia of Philosophy*. http://plato.stanford.edu/entries/models-science (2020).

Funkenstein, Amos. 1986. *Theology and the Scientific Imagination from the Middle Ages to the Seventeenth Century*. Princeton, NJ: Princeton University Press.

Gabbey, Alan. 1971. Force and inertia in seventeenth-century dynamics. *Studies in History and Philosophy of Science Part A*, 2(1): 1–67.

Gabbey, Alan. 1992. Newton's Mathematical Principles of Natural Philosophy: A Treatise on "Mechanics." In *The Investigation of Difficult Things*. P. M. Harman and A. E. Shapiro (eds.), 305–322. Cambridge, UK: Cambridge University Press.

Galileo Project (n.d.). *Pendulum Clock*. http://galileo.rice.edu/sci/instruments/pendulum.html

Garber, Daniel. 1987. How God causes motion: Descartes, divine sustenance, and occasionalism. *The Journal of Philosophy*, 84(10): 567–580.

Garber, Daniel. 2000. A different Descartes: Descartes and the programme for a mathematical physics in his correspondence. In *Descartes' Natural Philosophy*. S. Gaukroger, J. Schuster, and J. Sutton (eds.), 113–130. New York: Routledge.

Garrett, D. 1994. Spinoza's theory of metaphysical individuation. In *Individuation and Identity in Early Modern Philosophy. Descartes to Kant*. Kenneth F. Barber and Jorge J. E Gracia (eds.), 73–101. Albany: SUNY Press.

Glacken, Clarence J. 1967. *Traces on the Rhodian shore: Nature and Culture in Western Thought from Ancient Times to the End of the Eighteenth Century*. Vol. 170. Berkeley: University of California Press.

Goldish, M. 1999. Newton's "Of the Church": Its contents and implications. In *Newton and Religion: Context, Nature, and Influence*. J. E. Force and R. Popkin (eds.), 145–164. Dordrecht, The Netherlands: Kluwer.

Gorham, Geoffrey. 2007. Descartes on time and duration. *Early Science and Medicine*, 12: 28–54.

Gorham, Geoffrey. 2011a. Newton on God's relation to space and time: The Cartesian framework. *Archiv für Geschichte der Philosophie*, 93(3): 281–320.

Gorham, Geoffrey. 2011b How Newton solved the mind-body problem. *History of Philosophy Quarterly*, 28(1): 21–44.

Gorham, Geoffrey. 2012. "The twin-brother of space": Spatial analogy in the emergence of absolute time. *Intellectual History Review*, 22(1): 23–39.

Grant, E. 1981. *Much Ado about Nothing: Theories of Space and Vacuum from the Middle Ages to the Scientific Revolution*. Cambridge, UK: Cambridge University Press.

Grene, M. 1999 (1985). *Descartes*. Indianapolis: Hackett Press.

Guerlac, H., and M. C. Jacob. 1969. Bentley, Newton, and Providence. The Boyle lectures once more. *Journal of the History of Ideas*, 30(3): 307–318.

Guicciardini, N. 1999. *Reading the* Principia: *The Debate on Newton's Mathematical Methods for Natural Philosophy from 1687 to 1736*. Cambridge, UK: Cambridge University Press.

Hall, A. Rupert. 1952. *Ballistics in the Seventeenth Century*. Cambridge, UK: Cambridge University Press.

Halley, Edmond. 1705. *A Synopsis of the Astronomy of Comets*. Translated from the original. Oxford: John Senex.

Harper, William L. 2011. *Isaac Newton's Scientific Method: Turning Data into Evidence about Gravity and Cosmology*. Oxford: Oxford University Press.

Harper, W. 2012. Measurement and method: Some remarks on Newton, Huygens and Euler on natural philosophy. In *Interpreting Newton*. A. Janiak and E. Schliesser (eds.), Cambridge, UK: Cambridge University Press.

Harper, William, and George E. Smith. 1995. Newton's new way of inquiry. In *The creation of ideas in physics*. Edited by Jarrett Leplin, 113–166. Dordrecht, The Netherlands: Springer.

Harris, John. 1708. *Lexicon Technicum*. 2nd edition. London (n.p.).

Harrison, Peter. 2002. Voluntarism and early modern science. *History of Science*, 40(1): 63–89.

Harrison, P. 2004. Was Newton a voluntarist? In *Newton and Newtonianism*. J. E. Force and S. Hutton (eds.), Dordrecht, The Netherlands: Kluwer.

Harrison, Peter. 2007. *The Fall of Man and the Foundations of Science*. Cambridge, UK: Cambridge University Press.

Hattab, Helen. 2000. The problem of secondary causation in Descartes: A response to Des Chene. *Perspectives on Science*, 8(2): 93–118.

Hattab, Helen. 2007. Concurrence or divergence? Reconciling Descartes's physics with his metaphysics. *Journal of the History of Philosophy*, 45(1): 49–78.

Hazony, Yoram. 2014. Newtonian explanatory reduction and Hume's system of the sciences. In *Newton and Empiricism*. Edited by Z. Biener and E. Schliesser, 138–170. Oxford: Oxford University Press.

Henry, John. 1994. "Pray do not ascribe that notion to me": God and Newton's Gravity. *The Books of Nature and Scripture: Recent Essays on Natural Philosophy, Theology and Biblical Criticism in the Netherlands of Spinoza's Time and the British Isles of Newton's Time*. Edited by James E. Force and Richard H. Popkin, 123–147. Dordrecht, The Netherlands: Springer.

Henry, J. 1999. Isaac Newton and the problem of action at a distance. *Krisis*, 8–9: 30–46.

Henry, J. 2004. Metaphysics and the origins of modern science: Descartes and the importance of laws of nature. *Early Science and Medicine*, 73–114.

Henry, J. 2007. Isaac Newton y el problema de la acción a distancia. *Estudios de Filosofía*, 35: 189–226.

Henry, J. 2009. Voluntarist theology at the origins of modern science: A response to Peter Harrison. *History of Science*, 47: 79–113.

Henry, J. 2011. Gravity and De gravitatione: The development of Newton's ideas on action at a distance. *Studies in History and Philosophy of Science Part A*, 42(1): 11–27.

Henry, John. 2013. The reception of Cartesianism. In *The Oxford Handbook of British Philosophy in the Seventeenth Century*. Peter Anstey (ed.), Oxford: Oxford University Press.

Henry, John. 2014. Newton and action at a distance between bodies—A response to Andrew Janiak's "Three concepts of causation in Newton." *Studies in History and Philosophy of Science Part A*, 47: 91–97.

Henry, John. 2019 (forthcoming). Newton and action at a distance. The Oxford Handbook of Newton. Edited by Eric Schliesser and Chris Smeenk. Oxford: Oxford University Press. https://www.oxfordhandbooks.com/view/10.1093/oxfordhb/9780199930418.001.0001/oxfordhb-9780199930418-e-17.

Henry, John. 2020. Primary and secondary causation in Samuel Clarke's and Isaac Newton's theories of gravity. *Isis*, 111(3): 542–561.

Herivel, John. 1965. *The Background to Newton's Principia*. Oxford: Oxford University Press.

Hicks, Lewis Ezra. 1883. *A Critique of Design-arguments: A Historical Review and Free Examination of the Methods of Reasoning in Natural Theology*. New York: Scribner.

Hobbes, Thomas. 1656. *Six Lessons to the Professors Mathematiques One of Geometry the Other of Astronomy, in the Chaires Set Up by the Noble and Learned Sir Henry Savile in the University of Oxford*. London: Andrew Cook.

Hooker, R. 1888. *The Works of That Learned and Judicious Divine Mr. Richard Hooker with an Account of his Life and Death by Isaac Walton*. Vol. 1. Chapter: The First Book Concerning Laws and Their Several Kinds In General. (Arranged by the Rev. John Keble MA. 7th edition revised by the Very Rev. R. W. Church and the Rev. F. Paget). Oxford: Clarendon Press. http://oll.libertyfund.org/title/921/85481.

Hübner, Karolina. 2015. On the significance of formal causes in Spinoza's Metaphysics. *Archiv für Geschichte der Philosophie*, 97(2): 196–233.

Huggett, N. 2012. What did Newton mean by Absolute Motion? In *Interpreting Newton*. A. Janiak and E. Schliesser (eds.), Cambridge, UK: Cambridge University Press.

Hume, David. 1777. *An Enquiry Concerning Human Understanding, in Essays and Treatises on Several Subjects*. London: A. Millar. Retrieved from http://davidhume.org.

Hume, David. 1739. *A Treatise concerning Human Understanding*. Hume Texts Online. http://www.davidhume.org/

Hume, David. 2000 (1748). *An Enquiry concerning Human Understanding*. Edited by T. L. Beauchamp. 1st edition. Oxford: Clarendon Press.

Hume, David. 1983. *The History of England from the Invasion of Julius Caesar to the Revolution in 1688*. Foreword by William B. Todd. 6 vols. Indianapolis: Liberty Fund.

Hunter, Graeme. 2009. Cicero's neglected argument from design. *British Journal for the History of Philosophy*, 17(2): 235–245.

Hurlbutt, Robert H., III. 1965 (1985). *Hume, Newton, and the Design Argument*. Lincoln: University of Nebraska Press.

Hutton, Sarah. 1990. *Henry More*. Dordrecht, The Netherlands: Kluwer.

Huygens, Christiaan. 1669. Instructions concerning the use of pendulum-watches for finding the longitude at sea. *Philosophical Transactions of the Royal Society*, 4–47, 937. http://adcs.home.xs4all.nl/Huygens/06/kort-E.html

Huygens, C. 1698 (1888–1950). Cosmotheoros. In *Oeuvres completes de Christiaan Huygens*. Vol. 21. D. Bierens de Haan, L. Bosscha, and D. J. Korteweg (eds.), 653–842. The Hague: M. Nijhoff.

Huygens, Christiaan. 1690. *Traite' de la Lumiere* [Treatise on Light]. In *The Scientific Background to Modern Philosophy*. Edited and translated by M. Matthews, 124–132. Indianapolis and Cambridge, MA: Hackett.

Huygens, Christiaan. 1944. *Oeuvres completes*. Volume 21, *Cosmologie*. J. A. Vollgraff, (ed.), The Hague: Martinus Nijhoff, 1944.

Internet Encyclopedia of Philosophy. 2008. Emanation. http://www.iep.utm.edu/e/emanatio.htm. Accessed on October 24, 2008.

Israel, Jonathan. 2002. *Radical Enlightenment: Philosophy and the Making of Modernity, 1650–1750*. Oxford: Oxford University Press.

Israel, Jonathan. 2009. *A Revolution of the Mind: Radical Enlightenment and the Intellectual Origins of Modern Democracy*. Oxford: Oxford University Press.

Jacob, Alexander. 1991. *Henry More's Refutation of Spinoza*. Hildesheim, Germany: Olms.

Jacob, Margaret Candee. 1969. John Toland and the Newtonian ideology. *Journal of the Warburg and Courtauld institutes*, 32: 307–331.

Jacob, Margaret. 1981. *The Radical Enlightenment*. London: Allen & Unwin.

Jalobeanu, Dana. 2007. Space, bodies and geometry: Some sources of Newton's Metaphysics. *Zeitsprünge, Forschungen zur Frühen Neuzeit*, 11: 81–113.

Jalobeanu, Dana. 2011. The Cartesians of the Royal Society: The debate over collisions and the nature of body (1668–1670). In *Vanishing Matter and the Laws of Motion*. D. Jalobeanu and P. R. Anstey (eds.), 103–129.

Jalobeanu, Dana, and P. R. Anstey (eds.). 2011. *Vanishing Matter and the Laws of Motion: Descartes and Beyond*. New York: Routledge.

Janiak, A. 2007. Newton and the reality of force *The Journal of the History of Philosophy*, 45(1): 127–147.

Janiak, A. 2008. *Newton as Philosopher*. Cambridge, UK: Cambridge University Press.

Janiak, Andrew. 2010. Substance and action in Descartes and Newton. *The Monist*, 93.4(2010): 657–677.

Janiak, A. 2012. Newton and Descartes: Theology and natural philosophy. *The Southern Journal of Philosophy*, 50(3): 414–435.

Janiak, A. 2013. Three concepts of causation in Newton. *Studies in History and Philosophy of Science Part A*, 44(3): 396–407.

Janiak, Andrew. 2015. Mathematics and infinity in Descartes and Newton. *Mathematizing Space*. 209–230. Cham, Switzerland: Birkhauser.

Janiak, Andrew, and Eric Schliesser (eds.). 2012. *Interpreting Newton: Critical Essays*. Cambridge, UK: Cambridge University Press.

Jantzen, Bernard C. 2014. *An Introduction to Design Arguments* Cambridge, UK: Cambridge University Press.

Jones, Alexander. 2017. *A Portable Cosmos: Revealing the Antikythera Mechanism, Scientific Wonder of the Ancient World*. Oxford: Oxford University Press.

Jorink, Eric. 2009. Honouring Sir Isaac, or, Exorcising the Ghost of Spinoza: Some Remarks on the Success of Newton in the Dutch Republic. In *Future Perspectives on Newton Scholarship and the Newtonian Legacy in Eighteenth-Century Science and Philosophy*. Steffen Ducheyne (ed.), 23–33. Brussels: KVAB.

Joy, L. 2006 Scientific explanation: from formal causes to laws of nature. In *The Cambridge History of Science. Volume 3: Early Modern Science*. K. Park and L. Daston (eds.), 70–105. Cambridge, UK: Cambridge University Press.

Kant, I. 1977 (1960). *Allgemeine Naturgeschichte und Theorie des Himmels: W. Weischedel* (ed.), Theorie-Werkausgabe Immanuel Kant: *Vorkritische Schriften bis 1768*. Volume 1. Frankfurt Am Main, Suhrkampf.

Kant, I. 2008. *Universal Natural History and Theory of the Heavens*. Arlington, VA: Richer Resources Publications. http://records.viu.ca/~johnstoi/kant/kant2e.htm.

Kepler, J. 2003. *Kepler's Somnium: The Dream, or Posthumous Work on Lunar Astronomy*. Edited and translated by E. Rosen. Mineola: Dover Publications.

Kerszberg, P. 2012. Deduction versus discourse: Newton and the cosmic phenomena. *Foundations of Science*, 18: 529–544. http://dx.doi.org/10.1007/s10699-011-9283-2.

Kidd, I. G. 1988. *Posidonius II: The commentary: (i) Testimonia and Fragments 1–149*. Cambridge, UK: Cambridge University Press.

Kochiras, Hylarie. 2008. Force, matter, and metaphysics in Newton's natural philosophy. PhD diss., University of North Carolina, Chapel Hill.

Kochiras, H. 2009. Gravity and Newton's substance counting problem. *Studies in History and Philosophy of Science*, 40: 267–280.

Kochiras, Hylarie. 2013. The mechanical philosophy and Newton's mechanical force. *Philosophy of Science*, 80(4): 557–578.

Koyré, Alexandre. 1950. The significance of the Newtonian synthesis. *The Journal of General Education*, 4.4: 256–268.

Koyré, A. 1957. *From the Closed World to the Infinite Universe*. Baltimore: The Johns Hopkins University Press.

Koyré, Alexandre, and I. Bernard Cohen. 1961. The Case of the Missing Tanquam: Leibniz, Newton and Clarke. *Isis*, 52(4): 555–566.

Kristensen, L. K., and K. M. Pedersen. 2012. Roemer, Jupiter's satellites, and the velocity of light. *Centaurus*, 54: 4–38.

Kuhn, Thomas. 1977. *The Essential Tension: Selected Studies in Scientific Tradition and Change*. Chicago: University of Chicago Press.

Lærke, Mogens. 2013. The anthropological analogy and the constitution of historical perspectivism. In *Philosophy and Its History: Aims and Methods in the Study of Early Modern Philosophy*, Mogens Laerke, Justin E. H. Smith, and Eric Schliesser (eds.), 7–29. Oxford: Oxford University Press.

Lange, M. 2009. *Laws and Lawmakers*. Oxford: Oxford University Press.

Leask, Ian. 2012. Unholy force: Toland's Leibnizian "Consummation" of Spinozism. *British Journal for the History of Philosophy*, 20(3): 499–537.

Lee, R. A., Jr. 2006. The Cartesian resources for Descartes' concept of *causa sui*. *Oxford Studies in Early Modern Philosophy*, 3: 91–118.

Leibniz, G. W., and S. Clarke. 2000. In *Correspondence*. R. Ariew (ed.), Indianapolis: Hackett Publishing.

Leibniz, G., & Clarke, S. 1717. *A collection of papers, which passed between the late learned Mr. Leibnitz, and Dr. Clarke, in the years 1715 and 1716: Relating to the principles of natural philosophy and religion. With an appendix. To which are added, Letters to Dr. Clarke concerning liberty and necessity; from a gentleman of the University of Cambridge: with the Doctor's answers to them. Also Remarks upon a book, entitled, A philosophical enquiry concerning human liberty*. Retrieved from http://books.google.com/books/about/A_collection_of_papers_which_passed_betw.html?id=_RUHAAAAQAAJ&redir_esc=y.

Martin Lin, 2004. "Spinoza's Metaphysics of Desire: The Demonstration of IIIP6." *Archiv für Geschichte der Philosophie* 86.1 (2004), 21–55.

Lüthy, C. et al. (eds.). 2001. *Late Medieval and Early Modern Corpuscular Matter Theories*. Dordrecht, The Netherlands: Brill.

MacIntosh, J. J. 1991. Robert Boyle on Epicurean atheism and atomism. In *Atoms, Pneuma, and Tranquility: Epicurean and Stoic Themes in European Thought*. M. J. Osler (ed.), 197–220. Cambridge, UK: Cambridge University Press.

Maclaurin, Colin. 1748. *An Account of Sir Isaac Newton's philosophy*. Edinburgh: Patrick Murdoch.

Maglo, K. 2003. The reception of Newton's gravitational theory by Huygens, Varignon, and Maupertuis: How normal science may be revolutionary. *Perspectives on Science*, 11(2): 135–169.

Mancosu, P. 1999. *Philosophy of Mathematics and Mathematical Practice in the Seventeenth Century*. Oxford: Oxford University Press.

Manning, Richard. 2012. Spinoza's physical theory. In *The Stanford Encyclopedia of Philosophy* (Spring edition). Edited by Edward N. Zalta. http://plato.stanford.edu/archives/spr2012/entries/spinoza-physics.

Manson, Neil A. (ed.). 2003. *God and Design: The Teleological Argument and Modern Science*. London: Routledge.

Marchant, Jo. 2009. *Decoding the Heavens: Solving the Mystery of the World's First Computer*. London: Windmill Books.

Martin, C. 2003. D'un épicurisme "discret": Pour une lecture lucrécienne des entretiens sur la pluralité des mondes de Fontenelle. *L'épicurisme des lumières*, 35: 55–73.

Massimi, M. 2011. Kant's dynamical theory of matter in 1755, and its debt to speculative Newtonian experimentalism *Studies in History and Philosophy of Science*, 42: 525–543.

Massimi, Michela. 2014. "Prescribing Laws to Nature. Part I. Newton, the pre-Critical Kant, and Three Problems about the Lawfulness of Nature." *Kant-Studien*, 105(4): 491–508.

Maudlin, Tim. 2007. *The Metaphysics within Physics*. Oxford: Oxford University Press.

Maxwell, J. C. 1890. *The Scientific Papers of James Clerk Maxwell*, 2 vols. Edited by W. D. Niven. Cambridge, UK: Cambridge University Press.

Maxwell, James Clerk. 2010. *Matter and Motion*. Cambridge, UK: Cambridge University Press.

McGuire, J. E. 1968. Force, active principles, and Newton's invisible realm. *Ambix*, 15: 154–208.

McGuire, J. E. 1970a. Atoms and the 'Analogy of Nature': Newton's Third Rule of philosophizing. *Studies in History and Philosophy of Science*, 1(1): 3–58.

McGuire, J. E. 1970b Newton's "Principles of Philosophy": An intended Preface for the 1704 *Opticks* and a related draft fragment, *British Journal for the History of Science*, 5: 178–186.

McGuire, J. E. 1978. Existence, actuality and necessity: Newton on space and time. *Annals of Science*, 35(5): 463–508.

McGuire, J. E. 1995. *Tradition and innovation: Newton's metaphysics of nature* Dordrecht, The Netherlands: Kluwer.

McGuire, J. E. 2007. A dialogue with Descartes: Newton's ontology of true and immutable natures. *Journal of the History of Philosophy*, 45(1): 103–125.

McGuire, J. E. 2011. Ideas and texts: Newton and the intellectual history of science. *Sartoniana*, 24: 37–48. http://www.sartonchair.ugent.be/en/journal/archive.

McGuire, J. E., and E. Slowik. 2013. Newton's ontology of omnipresence and infinite space. *Oxford Studies in Early Modern Philosophy*, 6: 279–308.

McGuire, J. E., and P. M. Rattansi. 1966. Newton and the pipes of Pan. *Notes and Records of the Royal Society of London*, 21(2): 108–143.

McMullin, E. 2001. The impact of Newton's *Principia* on the philosophy of science. *Philosophy of Science*, 68: 279–310.

Melamed, Yitzhak Y. 2000. On the exact science of Nonbeings: Spinoza's view of mathematics. *Iyyun: The Jerusalem Philosophical Quarterly*/עיון: פילוסופי רבעון, (2000): 3–22.

Melamed, Yitzhak Y. 2006. Inherence and the immanent cause in Spinoza. *Leibniz Review*, 16: 43–52.

Melamed, Y. Y. 2010. Acosmism or weak individuals? Hegel, Spinoza, and the reality of the finite. *Journal of the History of Philosophy*, 48(1): 77–92.

Melamed, Yitzhak Y., and Martin Lin. 2020 (2016). Principle of Sufficient Reason. The Stanford Encyclopedia of Philosophy (Spring 2020 Edition). Edited by Edward N. Zalta. <https://plato.stanford.edu/archives/spr2020/entries/sufficient-reason/>.

Mercer, C. 2001. *Leibniz's Metaphysics: Its Origins and Development*. Cambridge, UK: Cambridge University Press.

Miller, D. M. 2009. Qualities, properties, and laws in Newton's method of induction. *Philosophy of Science*, 76: 1052–1063.

Montes, Leonidas. 2003. Smith and Newton: some methodological issues concerning general economic equilibrium theory. *Cambridge Journal of Economics*, 27(5): 723–747.

More, Thomas. 1662. *An Antidote Against Atheism*. 3rd edition. London: James Fletcher.

Moreau, Pierre-François. 2014. *Spinoza et le spinozisme*: Que sais-je? Paris: Presses universitaires de France.

Nadler, Stephen. 2006. *Spinoza's Ethics: An Introduction*. Cambridge, UK: Cambridge University Press.

Nadler, S. 2010 Benedictus Pantheismus. *Insiders and Outsiders in Seventeenth-Century Philosophy*. In G. A. J. Rogers, Tom Sorrell, and Jill Kraye (eds.), 240. London: Routledge, 2010.

Nelson, Alan. 2017. Descartes on the limited usefulness of mathematics. *Synthese*: 1–22.

Newman, L. (ed.). 2007. *The Cambridge Companion to Locke's Essay Concerning Human Understanding*. Cambridge, UK: Cambridge University Press.

Newton, I. 1728a *De mundi systemate liber*. London: Tonson. http://books.google.com/books?id=e44_AAAAcAAJ&printsec=frontcover   &hl=nl&source=gbs_v2_summary_r&cad=0#v=onepage&q&f=false.

Newton, I. 1728b. *A Treatise of the System of the World*. (Anonymous Trans.). London: Fayram.

Newton, I. 1730, fourth corrected edition, *Opticks: Or a Treatise of the Reflections, Refractions, Inflections & Colours of Light*. London: William Innys.

Newton, I. 1731. *A Treatise of the System of the World*. <http://books.google.com/books?id=DXE9AAAAcAAJ&dq=Newton+system+of+the+world&hl=nl&source=gbs_navlinks_s

Newton, I. 1952 (1730). *Opticks: Or a treatise of the reflections, refractions, inflections & colours of light*. Edited by I. Bernard Cohen (ed.) based on the 4th edition. London and New York: Dover Publications.

Newton, Isaac. 1969. *The System of the World*. Edited by I. B. Cohen. London: Dawsons of Pall Mall.

Newton, I. 1978. In *Unpublished scientific papers of Isaac Newton: A selection from the Portsmouth Collection in the University Library*. A. R. Hall and M. B. Hall (eds.), Cambridge, UK: Cambridge University Press.

Newton, I. 1979 (1704). *Opticks: Or a Treatise of the Reflections Inflections and Colours of Light*. New York: Dover Publications.

Newton, Isaac. 1989. *The Preliminary Manuscripts for Isaac Newton's 1687 Principia, 1684–1685.* Cambridge, UK: Cambridge University Press.

Newton I. 1999. *The Principia: Mathematical Principles of Natural Philosophy.* Translated by I. Bernard Cohen and Anne Whitman. Berkeley: University of California Press.

Newton, Isaac. 2004. *Isaac Newton: Philosophical Writings.* Edited by A. Janiak. Cambridge: Cambridge University Press.

Newton, Isaac. Forthcoming. *De Motu Corporum, Liber Secundus, ULC.* Add. 3990, folio 13. Translated by George E. Smith and Anne Whitman with commentary by George Smith and Samia Hesni.

Oakley, F. 1961. Christian theology and the Newtonian science: The rise of the concept of the laws of nature. *Church History,* 30: 433–457.

Osler, Margaret J. 1996. From immanent natures to nature as artifice: The reinterpretation of final causes in seventeenth-century natural philosophy. *The Monist,* 79(3): 388–407.

Ott, Walter. 2009. *Causation and Laws of Nature in Early Modern Philosophy.* Oxford: Oxford University Press.

Paley, William. 1809 [1802]. *Natural theology: Or, evidence of the existence and attributes of the deity.* 12th edition. London: J. Faulder.

Park, K., and L. Daston (eds.). 2006. *The Cambridge History of Science.* Volume 3: Early Modern Science. Cambridge, UK: Cambridge University Press.

Parker, Adwait A. 2019. Mathematical and Physical Space in Kantian Idealism. PhD diss., Stanford University.

Parker, Adwait A. 2020. Newton on active and passive quantities of matter. *Studies in History and Philosophy of Science Part A,* 84: 1–11. https://doi.org/10.1016/j.shpsa.2020.03.006.

Peterman, Alison. 2015. Spinoza on extension. *Philosopher's Imprint* 15.14.

Purrington, R. D. 2009. *The First Professional Scientist: Robert Hooke and the Royal Society of London* Dordrecht, The Netherlands: Springer.

Ratzsch, Del. 2014. Teleological arguments for God's existence. *The Stanford Encyclopedia of Philosophy* (Fall 2014 Edition). Edited by Edward N. Zalta. <http://plato.stanford.edu/archives/fall2014/entries/teleological-arguments/>

Ratzsch, Del, and Jeffrey Koperski. 2020. Teleological arguments for God's existence. *The Stanford Encyclopedia of Philosophy* (Summer 2020 Edition). Edited by Edward N. Zalta. <https://plato.stanford.edu/archives/sum2020/entries/teleological-arguments/>.

Riskin, Jessica. 2002. *Science in the Age of Sensibility: The Sentimental Empiricists of the French Enlightenment.* Chicago: University of Chicago Press.

Russell, Bertrand. 1986. On scientific method in philosophy. In *The Collected Papers of Bertrand Russell. Volume 8: The Philosophy of Logical Atomism and Other Essays, 1914–19.* Edited by John G. Slater, 55–73. London: Routledge.

Russell, Paul. 2008. *The Riddle of Hume's Treatise: Skepticism, Naturalism, and Irreligion.* Oxford: Oxford University Press.

Rynasiewicz, Robert. 2011. Newton's Views on Space, Time, and Motion. In *The Stanford Encyclopedia of Philosophy* (Fall 2011 Edition). Edited by Edward N. Zalta. [http://plato.stanford.edu/archives/fall2011/entries/newton-stm/]

Savan, David. 1986. Spinoza: Scientist and theorist of scientific method. In *Spinoza and the sciences.* Marjorie Glicksman Grene and Debra Nails (eds.), 95–123. Dordrecht, The Netherlands: Kluwer.

Schaffer, S. 1978. The phoenix of nature: Fire and evolutionary cosmology in Wright and Kant *Journal for the History of Astronomy,* 9: 180–200.

Schliesser, Eric, and George E. Smith (forthcoming). Huygens's 1688 Report to the Directors of the Dutch East India Company on the Measurement of Longitude at Sea and the Evidence it Offered Against Universal Gravity. Archive for the History of Exact Sciences.

Schliesser, E. 2005a. On the origin of modern naturalism: The significance of Berkeley's response to a Newtonian indispensability argument. *Philosophica*, 76: 45–66.

Schliesser, E. 2005b. Realism in the face of scientific revolutions: Adam Smith on Newton's "proof" of Copernicanism. *British Journal for the History of Philosophy*, 13(4): 697–732.

Schliesser, Eric. 2007. Two definitions of "cause," Newton, and the significance of the Humean distinction between natural and philosophical relations. *Journal of Scottish Philosophy*, 5(1): 83–101.

Schliesser, E. 2008. Hume's Newtonianism and anti-Newtonianism. In The Stanford Encyclopedia of Philosophy (Winter Edition) Edited by Edward N. Zalta, forthcoming. http://plato.stanford.edu/ archives/win2008/entries/hume-newton/.

Schliesser, Eric. 2010a. Spinoza's *conatus* as an essence-preserving, attribute-neutral immanent cause: Toward a new interpretation of attributes and modes. In *Causation and Modern Philosophy*. Keith Allen and Tom Stoneham (eds.), 65–86. London: Routledge.

Schliesser, E. 2010b. Book review of Epicureanism at the origins of modernity by Catherine Wilson. *Mind*, 119(474): 535–539.

Schliesser, Eric. 2011a. Newton's challenge to philosophy: A programmatic essay. *HOPOS: The Journal of the International Society for the History of Philosophy of Science*, 1: 101–128.

Schliesser, Eric. 2011c. Angels and philosophers: Toward a new interpretation of Spinozistic common notions. Proceedings of the Aristotelian Society 111: 497–518.

Schliesser, Eric. 2012a. The Newtonian refutation of Spinoza: Newton's challenge and the Socratic problem. In *Interpreting Newton*. Andrew Janiak and Eric Schliesser (eds.), 299–319. Cambridge: Cambridge University Press.

Schliesser, Eric. 2013. The methodological dimension of the Newtonian revolution. *Metascience*, 22(2): 329–333.

Schliesser, Eric. 2015a Introduction: On sympathy. In *Sympathy: A History*. E. Schliesser (ed.), 3–14. Oxford: Oxford University Press.

Schliesser, Eric. 2015b Review of Omri Boehm's Kant's Critique of Spinoza. *Graduate Faculty Philosophy Journal*, 36(2): 463–483.

Schliesser, E. 2017. Spinoza and the philosophy of science: Mathematics, motion, and being. In *Oxford Handbook of Spinoza*. M. D. Rocca (ed.), Oxford: Oxford University Press.

Schliesser, Eric. 2018. A genealogy of modernity and Dennett's strange inversion of reasoning. *Teorema: Revista internacional de filosofía*, 37(3): 171–180.

Schliesser, Eric. 2018a. Darwin on Savages (II): The Posidonian Argument, Revisited. Digressions and Impressions. August 29, 2018. https://digressionsnimpressions. typepad.com/digressionsimpressions/2018/08/darwin-on-savages-ii-and-the-posidonian-argument.html.

Schliesser, Eric. 2018b. Huxley on Wider Teleology. Digressions and Impressions. October 18, 2018. https://digressionsnimpressions.typepad.com/digressionsimpressions/2018/10/huxley-on-teleology.html.

Schliesser, Eric. 2018c. Four methods of empirical inquiry in the aftermath of Newton's Challenge. In *What Does it Mean to be an Empiricist?* Boston Studies in the Philosophy and History of Science 331. S. Bodenmann and A. L. Rey (eds.), 15–30. Dordrecht, The Netherlands: Springer.

Schliesser, Eric. 2020. Does Berkeley's immaterialism support Toland's Spinozism? The Posidonian argument and the eleventh objection. *Royal Institute of Philosophy Supplements*, 88: 33–71.

Schmaltz, Tad. 1999. Spinoza on the vacuum. *Archiv für Geschichte der Philosophie*, 81: 174–205.

Schneider, D. 2014. Spinoza's PSR as a principle of clear and distinct representation. *Pacific Philosophical Quarterly*, 95(1): 109–129.

Schönfeld, M. 2000. *The Philosophy of the Young Kant: The Precritical Project*. Oxford University Press: Oxford.

Sedley, David. 2007. *Creationism and Its Critics in Antiquity* Berkeley: University of California Press.

Sessions, William Lad. 2002. *Reading Hume's Dialogues: A Veneration for True Religion*. Bloomington: Indiana University Press.

Shapiro, Alan E. 1989. Huygens' *Traité de la lumière* and Newton's *Opticks*: Pursuing and eschewing hypotheses. *Notes and Records of the Royal Society of London*, 43(2): 223–247.

Shapiro, A. 2004. Newton's experimental philosophy. *Early Science and Medicine*, 9(3): 185–217.

Simons, P. 1985. Coincidence of things of a kind. *Mind*, 70–75.

Slowik, E. 2008. *Newton's metaphysics of space: A "Tertium Quid" betwixt Substantivalism and Relationalism, or merely a "God of the (Relational Mechanical) Gaps?"* PhilSci Archive. http://philsci-archive.pitt.edu/archive/00004185/.

Slowik, E. 2009. Newton's metaphysics of space: A Tertium Quid betwixt substantivalism and relationalism, or merely a "God of the (relational mechanical) gaps?" *Perspectives on Science*, 17(4): 429–456.

Slowik, E. 2013. Newton's neo-platonic ontology of space. *Foundations of Science*, 18(3): 419–448.

Smeenk, Chris. 2016. Philosophical geometers and geometrical philosophers. In *The Language of Nature: Reassessing the Mathematization of Natural Philosophy in the 17th century*. Edited by Geoffrey Gorham et al. Minneapolis: University of Minnesota Press. https://manifold.umn.edu/read/0b39d3d8-1898-492e-a954-e8a7114eb985/section/9407db1c-0d4c-4034-b50c-be2bc499915e.

Smeenk, Chris, and Eric Schliesser. 2013. Newton's *Principia*. In *Oxford Handbook of the History of Physics*, J. Buchwald and R. Fox (eds.), 109–165. Oxford: Oxford University Press.

Smeenk, Chris, and George E. Smith unpublished ms. "Newton on Constrained Motion."

Smith, Adam. (1982). *Essays on philosophical subjects* W. P. D. Wightman, J. C. Bryce (eds.), Glasgow edition of the works and correspondence of Adam Smith, Vol. III, Liberty Fund, Indianapolis. http://oll.libertyfund.org/title/201/56025/916354.

Smith, G. E. 2001 Comments on Ernan McMullin's "The impact of Newton's *Principia* on the philosophy of science." *Philosophy of Science*, 68: 327–338.

Smith, G. E. 2001. The Newtonian style in book II of the *Principia*. In *Isaac Newton's Natural Philosophy*. J. Z. Buchwald and I. B. Cohen (eds.), 249–298. Cambridge, MA: MIT Press.

Smith, G. E. 2002 Newton's Methodology in I. B. Cohen and G. E. Smith (eds.), 138–173.

Smith, George. 2008. "Isaac Newton." The Stanford Encyclopedia of Philosophy (Fall 2008 Edition), Edited by Edward N. Zalta. http://plato.stanford.edu/archives/fall2008/entries/newton/.

Smith, George E. 2014. Closing the loop: Testing Newtonian gravity, then and now. In _Newton and Empiricism_, Biener and Schliesser (eds.), 262–351. Oxford: Oxford University Press.

Smith, George E. 2019. Newton's numerator in 1685: A year of gestation. _Studies in History and Philosophy of Science Part B: Studies in History and Philosophy of Modern Physics_, 68: 163–177.

Smith. George. 2020. Experiments in the _Principia_. In _The Oxford Handbook of Newton_. Eric Schliesser and Chris Smeenk (eds.), Oxford: Oxford University Press. DOI: 10.1093/oxfordhb/9780199930418.013.36

Smith, George. "Indirect (i.e., derived) measurement and Evidence" in honor of Patrick Suppes' 90th birthday (Stanford). Unpublished.

Smith, Justin E. H. 2013. The history of philosophy as past and as process. In _Philosophy and Its History: Aims and Methods in the Study of Early Modern Philosophy_. Mogens Laerke, Justin E H Smith, and Eric Schliesser (eds.), 30–49. Oxford: Oxford University Press.

Snobelen, S. D. 1997. Caution, conscience and the Newtonian reformation: The public and private heresies of Newton, Clarke and Whiston. _Enlightenment and Dissent_, 16: 151–184.

Snobelen, S. D. 1999. Isaac Newton, Heretic: the strategies of a Nicodemite. _British Journal for the History of Science_, 32: 381–419.

Snobelen, Stephen D. 2001. "God of gods, and Lord of lords": The theology of Isaac Newton's General Scholium to the _Principia_. _Osiris_, 16: 169–208.

Solla Price, Derek de. 1974 Gears from the Greeks. The Antikythera mechanism: A calendar computer from ca. 80 BC. _Transactions of the American Philosophical Society_: 1–70.

Spinoza (ms.). _Theological Political Treatise_. Translated by E. Curley.

Spinoza, B. 2002. _Spinoza: Complete Works_. Edited and Translated by Samuel Shirley, with Introduction and Notes by Michael L. Morgan. Indianapolis: Hackett.

Spinoza, B. 2007. _Theological-Political Treatise_. Edited by J. Israel and Translated by M. Silverthorne. Cambridge, UK: Cambridge University Press.

Spinoza, Benedict de (1996). _Ethics_ Translated by Edwin Curley. London: Penguin.

Spinoza, B. 1994. _A Spinoza Reader: The Ethics and Other Works_. Edited and Translated by Edin Curley. Princeton, NJ: Princeton University Press.

Steenbakkers, Piet Een Vijandige Overname: Spinoza over Natura Naturans en Natura Naturata. In _Spinoza en Scholastiek_. Gunther Coppens (ed.), Leuven, Belgium: Acco.

Steenbergen, G. (ms). The role of measurement in Newton's De Gravitatione. (N.p.)

Stein, Howard. 1967a. Newtonian space-time. _The Texas Quarterly_ 10, 174–200. Reprinted in Palter, Robert (ed.) 1970. _The Annus Mirabilis of Sir Isaac Newton_. Cambridge, MA: MIT Press.

Stein, H. 1967b. Newtonian spacetime. _Texas Quarterly_, 10: 174–200.

Stein, H. 1970. On the notion of field in Newton, Maxwell, and beyond. In _Historical and Philosophical Perspectives of Science_. R. H. Stuewer (ed.), 264–287. Minneapolis: University of Minnesota Press.

Stein, Howard. 1977. Some philosophical prehistory of general relativity. _Minnesota Studies in the Philosophy of Science Minneapolis_, 8: 3–49.

Stein, Howard. 1990. "From the Phenomena of Motions to the Forces of Nature": Hypothesis or Deduction? _PSA: Proceedings of the Biennial Meeting of the Philosophy of Science Association_, 2: 209–222.

Stein, Howard. 2002. Newton's Metaphysics. In _Cambridge Companion to Isaac Newton_. I. B. Cohen and G. E. Smith (eds.), 256–307. Cambridge, UK: Cambridge University Press.

Steinle, Friedrich. 2002. Negotiating experiment, reason, and theology: The concept of laws of nature in the early Royal Society. In *Ideals and Cultures of Knowledge in Early Modern Europe*. W. Detel and C. Zittel (eds.), 197–212. Frankfurt: Akademie Verlag.

Strien, Marij van. 2014. On the origins and foundations of Laplacian determinism. *Studies in History and Philosophy of Science Part A*, 45: 24–31.

Stuart-Buttle, Tim. 2019. *From Moral Theology to Moral Philosophy: Cicero and Visions of Humanity from Locke to Hume*. Oxford: Oxford University Press.

Toland, John. 2013. *Letters to Serena* Edited by Ian Leask. Dublin: Four Courts Press.

Valentin, W. 2009. *Onwaarneembare hoeveelheid van materie*. BA thesis, Leiden University, Institute of Philosophy.

van Lunteren, F. 1991 *Framing hypotheses: conceptions of gravity in the 18th and 19th centuries*. PhD dissertation, Utrecht University.

Viljanen, Valtteri. 2008. On the derivation and meaning of Spinoza's *Conatus* doctrine. In *Oxford Studies in Early Modern Philosophy*, Volume 4. Daniel Garber and Steven Nadler (eds.), 89–112. Oxford: Oxford University Press.

Voltaire. 1822. *Oeuvres complètes de Voltaire*. Vol 37. Paris: Lequien.

Voltaire. 1843. *A Philosophical Dictionary*. Unknown translator. London: Dugdale.

Voltaire. 1927. *Essay on Newton* Translated by W. F. Fleming. In *The works of Voltaire; A Contemporary Version*. W. J. Fleming (ed.), 172–176. New York, Dingwall-Rock.

Watkins, Eric. 2013 The early Kant's (anti-) Newtonianism. *Studies in History and Philosophy of Science Part A*, 44(3): 429–437.

Westfall, Robert S. 1982. Isaac Newton's Theologia gentiles origines philosophicae. In *The Secular Mind: Transformations of Faith in Modern Europe*. W. Warren Wagar (ed.), 15–34. New York: Holmes and Meier.

Westfall, R. S. 1983. *Never at Rest*. Cambridge, UK: Cambridge University Press.

Whitehead, A. N. 1933. *Adventures of Ideas*. New York: Simon & Schuster.

Whewell, William. 1864. *Astronomy and General Physics: Considered with Reference to Natural Theology*. Cambridge, UK: Deighton, Bell.

Wigelsworth, Jeffrey R. 2003 Lockean Essences, Political Posturing, and John Toland's Reading of Isaac Newton's *Principia*. *Canadian Journal of History*, 38(3): 521–535.

Willmoth, F. 2012. Römer, Flamsteed, Cassini and the speed of light. *Centaurus*, 54: 39–57.

Wilson, Catherine. 2008. *Epicureanism at the Origins of Modernity*. Oxford: Oxford University Press.

Winsberg, Eric. 2010. *Science in the Age of Computer Simulation*. Chicago: University of Chicago Press.

Wolfe, Charles T. 2005. The materialist denial of monsters. In *Monsters and Philosophy*. C. T. Wolfe (ed.), 187–204. London: King's College Publications.

Wolfe, Charles. 2010a. Rethinking empiricism and materialism: The revisionist view. *Annales Philosophici*, 1: 101–113.

Wolfe, Charles. 2010b. Endowed molecules and emergent organization: The Maupertuis–Diderot debate. *Early Science and Medicine*, 15: 38–65.

Yenter, T. 2014. Clarke against Spinoza on the manifest diversity of the world. *British Journal for the History of Philosophy*, 22(2): 260–280.

Yenter, Timothy, and Ezio Vailati. (Winter 2020 Edition). Samuel Clarke. In *The Stanford Encyclopedia of Philosophy*, Edward N. Zalta (ed.), <https://plato.stanford.edu/archives/win2020/entries/clarke/>.

Yoder, Joella. 1988. *Unrolling Time: Christiaan Huygens and the Mathematization of Nature*. Cambridge, UK: Cambridge University Press.

Index Locorum

For the benefit of digital users, indexed terms that span two pages (e.g., 52–53) may, on occasion, appear on only one of those pages.

Newton
Principia
Halley's ode to Newton, 80–81, 86–87, 90–
    91, 98–100, 100n.25, 101, 103
Cotes's preface, 21n.12, 24–25, 39, 43n.16,
    58–59, 90, 92–93, 103, 164–65, 252–
    53, 257–58
Newton's preface, 5, 35–36, 51–52, 162–63,
    175–90, 194–95, 196, 213
definitions, 52, 121
    def. 1, 73
    def. 3, 21, 72–73, 74n.34, 76
    def. 6, 71–72
    def. 7, 71n.26
    def. 8, 26–27, 36, 52–53, 71–72, 95–96
    Scholium to the definitions, 5, 13, 91–
        92, 102–3, 164n.13, 172, 174, 175–77,
        179, 181, 182, 183–84, 185, 186, 188–
        90, 193, 195, 251, 252, 254
axioms, or the laws of motion, 23, 37, 52,
    121, 149, 157–58, 162–63, 171, 175
    law 1, 159, 164n.13, 178n.10
    law 2, 16, 46, 53–54, 58, 159
    law 3, 12–13, 16–17, 46, 51, 96, 162,
        164n.13
    corol. 5, 164n.13, 177
    corol. 6, 177
    Scholium, 52, 160–62, 253
Book I, 50, 51, 52–53, 195
sect. I.1–I.3, 90–91
sect. I.6
    prop. 31, 173
sect. I.10, 52, 161n.7
    prop. 52, 173
sect. I.11
    intr., 51
    prop. 58–61, 51
    schol. 25, 28–29, 41, 50, 95, 212–13

sect. I.14
    prop. 96, schol.–prop. 98, schol., 51
Book II, 51, 84, 89–90, 195
sect II.1
    Prop. 1, Corol., 196n.6
sect II.2
    Prop. 10, 176n.7
sect. II.6, 173
sect. II.8, 51
sect. II.9, 51
sect. II.9, Schol., 125
Book III, 10, 46, 50, 90–91, 195
    intr., 10, 36–37, 92, 97–98
    rules, 78, 89–90, 166–67, 196–97
        rule 1, 4n.6, 19, 102n.28, 208n.19
        rule 2, 4n.6, 102n.28, 208n.19
        rule 3, 12–13, 19, 20–21, 26–28, 36,
            37, 70, 70n.25, 77, 96, 165–67,
            171, 193–94
        rule 4, 38, 44, 89–90, 98n.22, 165–
            67, 171
    phenomena, 89–90
    propositions
        prop. 6, Corol. 2, 43n.17
        prop. 6, Corol. 3, 43, 126n.21
        prop. 8, 96–97
        prop. 10, 92, 125
        prop. 13, 61
        lemma 4–Prop. 42, 80–81, 90–91, 99–
            100, 101, 243
        lemma. 4, Corol. 3, 92, 125
        prop. 41, 44n.18, 99–100
        prop. 42, 99n.23, 100n.26
    General Scholium, 5n.7, 6, 10–12, 14, 16–
        17, 16n.8, 21, 31–32, 33, 34, 35n.7,
        35n.8, 36, 39, 40–42, 56, 58, 60–67,
        80–81, 84–86, 85n.7, 89, 90, 92–93,
        95, 96, 97, 100, 103, 104–7, 109–10,

Name Index

For the benefit of digital users, indexed terms that span two pages (e.g., 52–53) may, on occasion, appear on only one of those pages.

Subject Index

For the benefit of digital users, indexed terms that span two pages (e.g., 52–53) may, on occasion, appear on only one of those pages.

abstraction, 52, 181, 230
acceleration(s)
  measurement(s) of, 175, 178
acting
  per substantiam, 40–41
  per virtutem, 40–41
action
  cause(s) of (*see* cause[s])
  gravitational, 16–17, 53, 131
  local vs. distant, 23, 25 (*see also* action at a distance; principle of local causation [Kochiras])
  magnetic, 53
  "shared action," 27–28
  source of, 154–55 (*see also* mind[s]; substance[s])
  surface, 14
  through contact, 23, 24, 41, 88
  twofold and single, 15, 16
action at a distance, 16–17, 21, 23, 24–26, 36–37, 38–42, 46, 56, 58–59, 61, 62, 80–81, 90–91, 95, 101, 117–18, 213n.33, 235
  endorsement of/commitment to, 41, 90–91, 95
  Epicurean position (*see* Epicureanism [Epicurean])
  and matter, 23, 55–56, 117–18
  rejection/ruling out of, 26, 39–40, 88
actor's categories, 5–6, 6n.9 *See also* anachronism
actuality, 70–71, 155n.26 *See also* modality
aesthetics (aesthetic), 227, 227n.21, 245–46. *See also* beauty (beautiful)
agency
  God's, 66–67n.21, 67, 68–69, 235
agnosticism (agnostic), 42, 205

about the activity/passivity of matter, 38–39, 54, 57–58, 218
about the cause of gravitational force, 4, 46, 95, 213n.33, 218
about the causes of all forces, 4, 95, 213n.33
about the interaction between God and the world, 6
about Newton's views, 12
alchemy (alchemical), 11, 12, 13–14, 43, 93–94
anachronism (anachronistic), 2, 7, 211, 240, 243. *See also* actor's categories
analytic philosophy, 132
analytic/synthetic distinction, 4
anatomy, 241–42
ancients (ancient), 48, 55–56, 60–61, 83, 88, 174, 206–7, 226–27, 228, 247, 250
anthropomorphism (anthropomorphic), 150, 205, 206
anti-anthropomorphism, 109–10 (*see also* Kant, Immanuel; Spinozism [Spinozist])
  un-anthropomorphic, 150–51
anti-anthropocentrism, 100, 107–8
Antikythera Mechanism, 225–26, 226n.11, 226n.12 *See also* planetarium(s)
anti-mathematicism, 126n.*, 211–12, 212n.30, 212n.31
  global anti-mathematicist strategy, 211–13 (*see also* Spinoza, Benedict de; Toland, John)
Arianism, 10–11
Aristotelianism (Aristotelian), 170–71, 210n.24, 228–29n.24, 239. *See also* Scholasticism